13.00

Student Solutions Manual to Accompany

Chemistry

& Chemical Reactivity

SECOND EDITION

Kotz & Purcell

Alton J. Banks
Southwest Texas State University

Saunders College Publishing
Philadelphia   Fort Worth   Chicago   San Francisco
London   Tokyo   Sydney

Kotz/Purcell: Student Solutions Manual for
    CHEMISTRY AND CHEMICAL REACTIVITY, 2/E

ISBN # 0-03-053618-9

123 115 987654321

To the student:

The skills involved in solving chemistry problems are acquired only by discovering the paths which connect the available data to the desired piece(s) of information. These paths are found by repeated efforts. Therefore, this **Solutions Manual** will benefit you most if used as a reference--after you have attempted to solve a problem.

The selected Study Questions have been chosen by the authors of your text to allow you to discover the range and depth of your understanding of chemical concepts. The importance of mastering the "basics" cannot be overemphasized. You will find that the text, **CHEMISTRY AND CHEMICAL REACTIVITY**, has a wealth of study questions to assist you in your study of the science we call Chemistry.

Many of the questions contained in your book--and this manual--have multiple parts. In many cases, comments have been added to aid you in following the "path" from gathering available data and applicable unity-factors to arriving at the desired "answer". In these "multiple-step" questions, you may get an answer which differs slightly from those given here. This may be a result of "rounding" intermediate answers. The procedure followed in this manual was to report each intermediate answer to the appropriate number of significant figures, and to calculate the "final" answer without intermediate rounding. In cases involving atomic and molecular weights, those quantities were expressed with one digit more than the number needed for the data provided.

A word of appreciation is due to several people. Thanks go to Dr. John C. Kotz for the many helpful conversations held during the development of this manual. Additionally, I thank Sandi Kiselica of Saunders College Publishing for her assistance in this work. I would also like to thank Professors David Koster from Southern Illinois University at Carbondale and William J. Vining from Hartwick College in Oneonta for their assistance as accuracy checkers. Finally, I would like to thank my wife, Dr. Catherine Hamrick Banks for her invaluable assistance in typing and proofreading this manuscript. Rarely is a person twice blessed with a patient wife and chemical colleague. Thanks also go to Jonathan and Jennifer for enduring the many aggravations associated with a father in the throes of completing a manuscript.

We have worked diligently to remove all errors from the manual, but remain certain that some have escaped the many inspections. I accept responsibility for those errors.

This manuscript was typed using Microsoft® Word on a Macintosh® computer. Microsoft® Excel was used for the spreadsheet solutions. Atomic and molecular masses were determined using the ChemIntosh™ Molecular Mass Calculator. Graphics were prepared with SuperPaint™ and ChemDraw™, and graphs drawn with Cricket Graph™. Many of the mathematical expressions were constructed with Expressionist™.

Alton J. Banks
Department of Chemistry
Southwest Texas State University
San Marcos, Texas 78666

Table of Contents

CHAPTER 1: The Tools Of Chemistry

1. Tell whether the properties are physical or chemical:

    a. color - physical - involves no change in chemical composition

    b. transformed into rust - chemical - involves a change in chemical composition

    c. explode - chemical - interaction with oxygen indicates a change in chemical composition

    d. density - physical - involves no change in chemical composition

    e. melts - physical - involves a change of state, but not of chemical composition

3. Homogeneous mixtures are those in which every portion of a sample has the same composition as every other portion. The photo shows a mixture which does not fit this description, hence this mixture is **heterogeneous.** Since the components of a mixture retain their physical and chemical properties, one could separate the iron chips and sand by dragging a bar magnet through the mixture. The iron chips would adhere to the magnet while the sand would not.

5. Qualitative information   Quantitative information

    a. purple , solid   1.25 g

    b. silvery , floats   0.025 g

    c. blue , colorless   25 mL , 50 mL

7. Express 19 cm in millimeters, meters, inches:

$$19 \text{ cm} \cdot \frac{10 \text{ mm}}{1 \text{ cm}} = 190 \text{ mm}$$
$$19 \text{ cm} \cdot \frac{1 \text{ m}}{100 \text{ cm}} = 0.19 \text{ m}$$
$$19 \text{ cm} \cdot \frac{1 \text{ in}}{2.54 \text{ cm}} = 7.5 \text{ in}$$

2 significant figures in 19, hence there are 2 significant figures in each answer.

9. Express the metric mile (1600. m) in miles :

$$1600. \text{ m} \cdot \frac{39.37 \text{ in}}{1 \text{ m}} \cdot \frac{1 \text{ ft}}{12 \text{ in}} \cdot \frac{1 \text{ mi}}{5280 \text{ ft}} = 0.9942 \text{ mi}$$

(4 significant figures are possible since the original data was given with four, and all other conversion factors either contain at least 4 significant figures or are exact conversions (e.g. 1 ft = 12 in).

11.   Express 130. km/hr in miles/hr :

Note that the units of the denominator (hr) are the same for both speeds, so no change is necessary. We need only convert distance units of kilometers to miles.

From the table at the back of the textbook : 1 km = 0.62137 mi

$$\frac{130.\ km}{1\ hr} \cdot \frac{0.62137\ mi}{1\ km} = \frac{80.8\ mi}{1\ hr} \quad \text{(3 significant figures in 130. km)}$$

13.   Express 36 ft 7 inches, and 12 feet in meters ; in centimeters:

First, convert 36 ft 7 inches to one set of units (e.g. inches):

$$36\ ft \cdot \frac{12\ in}{1\ ft} = 432\ in\ +\ 7\ inches = 439\ inches$$

$$439\ in \cdot \frac{1\ m}{39.37\ in} = 11.2\ m \quad \text{and}\quad 11.2\ m \cdot \frac{100\ cm}{1\ m} = 1120\ cm$$

and $$12\ ft \cdot \frac{12\ in}{1\ ft} \cdot \frac{1\ m}{39.37\ in} = 3.7\ m \quad \text{and}\quad 3.7\ m \cdot \frac{100\ cm}{1\ m} = 370\ cm$$

15.   Express the area of a 2.5 cm x 2.1 cm stamp in cm^2 ; in m^2 ; in in^2:

$$2.5\ cm \cdot 2.1\ cm = 5.3\ cm^2$$

$$5.3\ cm^2 \cdot \left(\frac{1\ m}{100\ cm}\right)^2 = 5.3\ x\ 10^{-4}\ m^2$$

$$5.3\ cm^2 \cdot \left(\frac{1\ in}{2.54\ cm}\right)^2 = 0.81\ in^2$$

17.   Express the volume of a Saab luggage compartment in cm^3 ; in L ; in m^3:

$$100.cm \cdot 100.\ cm \cdot 150.\ cm = 1.50\ x\ 10^6\ cm^3$$

$$1.50\ x\ 10^6\ cm^3 \cdot \frac{1\ L}{1000\ cm^3} = 1500\ L\ \text{or}\ 1.50\ x\ 10^3\ L$$

$$1.50\ x\ 10^6\ cm^3 \cdot \left(\frac{1\ m}{100\ cm}\right)^3 = 1.50\ m^3$$

19.   Express the mass of a U. S. quarter (5.63 g) in units of kilograms; in milligrams:

$$\frac{5.63 \text{ g}}{1 \text{ quarter}} \cdot \frac{1 \text{ kg}}{1 \times 10^3 \text{ g}} = \frac{5.63 \times 10^{-3} \text{ kg}}{\text{quarter}}$$

$$\frac{5.63 \text{ g}}{1 \text{ quarter}} \cdot \frac{1 \times 10^3 \text{ mg}}{1 \text{ g}} = \frac{5.63 \times 10^3 \text{ mg}}{\text{quarter}}$$

21.   Complete the following table of masses.

MILLIGRAMS	GRAMS	KILOGRAMS
693	0.693	6.93×10^{-4}
156	0.156	1.56×10^{-4}
2.23×10^6	2.23×10^3	2.23

23.   Express a weight of 60. pounds in g; in kg

$$60. \text{ pounds} \cdot \frac{454 \text{ g}}{1 \text{ pound}} = 2.7 \times 10^4 \text{ g}$$

$$2.7 \times 10^4 \text{ g} \cdot \frac{1 \text{ kg}}{1 \times 10^3 \text{ g}} = 2.7 \times 10^1 \text{ or } 27 \text{ kg}$$

25.   Express a weight of 3 pounds in g; in kg :

$$3 \text{ pounds} \cdot \frac{454 \text{ g}}{1 \text{ pound}} = 1 \times 10^3 \text{ g}$$

$$1 \times 10^3 \text{ g} \cdot \frac{1 \text{ kg}}{1 \times 10^3 \text{ g}} = 1 \text{ kg}$$

27.   Express 800. mL in cm^3; in liters(L); in m^3 :

$$\frac{800. \text{ mL}}{1 \text{ beaker}} \cdot \frac{1 \text{ cm}^3}{1 \text{ mL}} = \frac{800. \text{ cm}^3}{1 \text{ beaker}}$$

$$\frac{800. \text{ cm}^3}{1 \text{ beaker}} \cdot \frac{1 \text{ L}}{1000 \text{ cm}^3} = \frac{0.800 \text{ L}}{1 \text{ beaker}}$$

$$\frac{800. \text{ cm}^3}{1 \text{ beaker}} \cdot \frac{1 \text{ m}^3}{1 \times 10^6 \text{cm}^3} = \frac{8.00 \times 10^{-4} \text{ m}^3}{1 \text{ beaker}}$$

29.   Express the volume 23.67 mL in cm^3 ; in liters :

$$23.67 \text{ mL} \quad \bullet \quad \frac{1 \text{ cm}^3}{1 \text{ mL}} \; = \; 23.67 \text{ cm}^3$$

$$23.67 \text{ mL} \quad \bullet \quad \frac{1 \text{ L}}{1000 \text{ mL}} \; = \; 0.02367 \text{ L or } 2.367 \times 10^{-2} \text{ L}$$

31.   Given that the density of diamond is 3.51 g/cm^3, calculate the mass (in g) of a diamond with a volume of 0.0270 cm^3 :

$$\text{Since } D = \frac{M}{V} \quad \text{we calculate mass by multiplying } D \bullet V:$$

$$\frac{3.51 \text{ g}}{1 \text{ cm}^3} \bullet 0.0270 \text{ cm}^3 \; = \; 0.0948 \text{ g}$$

33.   Calculate the mass (in g) of 500. mL of water, given D = 0.997 g/cm^3 at 25 °C.

$$\text{Mass} = D \bullet V = \frac{0.997 \text{ g}}{1 \text{ cm}^3} \bullet 500. \text{ mL} \; = \; 499 \text{ g}$$

$$\text{Express this mass in pounds: } 499 \text{ g} \quad \bullet \quad \frac{1 \text{ lb}}{454 \text{ g}} \; = \; 1.10 \text{ lb}$$

35.   Calculate the density of olive oil if 1 cup (237 mL) has a mass of 205 g:

$$\text{Since Density} = \frac{\text{Mass}}{\text{Volume}} \quad \text{then} \quad \frac{205 \text{ g olive oil}}{237 \text{ mL olive oil}} \; = \; 0.865 \text{ g/mL}$$

37.   Calculate the mass of 1 cup (237 mL) of peanut oil which has a density of 0.92 g/cm^3:

$$\text{Mass} \quad = \quad \text{Density} \bullet \text{Volume}$$

$$= \quad 0.92 \text{ g/cm}^3 \bullet \frac{1 \text{ cm}^3}{1 \text{ mL}} \bullet 237 \text{ mL} = 220 \text{ g}$$

39.   Mass of sulfuric acid in 500. mL of battery acid solution which is 38.08 % H_2SO_4:

$$\text{Mass} \quad = \quad \text{Density} \bullet \text{Volume}$$

$$= \quad \frac{1.285 \text{ g}}{1 \text{ cm}^3} \bullet \frac{1 \text{ cm}^3}{1 \text{ mL}} \bullet 500. \text{ mL} = 643 \text{ g of acid solution}$$

However, only 38.08 % is sulfuric acid. So the mass of sulfuric acid is:

$$643 \text{ g of acid solution } \cdot \frac{38.08 \text{ g sulfuric acid}}{100 \text{ g acid solution}} = 245 \text{ g}$$

41.   Express 25 °C in both Fahrenheit degrees and in Kelvin:

$$°F = \frac{9}{5}(25) + 32 = 45 + 32 = 77 \text{ °F}$$

$$K = (25 \text{ °C} + 273) \text{ or } 298 \text{ K}$$

43.   Make the following temperature conversions:

°F	°C	K
a. 57	5/9 (57-32) = 14	14 + 273.15 = 287
b. 9/5(37) + 32 = 99	37	37 + 273.15 = 310
c. - 40	5/9 (-40-32) = -40	- 40 + 273.15 = 233

45.   The accepted value for a normal human temperature is 98.6 °F.
       On the Celsius scale this corresponds to:

$$°C = \frac{5}{9}(98.6 - 32) = 37 \text{ °C}$$

So the gallium should melt in your hand.

47.   Express -218 °C in °F; in K :

$$°F = \frac{9}{5}(-218) + 32 = -360 \text{ °F}$$

$$K = -218 + 273.15 = 55 \text{ K}$$

49.   Calculate the average of 10.3 g, 9.334 g, and 9.25 g:
       1 digit past decimal
          ↓
       10.3 g + 9.334 g + 9.25 g = 28.9 g (one digit past the decimal is allowed)

$$\text{Average} = \overset{\overset{\text{3 sf}}{\downarrow}}{\frac{28.9}{3}} = 9.63 \text{ g } \textbf{but} \text{ the weight, 10.3 g, is only known } \pm 0.1 \text{ g}$$

so we report the average as 9.6 g.

51.   Express the product of three numbers to the proper number of significant figures:

$$(0.000523) \cdot (0.0263) \cdot (263.28) = 0.00362$$

The answer is limited to three significant figures by the first and second terms.

53.   Express the result of the calculation to the proper number of significant figures:

$$(0.0345)\left(\frac{35.45 - 1.2}{1.000 \times 10^3}\right) = 1.18 \times 10^{-3}$$

55.   Solve for n and report the answer to the correct number of significant figures:

$$\overset{\overset{\text{3 sf}}{\downarrow}}{\underset{\underset{\text{4 sf}}{\uparrow}}{\frac{11.2}{760.0}}} \cdot \underset{\underset{\text{4 sf}}{\uparrow}}{123.4} = n \cdot \underset{\underset{\text{3 sf}}{\uparrow}}{0.0821} \cdot \underset{\underset{\text{4 sf}}{\uparrow}}{298.3}$$

$$\frac{\frac{11.2}{760.0} \times 123.4}{0.0821 \times 298.3} = 0.0742546 \text{ or } 0.0743 \text{ to the correct number of}$$

significant figures.

57.   Calculate the volume of a cylindrical saucepan with a radius of 83.5 mm and height of 9.5 cm:

$$\text{Volume} = \pi r^2 h$$

$$= 3.1415 \times \underset{\underset{\text{5 sf}}{\uparrow}}{(8.35 \text{ cm})^2} \times \underset{\underset{\text{2 sf}}{\uparrow}}{9.5 \text{ cm}} = 2.1 \times 10^3 \text{ cm}^3$$

(with 3 sf marked under 3.1415 area)

59.   Express 1.97 Å in nm; in pm:

$$1.97\text{Å} \cdot \frac{1 \times 10^{-10}\text{ m}}{1 \text{ Å}} \cdot \frac{1 \text{ nm}}{1 \times 10^{-9}\text{ m}} = 0.197 \text{ nm}$$

$$1.97\text{Å} \cdot \frac{1 \times 10^{-10}\text{ m}}{1 \text{ Å}} \cdot \frac{1 \text{ pm}}{1 \times 10^{-12}\text{ m}} = 197 \text{ pm}$$

61.   Express the volume of a cube with edge length 0.563 nm in nm^3 ; in cm^3 :

Since $v = 1 \cdot w \cdot h$ then $(0.563 \text{ nm})^3 = 0.178 \text{ nm}^3$

Express the volume in cm^3 :

$$0.178 \text{ nm}^3 \cdot \left(\frac{1 \text{ m}}{1 \times 10^9 \text{ nm}}\right)^3 \cdot \left(\frac{1 \times 10^2 \text{ cm}}{1 \text{ m}}\right)^3 = 1.78 \times 10^{-22} \text{ cm}^3$$

63.   Compare a troy ounce to an avoirdupois ounce:

A troy ounce = 31.103 g

Calculate the mass of an avoirdupois ounce in grams:

$$\frac{453.59237 \text{ g}}{1 \text{ lb}} \cdot \frac{1 \text{ lb}}{16 \text{ oz avoir}} = \frac{28.350 \text{ g}}{1 \text{ oz avoir}}$$

A troy ounce is larger than an avoirdupois ounce.

65.   Express 65 mi/hr in km/hr; in ft/sec:

$$\frac{65 \text{ mi}}{1 \text{ hr}} \cdot \frac{5280 \text{ ft}}{1 \text{ mi}} \cdot \frac{1 \text{ m}}{3.28 \text{ ft}} \cdot \frac{1 \text{ km}}{1000 \text{ m}} = \frac{1.0 \times 10^2 \text{ km}}{1 \text{ hr}}$$

$$\frac{65 \text{ mi}}{1 \text{ hr}} \cdot \frac{5280 \text{ ft}}{1 \text{ mi}} \cdot \frac{1 \text{ hr}}{3600 \text{ s}} = \frac{95 \text{ ft}}{1 \text{ s}}$$

67.   Express the density of tourmaline in kg/m^3:

$$\frac{3.26 \text{ g}}{1 \text{ cm}^3} \cdot \frac{1 \text{ kg}}{1 \times 10^3 \text{ g}} \cdot \frac{1 \times 10^6 \text{ cm}^3}{1 \text{ m}^3} = \frac{3.26 \times 10^3 \text{ kg}}{1 \text{ m}^3}$$

69.   Calculate the mass of lead and of tin in 1.0 lb (454 g) of solder which is 67 % lead and 33 % tin:

$$1.0 \text{ lb} \cdot \frac{454 \text{ g solder}}{1 \text{ lb solder}} \cdot \frac{67 \text{ g lead}}{100 \text{ g solder}} = 3.0 \times 10^2 \text{ g lead}$$

$$1.0 \text{ lb} \cdot \frac{454 \text{ g solder}}{1 \text{ lb solder}} \cdot \frac{33 \text{ g tin}}{100 \text{ g solder}} = 1.5 \times 10^2 \text{ g tin}$$

71.   Calculate the grams of sugar in 250 mL of a sugar solution which is 8.1 % sugar: (Given the density is 1.05 g/cm^3)

$$\text{Since Density} = \frac{\text{Mass}}{\text{Volume}} \quad \text{we calculate Mass} = \text{Density} \cdot \text{Volume}$$

$$\text{Mass} = \frac{1.05 \text{ g sugar solution}}{1 \text{ cm}^3 \text{ sugar solution}} \cdot \frac{1 \text{ cm}^3}{1 \text{ mL}} \cdot 250 \text{ mL sugar solution}$$

$$= 2.6 \times 10^2 \text{ g sugar solution}$$

$$\text{Mass of sugar} = 2.6 \times 10^2 \text{ g sugar solution} \cdot \frac{8.1 \text{ g sugar}}{100 \text{ g sugar solution}}$$

$$= 21 \text{ g sugar}$$

73.   Calculate the percent of protein, carbohydrate, and fat in 2.0 ounce sample of macaroni:

$$2.0 \text{ ounce} \cdot \frac{454 \text{ g}}{16 \text{ oz}} = 57 \text{ g macaroni}$$

$$\frac{8.0 \text{ g protein}}{57 \text{ g macaroni}} \cdot 100 = 14\% \text{ protein}$$

$$\frac{42 \text{ g carbohydrate}}{57 \text{ g macaroni}} \cdot 100 = 74\% \text{ carbohydrate}$$

$$\frac{1.0 \text{ g fat}}{57 \text{ g macaroni}} \cdot 100 = 1.8\% \text{ fat}$$

75.   Express 12.1 million metric tons of sulfur in kg; in pounds: (1 metric T =1000 kg exactly)

$$12.1 \times 10^6 \text{ metric T} \cdot \frac{1000 \text{ kg}}{1 \text{ metric T}} = 1.21 \times 10^{10} \text{ kg}$$

$$1.21 \times 10^{10} \text{ kg} \cdot \frac{1 \times 10^3 \text{ g}}{1 \text{ kg}} \cdot \frac{1 \text{ lb}}{454 \text{ g}} = 2.67 \times 10^{10} \text{ lb}$$

77.   Calculate the mass of a coin which is 2.2 cm in diameter and 3.0 mm thick. The density of gold is 19.3 g/cm^3 . Assume the coin is a cylinder, i.e. Volume = $\pi r^2 h$

Express the thickness in centimeters, express the diameter (d) as a radius (1/2 d) and calculate the volume:

$$\text{Volume} = 3.1415 \cdot \left(\frac{2.2 \text{ cm}}{2}\right)^2 \cdot 0.30 \text{ cm} = 1.1 \text{ cm}^3$$

Mass of the coin = Volume • Density

$$= 3.1415 \cdot \left(\frac{2.2 \text{ cm}}{2}\right)^2 \cdot 0.30 \text{ cm} \cdot \frac{19.3 \text{ g}}{1 \text{ cm}^3} = 22 \text{ g}$$

If the price of gold is $ 410 per troy ounce, how much is the coin worth?

$$22 \text{ g} \cdot \frac{1 \text{ troy oz}}{31.10 \text{ g}} \cdot \frac{\$ 410}{1 \text{ troy oz}} = \$ 290$$

79.   According to Archimedes' principle, the volume in the cylinder will rise in an amount equal to the volume of the nickel sample. Since we know the mass and density of the sample we can calculate the volume of the metal.

$$\text{Mass} \cdot \frac{1}{\text{Density}} = \text{Volume}$$

$$35.69 \text{ g Ni} \cdot \frac{1 \text{ cm}^3}{8.908 \text{ g Ni}} = 4.007 \text{ cm}^3$$

and 4.007 cm^3 = 4.007 mL

Hence the new volume of the level in the graduated cylinder will be 50.0 mL + 4.007 mL or 54.0 mL.

81.   Calculate the mass of 12 ounces of aluminum in grams and from that mass, the volume:

$$\text{Volume} = \frac{\text{Mass}}{\text{Density}} = \frac{12 \text{ oz} \cdot \frac{454 \text{ g}}{16 \text{ oz}}}{2.70 \text{ g/cm}^3} = 130 \text{ cm}^3$$

Volume = Area • Thickness

Express the area in units of cm^2, then calculate the thickness by dividing the volume by the area:

$$\text{Area} = 75 \text{ ft}^2 \cdot \left(\frac{12 \text{ in}}{1 \text{ ft}}\right)^2 \cdot \left(\frac{2.54 \text{ cm}}{1 \text{ in}}\right)^2 = 7.0 \times 10^4 \text{ cm}^2$$

then $\text{Volume} = \text{Area} \cdot \text{Thickness}$

$$130 \text{ cm}^3 = 7.0 \times 10^4 \text{ cm}^2 \cdot \text{Thickness}$$

and $\text{Thickness} = \dfrac{130 \text{ cm}^3}{7.0 \times 10^4 \text{ cm}^2} = 1.8 \times 10^{-3} \text{ cm}$ or $1.8 \times 10^{-2} \text{ mm}$

83.   The volume of water displaced, and hence the volume of amethyst = 50.8 mL - 45.0 mL or 5.8 mL. This sample of amethyst has a mass of 15.25 g, therefore its density is

$$D = \frac{15.25 \text{ g}}{5.8 \text{ mL}} = 2.6 \text{ g/mL}$$

Calculate the thickness of a piece of amethyst having a mass of 6.50 g, if the amethyst has length of 1.8 cm and width of 0.95 cm:

$$\text{Volume} = \frac{\text{Mass}}{\text{Density}} = \frac{6.50 \text{ g}}{2.6 \text{ g/mL}} = 2.5 \text{ mL or } 2.5 \text{ cm}^3 \text{ and since}$$

$\text{Volume} = \text{length} \cdot \text{width} \cdot \text{thickness}$

$2.5 \text{ cm}^3 = 1.8 \text{ cm} \cdot 0.95 \text{ cm} \cdot \text{thickness or}$

$$\left(\frac{2.5 \text{ cm}^3}{1.8 \text{ cm} \cdot 0.95 \text{ cm}}\right) = 1.4 \text{ cm}$$

85.   Calculate the volume of the "cylinder" of copper wire with a density of 8.94 g/cm^3 and a mass of 125 lb:

$$125 \text{ lb} \cdot \frac{454 \text{ g}}{1 \text{ lb}} \cdot \frac{1 \text{ cm}^3}{8.94 \text{ g}} = 6350 \text{ cm}^3$$

Since the volume of a cylinder is equal to $\pi r^2 h$, we can calculate the "length" of the wire (the height of the cylinder) since we are told that the diameter of the wire is 9.50 mm.

Expressing the diameter in centimeters and converting the diameter into a radius (1/2 the diameter) we obtain:

$$\text{Volume} = \pi\, r^2\, h$$

$$6350\ cm^3 = 3.1415 \cdot (0.475\ cm)^2 \cdot h$$

$$\text{then } h = \frac{6350\ cm^3}{3.1415 \cdot (0.475\ cm)^2} = 8.96 \times 10^3\ cm$$

Expressing this number in feet:

$$8.96 \times 10^3\ cm \cdot \frac{1\ in}{2.54\ cm} \cdot \frac{1\ ft}{12\ in} = 294\ ft$$

CHAPTER 2: Atoms and Elements

16. Names for the following elements:

    a. C - carbon
    b. Na - sodium
    c. Cl - chlorine
    d. P - phosphorus
    e. Mg - magnesium
    f. Ge - germanium

18. Symbols for the following elements:

    a. lithium - Li
    b. titanium - Ti
    c. iron - Fe
    d. silicon - Si
    e. gold - Au
    f. lead - Pb

20. Element 25 is manganese. The symbol is Mn.

22.

element	protons	neutrons	mass number
a. Be	4	5	9
b. Ti	22	26	48
c. Ga	31	39	70

24. Mass number (A) = no. of protons + no. of neutrons;

    Atomic number (Z) = no. of protons

    a. $^{23}_{11}Na$　　　b. $^{39}_{18}Ar$　　　c. $^{70}_{31}Ga$

26.

substance	protons	neutrons	electrons
a. calcium-40	20	20	20
b. tin- 119	50	69	50
c. plutonium-244	94	150	94

Note that the number of protons and electrons are equal for any neutral atom. The number of protons is always equal to the atomic number. The mass number equals the sum of the numbers of protons and neutrons.

28.

Symbol	^{45}Sc	^{33}S	^{17}O	^{56}Mn
Number of protons	21	16	8	25
Number of neutrons	24	17	9	31
Number of electrons in the neutral atom	21	16	8	25

30.   Isotopes of the same element have the same atomic number and different mass numbers.

     a. $^{20}_{10}X$ c. $^{22}_{10}X$, and   e. $^{21}_{10}X$ are isotopes of the element containing 10 protons.

This element is neon.

32.   The average atomic weight of gallium is: $(0.6016)(68.95) + (0.3984)(70.95) = 69.75$

34.   The average atomic weight of silicon is:
$(0.9223)(27.97693) + (0.0467)(28.97649) + (0.0310)(29.97376) = 28.1$

36.   If the abundance of isotope 121 is designated as **x**, then the abundance of isotope 123 may be designated as **1 - x**. The average atomic weight of Sb is reported as 121.75, so we can write $(120.90)(x) + (122.90)(1 - x) = 121.75$. Simplifying gives

$$120.90\ x + 122.90 - 122.90\ x = 121.75$$
$$-2.00\ x = -1.15$$
$$\text{and} \quad x = 0.575$$

So the relative abundance of isotope 121 is 57.5% and that of isotope 123 is 42.5%.

38.   The element with an atomic weight of 72.6 g/mol is **germanium** (Atomic number 32). Its symbol is **Ge**.

40.   Element 7 (atomic number) is **nitrogen**. Its symbol is N and its atomic weight approximately 14.01.

42.   The elements are hydrogen, lithium, sodium, potassium, rubidium, cesium, and francium. Group 1A of the Periodic Table is commonly referred to as the **alkali metals**. As the name implies, most of the elements of Group 1A are metals.

44.   The elements of the second period are lithium, beryllium, boron, carbon, nitrogen, oxygen, fluorine, and neon.

46.
Symbol	Name	Character	Group	Period
a. Ca	calcium	metal	2A	4
b. Cd	cadmium	metal	2B	5
c. Si	silicon	metalloid	4A	3
d. I	iodine	nonmetal	7A	5

48.   a. $127.08 \text{ g Cu} \cdot \dfrac{1 \text{ mol Cu}}{63.546 \text{ g Cu}} = 1.9998 \text{ mol Cu}$

b. $20.0 \text{ g Ca} \cdot \dfrac{1 \text{ mol Ca}}{40.08 \text{ g Ca}} = 0.499 \text{ mol Ca}$

c. $16.75 \text{ g Al} \cdot \dfrac{1 \text{ mol Al}}{26.982 \text{ g Al}} = 0.6208 \text{ mol Al}$

d. $0.012 \text{ g K} \cdot \dfrac{1 \text{ mol K}}{39.1 \text{ g K}} = 3.1 \times 10^{-4} \text{ mol K}$

e. $5.0 \text{ mg Am} \cdot \dfrac{1 \text{ g Am}}{1.0 \times 10^3 \text{ mg}} \cdot \dfrac{1 \text{ mol Am}}{243 \text{ g Am}} = 2.1 \times 10^{-5} \text{ mol Am}$

50.   One mole of sodium has a mass of 22.99 g (its atomic weight to four significant figures). So 50.4 g of sodium represents

$$50.4 \text{ g Na} \cdot \dfrac{1 \text{ mol Na}}{22.99 \text{ g Na}} = 2.19 \text{ moles Na}$$

52.   Calculate the number of grams of each substance in:

a. $0.10 \text{ mol Fe} \cdot \dfrac{55.8 \text{ g Fe}}{1 \text{ mol Fe}} = 5.6 \text{ g Fe}$

b. $2.31 \text{ mol Si} \cdot \dfrac{28.09 \text{ g Si}}{1 \text{ mol Si}} = 64.9 \text{ g Si}$

c. $0.0023 \text{ mol C} \cdot \dfrac{12.0 \text{ g C}}{1 \text{ mol C}} = 0.028 \text{ g C}$

d. $0.54 \text{ mol Na} \cdot \dfrac{23.0 \text{ g Na}}{1 \text{ mol Na}} = 12 \text{ g Na}$

54.   Calculate the number of moles of Al in a 12 oz package of aluminum foil.

$$12 \text{ oz Al} \cdot \frac{28.35 \text{ g Al}}{1 \text{ oz Al}} \cdot \frac{1 \text{ mole Al}}{27.0 \text{ g Al}} = 13 \text{ moles Al}$$

56.   Calculate the number of grams of Ag which may be purchased with \$1.

$$\frac{0.055 \text{ mol Ag}}{1 \text{ dollar}} \cdot \frac{108 \text{ g Ag}}{1 \text{ mol Ag}} = \frac{5.9 \text{ g Ag}}{1 \text{ dollar}}$$

58.   To purchase 1.00 mol of gold:

$$1.00 \text{ mol Au} \cdot \frac{197.0 \text{ g Au}}{1 \text{ mol Au}} \cdot \frac{1 \text{ troy oz Au}}{31.1 \text{ g Au}} \cdot \frac{\$ 401.25}{1 \text{ troy oz Au}} = \$ 2540$$

60.   How many moles of Pt in 15.0 troy ounces ?

$$15.0 \text{ troy oz} \cdot \frac{31.1 \text{ g Pt}}{1 \text{ troy oz Pt}} \cdot \frac{1 \text{ mol Pt}}{195.1 \text{ g Pt}} = 2.39 \text{ mol Pt}$$

The density of Pt is 21.45 g/cm^3. What is the size of the block ?

$$15.0 \text{ troy oz} \cdot \frac{31.1 \text{ g Pt}}{1 \text{ troy oz Pt}} \cdot \frac{1 \text{ cm}^3}{21.45 \text{ g Pt}} = 21.7 \text{ cm}^3$$

62.   The number of atoms in 1.69 moles of aluminum:

$$1.69 \text{ mol Al} \cdot \frac{6.0221 \times 10^{23} \text{ atoms Al}}{1 \text{ mol Al}} = 1.02 \times 10^{24} \text{ atoms Al}$$

64.   The number of atoms of chromium in 35.67 g Cr :

$$35.67 \text{ g Cr} \cdot \frac{1 \text{ mol Cr}}{51.996 \text{ g Cr}} \cdot \frac{6.0221 \times 10^{23} \text{ atoms Cr}}{1 \text{ mol Cr}} = 4.131 \times 10^{23} \text{ atoms Cr}$$

66.   The mass of one atom of copper:

$$\frac{63.546 \text{ g Cu}}{1 \text{ mol Cu}} \cdot \frac{1 \text{ mol Cu}}{6.0221 \times 10^{23} \text{ atoms Cu}} = 1.0552 \times 10^{-22} \text{ g Cu/atom}$$

68.   The number of atoms of Al in a piece of foil with a volume of 2.00 cm^3.

$$2.00 \text{ cm}^3 \text{ Al} \cdot \frac{2.702 \text{ g Al}}{1 \text{ cm}^3 \text{ Al}} \cdot \frac{1 \text{ mol Al}}{26.98 \text{ g Al}} \cdot \frac{6.0221 \times 10^{23} \text{ atoms Al}}{1 \text{ mol Al}}$$

$$= 1.21 \times 10^{23} \text{ atoms Al}$$

70.   The atomic weight of arsenic (74.9216) indicates that the mass number for the stable isotope of arsenic is 75. The atomic number for arsenic is 33 (number of protons), hence the number of neutrons in an atom of arsenic is (75-33) or 42.

72. $454 \text{ g gunpowder} \cdot \dfrac{10.09 \text{ g S}}{100 \text{ g gunpowder}}$ $= 45.8 \text{ g S}$

       $454 \text{ g gunpowder} \cdot \dfrac{10.09 \text{ g S}}{100 \text{ g gunpowder}} \cdot \dfrac{1 \text{ mol S}}{32.06 \text{ g S}}$ $= 1.43 \text{ mol S}$

       $454 \text{ g gunpowder} \cdot \dfrac{14.29 \text{ g C}}{100 \text{ g gunpowder}}$ $= 64.9 \text{ g C}$

       $454 \text{ g gunpowder} \cdot \dfrac{14.29 \text{ g C}}{100 \text{ g gunpowder}} \cdot \dfrac{1 \text{ mol C}}{12.01 \text{ g C}}$ $= 5.40 \text{ mol C}$

74.   Calculate the mass of 7.83×10^{25} atoms of Fe in pounds:

$$7.83 \times 10^{25} \text{ atoms Fe} \cdot \frac{1 \text{ mol Fe}}{6.0221 \times 10^{23} \text{ atoms Fe}} \cdot \frac{55.85 \text{ g Fe}}{1 \text{ mol Fe}} \cdot \frac{1 \text{ lb Fe}}{454 \text{ g Fe}}$$

$$= 16.0 \text{ lb Fe}$$

76.   Calculate the volume of the foil: V = length x width x thickness.
       Since the density is given with volume units of cm^3, express all measurements in cm. Then
       V x D = mass.

$$0.0254 \text{ cm} \cdot 5.00 \text{ cm} \cdot 5.00 \text{ cm} \cdot \frac{21.45 \text{ g Pt}}{1 \text{ cm}^3} \cdot \frac{1 \text{ mol Pt}}{195.08 \text{ g Pt}} = 0.0698 \text{ mol Pt}$$

78.   a. Calculate the number of grams of Ag and Au per cubic mile:

$$\frac{45 \text{ tons Ag}}{1 \text{ cubic mile}} \cdot \frac{2000 \text{ lb Ag}}{1 \text{ ton Ag}} \cdot \frac{454 \text{ g Ag}}{1 \text{ lb Ag}} = \frac{4.1 \times 10^7 \text{ g Ag}}{1 \text{ cubic mile}}$$

$$\frac{9.0 \text{ lb Au}}{1 \text{ cubic mile}} \cdot \frac{454 \text{ g Au}}{1 \text{ lb Au}} = \frac{4.1 \times 10^3 \text{ g Au}}{1 \text{ cubic mile}}$$

b. Calculate the number of moles of Ag and Au per cubic mile:

$$\frac{45 \text{ tons Ag}}{1 \text{ cubic mile}} \cdot \frac{2000 \text{ lb Ag}}{1 \text{ ton Ag}} \cdot \frac{454 \text{ g Ag}}{1 \text{ lb Ag}} \cdot \frac{1 \text{ mol Ag}}{108 \text{ g Ag}} = 3.8 \times 10^5 \text{ mol Ag}$$

$$\frac{9.0 \text{ lb Au}}{1 \text{ cubic mile}} \cdot \frac{454 \text{ g Au}}{1 \text{ lb Au}} \cdot \frac{1 \text{ mol Au}}{197 \text{ g Au}} = 21 \text{ mol Au}$$

80. Volume of can = 2 in x 2 in x 5 in = 20 in^3

Volume of one mole of cans $= \dfrac{20 \text{ in}^3}{\text{can}} \cdot \dfrac{6.0221 \times 10^{23} \text{ can}}{\text{mol cans}}$

$$= 1 \times 10^{25} \text{ in}^3/\text{mol cans}$$

Express the land area of the U.S. (3.6 x 10^6 mi^2) in units of in^2:

$$3.6 \times 10^6 \text{ mi}^2 \cdot \left(\frac{5280 \text{ ft}}{1 \text{ mi}}\right)^2 \cdot \left(\frac{12 \text{ in}}{1 \text{ ft}}\right)^2 = 1.4 \times 10^{16} \text{ in}^2$$

We can calculate the depth of the stack since we know the area (1.4 x 10^{16} in^2) and the volume of one mole of cans (1 x 10^{25} in^3/mol).

Depth of stack $= \dfrac{\text{volume}}{\text{area}} = \dfrac{1 \times 10^{25} \text{ in}^3 /\text{mol cans}}{1.4 \times 10^{16} \text{ in}^2} = 8 \times 10^8$ in/mol cans

Converting to larger units (miles) we get:

$$8 \times 10^8 \text{ in/mol} \cdot \frac{1 \text{ ft}}{12 \text{ in}} \cdot \frac{1 \text{ mi}}{5280 \text{ ft}} = 10000 \text{ miles}$$

Note the limiting factor in significant figures arises from the dimensions of the can (1 significant figure).

82. To determine the volume of Aluminum needed for this "cylinder" of Al, we use the formula **Volume = $\pi r^2 h$** where **h** corresponds to the "thickness" of the Al coating, and **r** the radius. In order to determine the mass of Al needed, we will need to use the density of Al which is given in units of g/cm^3. Begin by converting the radius and thickness to units of cm: [With a diameter of 6.0 inches, the radius is 3.0 in]

$$3.0 \text{ in} \cdot \frac{2.54 \text{ cm}}{1 \text{ in}} = 7.6 \text{ cm}$$

17

and expressing the thickness in centimeters:

$$0.015 \text{ mm} \cdot \frac{1 \text{ cm}}{10 \text{ mm}} = 0.0015 \text{ cm}$$ so the volume of this coating will be:

$$\text{Volume} = 3.1415927 \cdot (7.6 \text{ cm})^2 \cdot 0.0015 \text{ cm} = 0.27 \text{ cm}^3$$

The density of aluminum is given as 2.702 g/cm^3. From this we can calculate the mass of aluminum with the density relationship $(D = \frac{M}{V})$

The mass of Al = 0.27 cm^3 • 2.702 g/cm^3 = 0.74 g Al

To determine the number of Al atoms, we make use of Avogadro's number:
Dividing the mass of Al by the molar mass of Al (26.98 g/mol) and then multiplying by Avogadro's number yields:

$$0.74 \text{ g Al} \cdot \frac{1 \text{ mol Al}}{26.98 \text{ g Al}} \cdot \frac{6.0221 \times 10^{23} \text{ Al atoms}}{1 \text{ mol Al}}$$

$$= 1.7 \times 10^{22} \text{ Al atoms}$$

84.   Converting the length of the side of the cube into centimeters and calculating the volume of the cube yields:

$$\left(543.1 \text{ pm} \cdot \frac{1 \times 10^{-10} \text{ cm}}{1 \text{ pm}}\right)^3 = \text{volume of cube in cm}^3$$

$$= 1.602 \times 10^{-22} \text{ cm}^3$$

Since we know that there are 8 Si atoms in this volume, we can calculate the number of grams of Si corresponding to 8 Si atoms. Given that the mass of 1 mole of silicon has a mass of 28.0854 g Si, we should be able to calculate the number of Si atoms in one mole of silicon.

$$\frac{8 \text{ Si atoms}}{1.602 \times 10^{-22} \text{ cm}^3} \cdot \frac{1 \text{ cm}^3}{2.328994 \text{ g Si}} \cdot \frac{28.0854 \text{ g Si}}{1 \text{ mol Si}} = \frac{6.022 \times 10^{23} \text{ Si atoms}}{1 \text{ mol Si}}$$

85.   Crossword puzzle:

Clues:

1-2 A metal used in ancient times: tin (Sn)

3-4 A metal that burns in air and is found in Group 5A: bismuth (Bi)

1-3 A metalloid: antimony (Sb)

2-4 A metal used in U.S. coins: nickel (Ni)

1. A colorful nonmetal: sulfur (S)

2. A colorless gaseous nonmetal: nitrogen (N)

3. An element that makes fireworks green: boron (B)

4. An element that has medicinal uses: iodine (I)

1-4 An element used in electronics: silicon (Si)

2-3 A metal used with Zr to make wires for superconducting magnets: niobium (Nb)

Using these solutions, the following letters fit in the boxes:

Box 1: S   Box 2: N   Box 3: B   Box 4: I

CHAPTER 3: Compounds and Molecules

9.   a. TiO_2 b. B_4H_{10} c. AlC_3H_9 or since the methyl group
                                               is often written as CH_3: $Al(CH_3)_3$

11.   The total number of atoms of each kind in the following molecules:

      a. CaC_2O_4 Ca = 1 C = 2 O = 4
      b. $C_6H_5CHCH_2$ C = 8 H = 8
      c. $(NH_4)_2SO_4$ N = 2 H = 8 S = 1 O = 4
      d. $Pt(NH_3)_2Cl_2$ Pt = 1 N = 2 H = 6 Cl = 2
      e. $K_4Fe(CN)_6$ K = 4 Fe = 1 C = 6 N = 6

13.   The symbol (including the charge) for each of the following ions:
      a. barium ion Ba^{2+} e. carbonate ion CO_3^{2-}
      b. aluminum ion Al^{3+} f. iodide ion I^-
      c. oxide ion O^{2-} g. phosphate ion PO_4^{3-}
      d. cobalt (III) ion Co^{3+} h. ammonium ion NH_4^+

15.   Formulas for:
      a. potassium iodide, KI   d. magnesium sulfide, MgS
      b. calcium oxide, CaO   e. sodium carbonate, Na_2CO_3
      c. sodium bromide, NaBr   f. ammonium nitrate, NH_4NO_3

17.   Formula for each of the following ionic compounds:
      a. sodium acetate, $NaCH_3COO$ d. titanium (IV) bromide, $TiBr_4$
      b. silver perchlorate, $AgClO_4$ e. sodium nitrate, $NaNO_3$
      c. potassium chlorate, $KClO_3$ f. manganese (II) sulfate, $MnSO_4$

19.

Compound	Cation			Anion		
Name	Formula	Charge	No.	Formula	Charge	No.
a. Sodium hydrogen carbonate	Na	1+	1	HCO_3	1-	1
b. calcium phosphate	Ca	2+	3	PO_4	3-	2
c. ammonium bromide	NH_4	1+	1	Br	1-	1

Compound	Cation			Anion		
Name	Formula	Charge	No.	Formula	Charge	No.
d. potassium perchlorate	K	1+	1	ClO_4	1-	1
e. sodium cyanide	Na	1+	1	CN	1-	1
f. copper(II) sulfate	Cu	2+	1	SO_4	2-	1
g. potassium permanganate	K	1+	1	MnO_4	1-	1
h. zinc oxide	Zn	2+	1	O	2-	1

21.   Formulas for ionic compounds are written by assembling positively charged ions (cations) and negatively charged ions (anions) in numbers which will provide a charge (on the ion pair) of zero. If we were to assemble it in a formula, it could be written :

|(charge on cation)(number of cations)| = |(charge on anion)(number of anions)|

where the vertical lines (| |) indicate absolute magnitude. The task in writing the formula would then be to determine the smallest number of cations (anions) which would make the equation valid.

In the case of sodium carbonate we could write : $|(+1) (x)| = |(-2) (y)|$
since +1 is the charge for a sodium ion, and -2 the charge for a carbonate ion. We then determine the smallest values of x and y which satisfy the equation--in this case $x = 2$ and $y = 1$. The formula for sodium carbonate then has a subscript of 2 for the sodium ion and an (unwritten) subscript of 1 for the carbonate ion.

Na_2CO_3	sodium carbonate	$SrCO_3$	strontium carbonate
NaI	sodium iodide	SrI_2	strontium iodide
$NaNO_3$	sodium nitrate	$Sr(NO_3)_2$	strontium nitrate

$(NH_4)_2CO_3$	ammonium carbonate
NH_4I	ammonium iodide
NH_4NO_3	ammonium nitrate

23.   Formula for each of the following binary nonmetal compounds:

a. boron tribromide, BBr_3 c. hydrogen iodide, HI

b. xenon difluoride, XeF_2 d. disulfur dichloride, S_2Cl_2

25.   Molar mass of the following: (with atomic weights expressed to 4 significant figures)

a. Fe_2O_3 $(2)(55.85) + (3)(16.00) = 159.70$

b. BF_3 $(1)(10.81) + (3)(19.00) = 67.81$

c. N_2O $(2)(14.01) + (1)(16.00) = 44.02$

d. $MnCl_2 \cdot 4\ H_2O$ $(1)(54.94) + (2)(35.45) + (8)(1.008) + (4)(16.00) = 197.90$

e. $C_6H_8O_6$ $(6)(12.01) + (8)(1.008) + (6)(16.00) = 176.12$

27.   Moles in 1.00 g of the following compounds:

	Molecular Weight	Number of Moles
a. CH_3OH	32.04	$1.00\ g \cdot \dfrac{1\ mol}{32.04\ g} = 3.12 \times 10^{-2}$
b. $COCl_2$	98.92	$1.00\ g \cdot \dfrac{1\ mol}{98.92\ g} = 1.01 \times 10^{-2}$
c. NH_4NO_3	80.04	$1.00\ g \cdot \dfrac{1\ mol}{80.04\ g} = 1.25 \times 10^{-2}$
d. $MgSO_4 \cdot 7\ H_2O$	246.47	$1.00\ g \cdot \dfrac{1\ mol}{246.47\ g} = 4.06 \times 10^{-3}$
e. $AgCH_3CO_2$	166.91	$1.00\ g \cdot \dfrac{1\ mol}{166.91\ g} = 5.99 \times 10^{-3}$

29.   The number of moles of vinyl chloride (VC) and molecules of VC in 1.00 lb of VC. (Molar mass of $C_2H_3Cl = 62.5$ g)

$$1.00\ lb\ VC \cdot \frac{454\ g\ VC}{1.00\ lb\ VC} \cdot \frac{1\ mol\ VC}{62.5\ g\ VC} = 7.26\ mol\ VC$$

$$\text{and } 7.26\ mol\ VC \cdot \frac{6.0221 \times 10^{23}\ molecules\ VC}{1\ mol\ VC} = 4.37 \times 10^{24}\ molecules\ VC$$

31.   To begin, express the mass of aspirin, sodium hydrogen carbonate, and aspirin in grams. Convert those masses to moles, using the molar mass of each substance.

a. Moles of aspirin:

$$0.324\ g\ C_9H_8O_4 \cdot \frac{1\ mol\ C_9H_8O_4}{180.16\ g\ C_9H_8O_4} = 1.80 \times 10^{-3}\ mol\ C_9H_8O_4$$

Moles of sodium hydrogen carbonate:

$$1.904 \text{ g NaHCO}_3 \cdot \frac{1 \text{ mol NaHCO}_3}{84.007 \text{ g NaHCO}_3} = 2.266 \times 10^{-2} \text{ mol NaHCO}_3$$

Moles of citric acid:

$$1.000 \text{ g C}_6\text{H}_8\text{O}_7 \cdot \frac{1 \text{ mol C}_6\text{H}_8\text{O}_7}{192.13 \text{ g C}_6\text{H}_8\text{O}_7} = 5.205 \times 10^{-3} \text{ mol C}_6\text{H}_8\text{O}_7$$

b. Molecules of aspirin per tablet:

$$1.80 \times 10^{-3} \text{ mol C}_9\text{H}_8\text{O}_4 \cdot \frac{6.022 \times 10^{23} \text{ molecules aspirin}}{1 \text{ mol aspirin}}$$

$$= 1.08 \times 10^{21} \text{ molecules aspirin}$$

33.   The number of moles of SO_3 in 1.00 lb of the compound:

$$1.00 \text{ lb SO}_3 \cdot \frac{454 \text{ g SO}_3}{1.00 \text{ lb SO}_3} \cdot \frac{1 \text{ mol SO}_3}{80.06 \text{ g SO}_3} = 5.67 \text{ mol SO}_3$$

The number of molecules of SO_3 :

$$5.67 \text{ mol SO}_3 \cdot \frac{6.022 \times 10^{23} \text{ molecules SO}_3}{1 \text{ mol SO}_3} = 3.41 \times 10^{24} \text{ molecules SO}_3$$

The number of sulfur atoms is easily calculable by noting that in each molecule of sulfur trioxide, there is one (1) atom of sulfur.

$$3.41 \times 10^{24} \text{ molecules SO}_3 \cdot \frac{1 \text{ atom S}}{1 \text{ molecule SO}_3} = 3.41 \times 10^{24} \text{ atoms S}$$

Similarly in each molecule of sulfur trioxide, there are three (3) atoms of oxygen.

$$3.41 \times 10^{24} \text{ molecules SO}_3 \cdot \frac{3 \text{ atoms O}}{1 \text{ molecule SO}_3} = 1.02 \times 10^{25} \text{ atoms O}$$

35.   The number of moles of vanillin, $C_8H_8O_3$, in 0.55 g vanillin:

The molar mass of $C_8H_8O_3$ is 152.14. Hence the number of moles is:

$$0.55 \text{ g C}_8\text{H}_8\text{O}_3 \cdot \frac{1 \text{ mol C}_8\text{H}_8\text{O}_3}{152.14 \text{ g C}_8\text{H}_8\text{O}_3} = 3.6 \times 10^{-3} \text{ mol C}_8\text{H}_8\text{O}_3$$

The number of molecules is arrived at using Avogadro's number:

$$3.6 \times 10^{-3} \text{ mol } C_8H_8O_3 \cdot \frac{6.022 \times 10^{23} \text{ molecules } C_8H_8O_3}{1 \text{ mol } C_8H_8O_3} = 2.2 \times 10^{21} \text{ molecules}$$

The formula for vanillin tells us that for each molecule of vanillin there are eight (8) C atoms. Hence the number of carbon atoms is:

$$2.2 \times 10^{21} \text{ molecules } C_8H_8O_3 \cdot \frac{8 \text{ C atoms}}{1 \text{ molecule } C_8H_8O_3} = 1.7 \times 10^{22} \text{ C atoms}$$

37.   Molar mass and weight percent for: [4 significant figures]
       a. PbS: (1)(207.2) + (1)(32.06) = 239.3 g/mol

$$\%Pb = \frac{207.2 \text{ g Pb}}{239.3 \text{ g PbS}} \times 100 = 86.60 \%$$

$$\%S = 100.00 - 86.60 = 13.40 \%$$

       b. C_2H_6: (2)(12.01)+(6)(1.008) = 30.07 g/mol

$$\%C = \frac{24.02 \text{ g C}}{30.07 \text{ g } C_2H_6} \times 100 = 79.88\%$$

$$\%H = 100.00 - 79.88 = 20.12\%$$

       c. CH_3CO_2H : (2)(12.01)+(4)(1.008)+(2)(16.00) = 60.05 g/mol

$$\%C = \frac{24.02 \text{ g C}}{60.05 \text{ g } C_2H_4O_2} \times 100 = 40.00\%$$

$$\%H = \frac{4.032 \text{ g H}}{60.05 \text{ g } C_2H_4O_2} \times 100 = 6.714\%$$

$$\%O = 100.00 - 46.714 = 53.29\%$$

       d. NH_4NO_3: (2)(14.01) + (4)(1.008) + (3)(16.00) = 80.05 g/mol

$$\%H = \frac{4.032 \text{ g H}}{80.05 \text{ g } NH_4NO_3} \times 100 = 5.037 \%H$$

$$\%O = \frac{48.00 \text{ g O}}{80.05 \text{ g } NH_4NO_3} \times 100 = 59.96 \%O$$

$$\%N = 100.00 - (5.037 + 59.96) = 35.00 \%N$$

39.   a. Molar mass of acrylonitrile, H_2CCHCN (or C_3H_3N) :

$$(3)(12.01) + (3)(1.008) + (1)(14.01) = 53.06 \text{ g/mol}$$

   b. Weight percent of C, H, N:

$$\%C = \frac{36.03 \text{ g C}}{53.06 \text{ g } C_3H_3N} \times 100 = 67.90\%$$

$$\%H = \frac{3.024 \text{ g H}}{53.06 \text{ g } C_3H_3N} \times 100 = 5.70\%$$

$$\%N = 100.00 - 73.60 = 26.40\%$$

41.   Weight percent of C, H, Cl, O in hexachlorophene, $C_{13}H_6Cl_6O_2$:

$$(13)(12.01) + (6)(1.008) + (6)(35.45) + (2)(16.00) = 406.88 \text{ g/mol}$$

$$\%C = \frac{156.13 \text{ g C}}{406.88 \text{ g } C_{13}H_6Cl_6O_2} \times 100 = 38.37\%$$

$$\%H = \frac{6.048 \text{ g H}}{406.88 \text{ g } C_{13}H_6Cl_6O_2} \times 100 = 1.486\%$$

$$\%Cl = \frac{212.7 \text{ g Cl}}{406.88 \text{ g } C_{13}H_6Cl_6O_2} \times 100 = 52.28\%$$

$$\%O = 100.00 - 92.13 = 7.87\%$$

43.   The empirical formula (CHO) would have a mass of 29.02 g.
        Since the molar mass is 116.7 g/mol we can write

$$\frac{1 \text{ empirical formula}}{29.02 \text{ g maleic acid}} \cdot \frac{116.1 \text{ g maleic acid}}{1 \text{ mol maleic acid}} = \frac{4.0 \text{ empirical formulas}}{1 \text{ mol maleic acid}}$$

So the molecular formula contains 4 empirical formulas (4 x CHO) or $C_4H_4O_4$.

45.   The percent composition for acetylene tells us that 100 g of the compound would contain
        92.26 g C and 7.74 g H. From this we can calculate the number of moles of carbon and
        hydrogen present in a 100 g sample.

$$7.74 \text{ g H} \cdot \frac{1 \text{ mol H}}{1.008 \text{ g H}} = 7.68 \text{ mol H} \quad \text{and} \quad 92.26 \text{ g C} \cdot \frac{1 \text{ mol C}}{12.011 \text{ g C}} = 7.68 \text{ mol C}$$

These numbers indicate that the hydrogen and carbon are present in a 1:1 molar ratio, indicating that the empirical formula for acetylene is CH. The "formula mass" of CH is 13.02 g. Data from the problem indicate that its molar mass is 26.04 g/mol. Dividing the molar mass by the formula mass indicates that there are two empirical formula units in one molecular unit, hence the molecular formula for acetylene is C_2H_2.

47. Calculate the empirical formula of a nitrogen oxide which is 36.85 % N.

This compound must contain (100.00-36.85) 63.15 % O. With this knowledge we can calculate the molar ratio of atoms in 100 g of the compound.

$$63.15 \text{ g O} \cdot \frac{1 \text{ mol O}}{16.00 \text{ g O}} = 3.947 \text{ mol O}$$

$$36.85 \text{ g N} \cdot \frac{1 \text{ mol N}}{14.01 \text{ g N}} = 2.630 \text{ mol N}$$

Calculating the ratio of atoms:

$$\frac{3.947 \text{ mol O}}{2.630 \text{ mol N}} = \frac{1.5 \text{ mol O}}{1 \text{ mol N}}$$

The empirical formula of this oxide indicates that there are 1.5 O atoms for each N atom. Since we cannot have a fractional part of an atom we write the empirical formula to express this ratio-- N_2O_3 .

49. Empirical and molecular formula for acetic acid:

Calculate moles of each atom present in 100 g of the compound:

$$40.0 \text{ g C} \cdot \frac{1 \text{ mol C}}{12.01 \text{ g C}} = 3.33 \text{ mol C}$$

$$6.71 \text{ g H} \cdot \frac{1 \text{ mol H}}{1.008 \text{ g H}} = 6.66 \text{ mol H}$$

$$53.29 \text{ g O} \cdot \frac{1 \text{ mol O}}{15.999 \text{ g O}} = 3.33 \text{ mol O}$$

Calculate the ratio of atoms: $\quad \frac{6.66 \text{ mol H}}{3.33 \text{ mol C}} = 2 \qquad \frac{3.33 \text{ mol O}}{3.33 \text{ mol C}} = 1$

This ratio indicates that for each C atom in the molecule, there is **one** O atom and **two** H atoms. This informs us that the empirical formula is CH_2O. The molar mass is reported as 60.0 g/mol. If we add the atomic masses of the atoms in the empirical formula, we obtain a formula mass of 30.0. Now we can determine the number of empirical formulas per molecular formula.

$$\frac{60.0 \text{ g acetic acid}}{1 \text{ mol acetic acid}} \cdot \frac{1 \text{ empirical formula}}{30.0 \text{ g acetic acid}} = \frac{2 \text{ empirical formulas}}{1 \text{ mol acetic acid}}$$

So the molecular formula is $C_2H_4O_2$.

51.   Empirical and molecular formula for cacodyl :

$$22.88 \text{ g C} \cdot \frac{1 \text{ mol C}}{12.011 \text{ g C}} = 1.905 \text{ mol C}$$

$$5.76 \text{ g H} \cdot \frac{1 \text{ mol H}}{1.008 \text{ g H}} = 5.71 \text{ mol H}$$

$$71.36 \text{ g As} \cdot \frac{1 \text{ mol As}}{74.922 \text{ g As}} = 0.9525 \text{ mol As}$$

Expressing these values as ratios of each element to one element we obtain:

$$\frac{1.905 \text{ mol C}}{0.9525 \text{ mol As}} = \frac{2.000 \text{ mol C}}{1 \text{ mol As}}$$

$$\frac{5.71 \text{ mol H}}{0.9525 \text{ mol As}} = \frac{5.99 \text{ mol H}}{1 \text{ mol As}}$$

From these ratios we know that for each As atom in the molecule there are 2 C atoms and 6 H atoms, for an empirical formula of AsC_2H_6.
Adding the atomic weights of the empirical formula gives a mass of :
    empirical formula = $1(74.922) + 2(12.011) + 6(1.008) = 104.99$

$$\frac{210 \text{ g cacodyl}}{1 \text{ mol cacodyl}} \cdot \frac{1 \text{ empirical formula}}{104.99 \text{ g cacodyl}} = \frac{2 \text{ empirical formulas}}{1 \text{ mol cacodyl}}$$

Multiplying each element in the empirical formula by two yields a molecular formula of $As_2C_4H_{12}$.

53.   To calculate the empirical formula, begin by calculating the atomic ratio of C, H, and O in 100 g of vanillin.

$$63.15 \text{ g C} \cdot \frac{1 \text{ mol C}}{12.011 \text{ g C}} = 5.258 \text{ mol C}$$

$$5.30 \text{ g H} \cdot \frac{1 \text{ mol H}}{1.008 \text{ g H}} = 5.258 \text{ mol H}$$

To calculate the mass of oxygen: $100.00 - (63.15 + 5.30) = 31.55 \text{ g O}$

$$31.55 \text{ g O} \cdot \frac{1 \text{ mol O}}{15.999 \text{ g O}} = 1.972 \text{ mol O}$$

Calculate the atomic ratios:

$$\frac{5.258 \text{ mol C}}{1.972 \text{ mol O}} = 2.666 \qquad \frac{5.258 \text{ mol H}}{1.972 \text{ mol O}} = 2.666$$

These ratios imply an empirical formula which we might be tempted to write as $C_{2.666}H_{2.666}O$. Noting however that 2.666 may also be expressed as 2 & 2/3 or 8/3, we multiply each of the subscripts by three to obtain an empirical formula of $C_8H_8O_3$. The formula weight of this empirical formula is 152.1, implying that there is one empirical formula unit in a molecular unit of vanillin.

55.   Calculate the empirical formula of naphthalene by calculating the atomic ratios of carbon and hydrogen in 100 g of the compound.

$$93.71 \text{ g C} \cdot \frac{1 \text{ mol C}}{12.011 \text{ g C}} = 7.803 \text{ mol C}$$

$$6.29 \text{ g H} \cdot \frac{1 \text{ mol H}}{1.008 \text{ g H}} = 6.24 \text{ mol H}$$

Calculate the atomic ratio: $\frac{7.803 \text{ mol C}}{6.24 \text{ mol H}} = \frac{1.25 \text{ mol C}}{1 \text{ mol H}}$

The atomic ratio indicates that there are 5 C atoms for 4 H atoms (1.25:1). The empirical formula is then C_5H_4. The formula mass is 64.08. Given that the molar mass of the compound is 128 g/mol, there are two formula units per molecular unit, hence the molecular formula for naphthalene is $C_{10}H_8$.

57.   The mass loss which is noted when removing the water from the hydrated ruthenium salt gives us information about the number of moles of water removed.

(1.056 g hydrated salt - 0.838 g anhydrous salt) = 0.218 g water.

Calculating the number of moles of water:

$$0.218 \text{ g water } \cdot \frac{1 \text{ mol water}}{18.02 \text{ g water}} = 0.0121 \text{ mol water}$$

Calculating the number of moles of the anhydrous salt:

$$0.838 \text{ g RuCl}_3 \cdot \frac{1 \text{ mol RuCl}_3}{207.4 \text{ g RuCl}_3} = 0.00403 \text{ mol RuCl}_3$$

Calculating the ratio of water to anhydrous salt gives:

$$\frac{0.0121 \text{ mol water}}{0.00403 \text{ mol RuCl}_3} = \frac{3 \text{ mol water}}{1 \text{ mol RuCl}_3}$$

hence $RuCl_3$ is a trihydrate, and the value of $x = 3$.

59.   The amount of water present in the 1.687 g sample of epsom salt is:

(1.687 g hydrate - 0.824 g $MgSO_4$) = 0.863 g water

$$0.863 \text{ g water} \cdot \frac{1 \text{ mol water}}{18.02 \text{ g water}} = 0.0479 \text{ mol water}$$

Calculate the number of moles of the anhydrous salt:

$$0.824 \text{ g MgSO}_4 \cdot \frac{1 \text{ mol MgSO}_4}{120.4 \text{ g MgSO}_4} = 0.00684 \text{ mol MgSO}_4$$

Calculating the ratios of water to anhydrous salt gives:

$$\frac{0.0479 \text{ mol water}}{0.00684 \text{ mol MgSO}_4} = \frac{7 \text{ mol water}}{1 \text{ mol MgSO}_4}$$

so there are 7 molecules of water for each formula unit of $MgSO_4$.

61.   a. $NaHCO_3$ is called sodium hydrogen carbonate or sodium bicarbonate.

b. Moles of $NaHCO_3$:

$$908 \text{ g NaHCO}_3 \cdot \frac{1 \text{ mol NaHCO}_3}{84.01 \text{ g NaHCO}_3} = 10.8 \text{ mol NaHCO}_3$$

c. Moles and grams of oxygen atoms:

The formula for $NaHCO_3$ indicates that each mol of $NaHCO_3$ contains **three** moles of O atoms.

$$10.8 \text{ mol } NaHCO_3 \cdot \frac{3 \text{ mol O atoms}}{1 \text{ mol } NaHCO_3} = 32.4 \text{ mol O atoms}$$

The mass of oxygen atoms may be obtained by recognizing that for each mol of $NaHCO_3$ (84.01 g), there are (3 x 16.00) 48.00 g of oxygen atoms, hence:

$$908 \text{ g } NaHCO_3 \cdot \frac{48.00 \text{ g O}}{84.01 \text{ g } NaHCO_3} = 519 \text{ g O atoms.}$$

63.   Calculate:

a. the molar mass of CaF_2: $40.08 + 2(19.00) = 78.08$ g/mol

b. the number of moles of CaF_2 in 1.56 g of the compound:

$$1.56 \text{ g } CaF_2 \cdot \frac{1 \text{ mol } CaF_2}{78.08 \text{ g } CaF_2} = 2.00 \times 10^{-2} \text{ mol}$$

c. the number of grams of CaF_2 to contain 12.0 g of fluoride ion:

$$12.0 \text{ g F}^- \cdot \frac{78.08 \text{ g } CaF_2}{38.00 \text{ g F}^-} = 24.7 \text{ g } CaF_2$$

65.   Given the masses of sulfur and fluorine involved, we can calculate the molar composition of the compound:

$$1.256 \text{ g S} \cdot \frac{1 \text{ mol S}}{32.06 \text{ g S}} = 0.03918 \text{ mol S}$$

The mass of fluorine present is: 5.722 g compound - 1.256 g S = 4.466 g F

$$4.466 \text{ g F} \cdot \frac{1 \text{ mol F}}{19.00 \text{ g F}} = 0.2351 \text{ mol F}$$

Calculating atomic ratios:

$$\frac{0.2351 \text{ mol F}}{0.03918 \text{ mol S}} = \frac{6 \text{ mol F}}{1 \text{ mol S}} \quad \text{indicating that the value of } \mathbf{x} \text{ is 6.}$$

67.   Determine the empirical formula for the compound by first calculating the atomic ratios of sulfur and chlorine in 100 g of the compound.

$$100.00 \text{ g of compound} - 52.51 \text{ g Cl} = 47.49 \text{ g S}$$

$$52.51 \text{ g Cl} \cdot \frac{1 \text{ mol Cl}}{35.453 \text{ g Cl}} = 1.481 \text{ mol Cl}$$

$$47.49 \text{ g S} \cdot \frac{1 \text{ mol S}}{32.066 \text{ g S}} = 1.481 \text{ mol S}$$

The empirical formula for this compound is SCl, with a formula mass of 67.51. Dividing this value into the molar mass of 135.0 g/mol indicates that there are two formula units per molecular unit, hence the molecular formula for this compound is S_2Cl_2.

69.   Name the compounds appearing in the equation given:

Reactants: Products:

$Cd(NO_3)_2$ cadmium nitrate NH_4NO_3 ammonium nitrate

$(NH_4)_2S$ ammonium sulfide CdS cadmium sulfide

71.   The number of moles of bismuth subsalicylate in two tablets:

$$2 \text{ tablets} \cdot \frac{0.300 \text{ g } C_7H_5BiO_4}{1 \text{ tablet}} \cdot \frac{1 \text{ mol } C_7H_5BiO_4}{362.10 \text{ g } C_7H_5BiO_4} = 1.66 \times 10^{-3} \text{ mol}$$

$$2 \text{ tablets} \cdot \frac{0.300 \text{ g } C_7H_5BiO_4}{1 \text{ tablet}} \cdot \frac{208.98 \text{ g Bi}}{362.10 \text{ g } C_7H_5BiO_4} = 0.346 \text{ g Bi}$$

73.   Moles of strychnine in 0.75 g of the compound:

$$0.75 \text{ g } C_{21}H_{22}N_2O_2 \cdot \frac{1 \text{ mol } C_{21}H_{22}N_2O_2}{334.42 \text{ g } C_{21}H_{22}N_2O_2} = 2.2 \times 10^{-3} \text{ mol strychnine}$$

$$2.2 \times 10^{-3} \text{ mol strychnine} \cdot \frac{6.022 \times 10^{23} \text{ molecules}}{1 \text{ mol strychnine}} = 1.4 \times 10^{21} \text{ molecules}$$

75.   The number of moles of quinine in 2.0 g of quinine ($C_{20}H_{24}N_2O_2 \cdot 2 \text{ HCl}$):

$$2.0 \text{ g quinine} \cdot \frac{1 \text{ mol quinine}}{397.3 \text{ g quinine}} = 5.0 \times 10^{-3} \text{ mol quinine}$$

77.   The substance contains 1.00 mol S, 24.1 x 10^{23} atoms F, and 71.0 g Cl in 1.00 mole.
       Convert all these values to a common unit -- <u>moles.</u>

$$24.1 \times 10^{23} \text{ atoms F} \cdot \frac{1 \text{ mol F}}{6.022 \times 10^{23} \text{ atoms F}} = 4.00 \text{ mol F}$$

$$71.0 \text{ g Cl} \cdot \frac{1 \text{ mol Cl}}{35.45 \text{ g Cl}} = 2.00 \text{ mol Cl}$$

Thus the molecular formula is SF_4Cl_2.

79.   The mass of iron in one mole of hemoglobin may be calculated from the percent
       composition data:

$$\frac{0.33 \text{ g iron}}{100 \text{ g hemoglobin}} \cdot \frac{6.8 \times 10^4 \text{ g hemoglobin}}{1 \text{ mol hemoglobin}} = \frac{2.2 \times 10^2 \text{ g iron}}{1 \text{ mol hemoglobin}}$$

and multiplying this mass by iron's atomic weight gives:

$$\frac{2.2 \times 10^2 \text{ g iron}}{1 \text{ mol hemoglobin}} \cdot \frac{1 \text{ mol iron}}{55.847 \text{ g iron}} = \frac{4.0 \text{ mol iron}}{1 \text{ mol hemoglobin}}$$

To calculate the exact number of iron atoms in 1 hemoglobin molecule, we would multiply
both the numerator and denominator of the last fraction by Avogadro's number. Note that
the **ratio** of iron atoms to hemoglobin molecules is identical to the **ratio** of moles of iron
atoms to moles of hemoglobin molecules (**4**).

81.   The number of moles of dioxin ($C_{12}H_4Cl_4O_2$) in the 1 ounce dirt sample:
       The mass of dioxin in the sample is:

$$28.1 \text{ g dirt} \cdot \frac{1.0 \times 10^{-4} \text{ g dioxin}}{100 \text{ g dirt}} = 2.81 \times 10^{-5} \text{ g dioxin}$$

Converting to moles:

$$2.81 \times 10^{-5} \text{ g dioxin} \cdot \frac{1 \text{ mol dioxin}}{321.98 \text{ g dioxin}} = 8.8 \times 10^{-8} \text{ mol dioxin}$$

83. $$1.00 \text{ lb DDT} \cdot \frac{454 \text{ g DDT}}{1.00 \text{ lb DDT}} \cdot \frac{1 \text{ mol DDT}}{354.49 \text{ g DDT}} \cdot \frac{6.022 \times 10^{23} \text{ molecules DDT}}{1 \text{ mol DDT}}$$

$$= 7.71 \times 10^{23} \text{ molecules DDT}$$

The number of grams of atomic Cl in 1.00 lb:

Using the fraction which gives the mass of chlorine in 1 mol of DDT (5 Cl atoms per formula unit - or 5 x 35.453 g Cl) gives :

$$1.00 \text{ lb DDT} \cdot \frac{454 \text{ g DDT}}{1.00 \text{ lb DDT}} \cdot \frac{177.3 \text{ g Cl}}{354.49 \text{ g DDT}} = 227 \text{ g Cl}$$

85.   The mass of the compound (1.322 g) includes the mass of both copper and oxygen atoms. Given that the mass of copper is 1.056 g, the mass of oxygen is then:

$$1.322 \text{ g compound} - 1.056 \text{ g copper} = 0.266 \text{ g oxygen}$$

Calculating the number of moles of copper and oxygen atoms present gives:

$$1.056 \text{ g copper} \cdot \frac{1 \text{ mol copper}}{63.546 \text{ g copper}} = 0.01662 \text{ mol copper and}$$

$$0.266 \text{ g oxygen} \cdot \frac{1 \text{ mol oxygen}}{16.00 \text{ g oxygen}} = 0.0166 \text{ mol oxygen, hence the}$$

empirical formula of the copper oxide is CuO.

87.   Pyrite has the formula FeS_2. Its molar mass is 119.98 g/mol.

$$55.847 + 2(32.066) = 119.979$$

The portion of iron in the sample is then $\frac{55.847 \text{ g Fe}}{119.98 \text{ g FeS}_2}$.

Note that this is also **identical to the ratio** of (iron : FeS_2) when both masses are expressed in kilograms. To calculate the mass of iron in 15.8 kg of pyrite:

$$15.8 \text{ kg FeS}_2 \cdot \frac{55.847 \text{ kg Fe}}{119.98 \text{ kg FeS}_2} = 7.35 \text{ kg Fe.}$$

89.   The antimony metal contained in the ore is present in the form of Sb_2S_3. From the molar mass of this compound (339.70 g/mol) and the atomic mass of antimony (121.75 g/mol), we can express this relation as either:

$$\frac{2 \times 121.75 \text{ g Sb}}{339.70 \text{ g Sb}_2\text{S}_3} \quad \text{or} \quad \frac{339.70 \text{ g Sb}_2\text{S}_3}{2 \times 121.75 \text{ g Sb}}$$

The mass of antimony present in 1.00 lb (454 g) of the ore may then be calculated:

$$454 \text{ g ore} \cdot \frac{10.6 \text{ g Sb}}{100 \text{ g ore}} = 48.1 \text{ g Sb.}$$

Multiplying this mass by the second fraction shown above gives:

$$454 \text{ g ore} \cdot \frac{10.6 \text{ g Sb}}{100 \text{ g ore}} \cdot \frac{339.70 \text{ g Sb}_2\text{S}_3}{243.5 \text{ g Sb}} = 67.1 \text{ g Sb}_2\text{S}_3 \, .$$

91.   Some complexity arises from the fact that both copper and oxygen appear in the carbonate and hydroxide components of azurite. Perhaps the simplest approach involves calculating the number of moles of carbon (appears only in the carbonate) and of hydrogen (appears only in the hydroxide). The percent composition of the azurite is such that in 100 grams of ore there are:

100.00 g ore - (55.31 g copper + 0.58 g hydrogen + 6.97 g carbon) = 37.14 g oxygen

Calculating the number of moles of each element present yields:

6.97 g C = 0.580 mol C   55.31 g Cu = 0.8704 mol Cu

37.14 g O = 2.321 mol O   0.58 g H = 0.58 mol H

Since carbon appears only in the carbonate, the molar amounts of copper and oxygen present in this part of the ore are determined by the molar amount of carbon present and the stoichiometry of the copper carbonate.

For the carbonate portion [$CuCO_3$] of the ore: 0.580 mol C requires 0.580 mol Cu and 1.74 mol O. Similarly for the hydroxide portion of azurite, the molar amounts of copper and oxygen are determined by the amount of hydrogen present.

For the hydroxide portion [$Cu(OH)_2$] of the ore: 0.58 mol H requires 0.58 mol O and 0.29 mol Cu. Note the amount of Cu from $Cu(OH)_2$ is exactly half the amount of Cu from $CuCO_3$. Hence there must be twice as many moles of $CuCO_3$ present or x = 2. Hence the formula for the mineral azurite may be written: [$CuCO_3$]$_2$ • [$Cu(OH)_2$]$_1$

As a double check, use this formulation to calculate the percentage of copper. For this formulation, the %Cu = 55.31-- a percentage consistent with the analytical data given in this problem.

CHAPTER 4: Chemical Reactions: An Introduction

10.   Balancing equations can be a matter of "running in circles" if a reasonable methodology is not employed. While there isn't one "right place" to begin, generally you will suffer fewer complications if you begin the balancing process using a substance that contains the greatest number of elements or the largest subscript values. Noting that you must have at least that many atoms of each element involved, coefficients can be used to increase the inventory of each atom. In the next few questions, you will see one **emboldened** substance in each equation. This underlined substance is the one that I judge to be a "good" starting place. One last hint -- modify the coefficients of uncombined elements, i.e. those not in compounds, <u>after</u> you modify the coefficients for compounds containing those elements -- <u>not before</u>!

a. $4\, Al\,(s) + 3\, O_2\,(g) \rightarrow 2\, \mathbf{Al_2O_3}\,(s)$

1. Note the need for <u>at least</u> 2 Al and 3 O atoms.
2. Oxygen is diatomic -- we'll need an <u>even</u> number of oxygen atoms, so
        try : $2\, Al_2O_3$.
3. $3\, O_2$ would give 6 O atoms on both sides of the equation.
4. 4 Al would give 4 Al atoms on both sides of the equation.

b. $N_2\,(g) + 3\, H_2\,(g) \rightarrow 2\, \mathbf{NH_3}\,(g)$

1. A minimum of 3 H in NH_3 is required, but elemental hydrogen is
        diatomic, so an even number of H atoms is useful -- $2\, NH_3$.
2. 6 H atoms (on the right) indicates $3\, H_2$ (on the left).
3. N_2 is balanced with 2 N atoms on each side.

c. $2\, \mathbf{C_6H_6} + 15\, O_2\,(g) \rightarrow 6\, H_2O\,(g) + 12\, CO_2\,(g)$

1. A minimum of 6 C and 6 H is required.
2. $6\, CO_2$ furnishes 6 C and $3\, H_2O$ furnishes 6 H atoms.
3. $3\, H_2O$ and $6\, CO_2$ furnish a total of 15 O atoms, making the coefficient
        of $O_2 = 15/2$.
4. Multiply <u>all</u> coefficients by 2.

12.   a. UO_2 (s) + 4 HF (l) → **UF_4** (s) + 2 H_2O (l)

       b. **B_2O_3** (s) + 6 HF (l) → 2 BF_3 (g) + 3 H_2O (l)

       c. BF_3 (g) + 3 H_2O (l) → 3 HF (l) + **H_3BO_3** (s)

14.   a. **Na_2O_2** (s) + 2 H_2O (l) → 2 NaOH (aq) + H_2O_2 (l)

       b. 4 PH_3 (g) + 8 O_2 (g) → **P_4O_{10}** (s) + 6 H_2O (g)

       c. 2 **C_2H_3Cl** (l) + 5 O_2 (g) → 4 CO_2 (g) + 2 H_2O (g) + 2 HCl (g)

16.   a. H_2NCl (aq) + 2 NH_3 (g) → NH_4Cl (aq) + **N_2H_4** (aq)

       b. **$(CH_3)_2N_2H_2$** (l) + 2 N_2O_4 (l) → 3 N_2 (g) + 4 H_2O (g) + 2 CO_2 (g)

       c. CaC_2 (s) + 2 H_2O (l) → **$Ca(OH)_2$** (s) + C_2H_2 (g)

18.   In general, combustion of a metal in oxygen forms the oxide.
       a. 2 Mg (s) + O_2 (g) → 2 MgO (s)   magnesium oxide
       b. 2 Ca (s) + O_2 (g) → 2 CaO (s)   calcium oxide
       c. 4 In (s) + 3 O_2 (g) → 2 In_2O_3 (s)   indium (III) oxide

20.   Combination of a metal with a halogen results in the formation of a halide:
       a. 2 Na (s) + Cl_2 (g) → 2 NaCl (s)   sodium chloride
       b. Mg (s) + Br_2 (l) → $MgBr_2$ (s)   magnesium bromide
       c. 2 Al (s) + 3 F_2 (g) → 2 AlF_3 (s)   aluminum fluoride

22.   In general, complete combustion of a hydrocarbon, a compound consisting primarily of
       hydrogen and carbon, yields the oxides of carbon and hydrogen which are named carbon
       dioxide and water respectively.
       a. CH_4 (g) + 2 O_2 (g) → CO_2 (g) + 2 H_2O (l)
       b. 2 C_8H_{18} (l) + 25 O_2 (g) → 16 CO_2 (g) + 18 H_2O (l)
       c. C_2H_5OH (l) + 3 O_2 (g) → 2 CO_2 (g) + 3 H_2O (l)

24.   The simplest method for producing an oxide is to allow the element to react with oxygen.

   a. $2\,C\,(s)\ +\ O_2\,(g)\ \rightarrow\ 2\,CO\,(g)$

   b. $2\,Ni\,(s)\ +\ O_2\,(g)\ \rightarrow\ 2\,NiO\,(s)$

   c. $4\,Cr\,(s)\ +\ 3\,O_2\,(g)\ \rightarrow\ 2\,Cr_2O_3\,(s)$

26.   The decomposition of carbonates typically result in the formation of an oxide and carbon dioxide.

   a. $BeCO_3\,(s)\ +\ heat\ \rightarrow\ CO_2\,(g)\ +\ BeO(s)$ beryllium oxide

   b. $NiCO_3\,(s)\ +\ heat\ \rightarrow\ CO_2\,(g)\ +\ NiO(s)$ nickel(II) oxide

   c. $Al_2(CO_3)_3\,(s)\ +\ heat\ \rightarrow\ 3\,CO_2\,(g)\ +\ Al_2O_3(s)$ aluminum oxide

28.   Mass of H_2 required to react with 280. g of N_2 :

$$280.\ g\ N_2\ \cdot\ \frac{1\ mol\ N_2}{28.01\ g\ N_2}\ \cdot\ \frac{3\ mol\ H_2}{1\ mol\ N_2}\ \cdot\ \frac{2.016\ g\ H_2}{1\ mol\ H_2}\ =\ 60.5\ g\ H_2$$

Since we calculated the stoichiometric amount of hydrogen needed to exactly react with 280. g of nitrogen, and only one product is formed, the mass of product anticipated is:

   280. + 60.5 or 341 g of NH_3 .

30.   Mass of N_2O obtainable from 1.00×10^3 g of NH_4NO_3:

The balanced equation for the decomposition shows that for each mole of NH_4NO_3, 1 mol of N_2O and 2 moles of H_2O result. With these factors and the molar mass of NH_4NO_3 we can calculate:

$$1.00 \times 10^3\ g\ NH_4NO_3\ \cdot\ \frac{1\ mol\ \ NH_4NO_3}{80.04\ gNH_4NO_3}\ \cdot\ \frac{1\ mol\ N_2O}{1\ mol\ \ NH_4NO_3}\ \cdot\ \frac{44.01\ g\ N_2O}{1\ mol\ N_2O}$$

$$=\ 5.50 \times 10^2\ g\ N_2O$$

Mass of water obtainable :

$$1.00 \times 10^3\ g\ NH_4NO_3\ \cdot\ \frac{1\ mol\ \ NH_4NO_3}{80.04\ gNH_4NO_3}\ \cdot\ \frac{2\ mol\ H_2O}{1\ mol\ \ NH_4NO_3}\ \cdot\ \frac{18.02\ g\ H_2O}{1\ mol\ H_2O}$$

$$=\ 4.50 \times 10^2\ g\ H_2O$$

32.   Knowing the mass of a product (ammonia) we focus on the molar mass of ammonia and the ratio of moles of ammonia to moles of water:

$$CaCN_2 \text{ (s)} + 3\,H_2O \text{ (l)} \rightarrow CaCO_3 \text{ (s)} + 2\,NH_3 \text{ (g)}$$

$$68.0 \text{ g } NH_3 \cdot \frac{1 \text{ mol } NH_3}{17.03 \text{ g } NH_3} \cdot \frac{3 \text{ mol } H_2O}{2 \text{ mol } NH_3} \cdot \frac{18.02 \text{ g } H_2O}{1 \text{ mol } H_2O} = 108 \text{ g } H_2O$$

$$68.0 \text{ g } NH_3 \cdot \frac{1 \text{ mol } NH_3}{17.03 \text{ g } NH_3} \cdot \frac{1 \text{ mol } CaCN_2}{2 \text{ mol } NH_3} \cdot \frac{80.10 \text{ g } CaCN_2}{1 \text{ mol } CaCN_2} = 160. \text{ g } CaCN_2$$

34.   Begin by writing and balancing the equation.

$$3\,Fe \text{ (s)} + 2\,O_2 \text{ (g)} \rightarrow Fe_3O_4$$

Using the atomic masses of iron and oxygen and the ratio in which these elements react we can calculate the mass of oxygen which reacts with 12.58 g of iron.

$$12.58 \text{ g } Fe \cdot \frac{1 \text{ mol } Fe}{55.847 \text{ g } Fe} \cdot \frac{2 \text{ mol } O_2}{3 \text{ mol } Fe} \cdot \frac{32.0 \text{ g } O_2}{1 \text{ mol } O_2} = 4.81 \text{ g } O_2$$

As there is only one product, the mass of that product may be obtained by adding the masses of both reactants: (12.58 + 4.81) or 17.39 g

36.   The balanced equation for the preparation of the platinum gives a 3:3 mole ratio for the platinum salt and platinum and a 3:16 mol ratio for the platinum salt and HCl.

    i.e.   3 moles platinum salt $\rightarrow$ 3 moles Pt and

           3 moles platinum salt $\rightarrow$ 16 moles HCl

$$34.6 \text{ g } (NH_4)_2PtCl_6 \cdot \frac{1 \text{ mol } (NH_4)_2PtCl_6}{443.9 \text{ g } (NH_4)_2PtCl_6} \cdot \frac{3 \text{ mol } Pt}{3 \text{ mol } (NH_4)_2PtCl_6} \cdot \frac{195.1 \text{ g } Pt}{1 \text{ mol } Pt}$$

$$= 15.2 \text{ g } Pt$$

Calculate the mass of HCl anticipated:

$$34.6 \text{ g } (NH_4)_2PtCl_6 \cdot \frac{1 \text{ mol } (NH_4)_2PtCl_6}{443.9 \text{ g } (NH_4)_2PtCl_6} \cdot \frac{16 \text{ mol } HCl}{3 \text{ mol } (NH_4)_2PtCl_6} \cdot \frac{36.46 \text{ g } HCl}{1 \text{ mol } HCl}$$

$$= 15.2 \text{ g } HCl$$

38. Fe_2O_3 (s) + 3 CO (g) → 2 Fe (s) + 3 CO_2 (g)

a. Mass of CO required to consume 365 g of Fe_2O_3:

$$365 \text{ g } Fe_2O_3 \cdot \frac{1 \text{ mol } Fe_2O_3}{159.7 \text{g } Fe_2O_3} \cdot \frac{3 \text{ mol CO}}{1 \text{ mol } Fe_2O_3} \cdot \frac{28.01 \text{ g CO}}{1 \text{ mol CO}} = 192 \text{ g CO}$$

b. Mass of Fe_2O_3 to produce 27.9 kg of iron:

$$27.9 \text{ kg Fe} \cdot \frac{1.00 \times 10^3 \text{ g Fe}}{1 \text{ kg Fe}} \cdot \frac{1 \text{ mol Fe}}{55.847 \text{ g Fe}} \cdot \frac{1 \text{ mol } Fe_2O_3}{2 \text{ mol Fe}} \cdot$$

$$\frac{159.7 \text{g } Fe_2O_3}{1 \text{ mol } Fe_2O_3} \cdot \frac{1 \text{ kg } Fe_2O_3}{1.00 \times 10^3 \text{g } Fe_2O_3} = 39.9 \text{ kg } Fe_2O_3$$

Note that the conversion of kg of Fe to g of Fe is unnecessary since the units of the answer were also requested in kg. [A time-saving hint!]

40.   PbS (s) + 4 H_2O_2 (aq) → $PbSO_4$ (s) + 4 H_2O (l)

a. $0.24 \text{ g PbS} \cdot \frac{1 \text{ mol PbS}}{239 \text{ g PbS}} \cdot \frac{4 \text{ mol } H_2O_2}{1 \text{ mol PbS}} \cdot \frac{34.0 \text{ g } H_2O_2}{1 \text{ mol } H_2O_2} = 0.14 \text{ g } H_2O_2$

b. $0.072 \text{ g } H_2O \cdot \frac{1 \text{ mol } H_2O}{18.0 \text{ g } H_2O} \cdot \frac{1 \text{ mol } PbSO_4}{4 \text{ mol } H_2O} \cdot \frac{303 \text{ g } PbSO_4}{1 \text{ mol } PbSO_4}$

$$= 0.30 \text{ g } PbSO_4$$

42.   The limiting reagent can be determined by calculating the mole-available and mole-needed ratios for the equation:

$$2 NH_3 \text{ (g)} + 5 F_2 \text{ (g)} → N_2F_4 \text{ (g)} + 6 \text{ HF (g)}$$

Moles of each reactant present:

$$4.00 \text{ g } NH_3 \cdot \frac{1 \text{ mol } NH_3}{17.03 \text{ g } NH_3} = 0.235 \text{ mol } NH_3$$

$$14.0 \text{ g } F_2 \cdot \frac{1 \text{ mol } F_2}{38.00 \text{ g } F_2} = 0.368 \text{ mol } F_2$$

moles-required ratio: $\dfrac{5 \text{ mol } F_2}{2 \text{ mol } NH_3} = \dfrac{2.5 \text{ mol } F_2}{1 \text{ mol } NH_3}$

moles-available ratio: $\dfrac{0.368 \text{ mol } F_2}{0.235 \text{ mol } NH_3} = \dfrac{1.57 \text{ mol } F_2}{1.00 \text{ mol } NH_3}$

Since we require 2.5 mol of fluorine per mole of ammonia, but we only have 1.57 mol of fluorine per mole of ammonia, **F_2 is the limiting reagent**; and **NH_3 is present in excess.**

To determine how much excess ammonia is present, calculate the amount of NH_3 needed to react with 0.368 mol of F_2:

$$0.368 \text{ mol } F_2 \cdot \dfrac{2 \text{ mol } NH_3}{5 \text{ mol } F_2} = 0.147 \text{ mol } NH_3 \text{ (needed)}$$

$$\text{Excess } NH_3 = (0.235 - 0.147) \text{ mol } NH_3 \cdot \dfrac{17.0 \text{ g } NH_3}{1 \text{ mol } NH_3} = 1.49 \text{ g } NH_3$$

To determine the mass of N_2F_4 obtainable:

$$0.368 \text{ mol } F_2 \cdot \dfrac{1 \text{ mol } N_2F_4}{5 \text{ mol } F_2} \cdot \dfrac{104.0 \text{ g } N_2F_4}{1 \text{ mol } N_2F_4} = 7.66 \text{ g } N_2F_4$$

Similarly to determine the mass of HF produced:

$$0.368 \text{ mol } F_2 \cdot \dfrac{6 \text{ mol } HF}{5 \text{ mol } F_2} \cdot \dfrac{20.01 \text{ g } HF}{1 \text{ mol } HF} = 8.84 \text{ g } HF$$

44.   The limiting reagent can be determined by calculating the moles-available and moles-required ratios for the equation:

$$CO \text{ (g)} + 2 H_2 \text{ (g)} \rightarrow CH_3OH \text{ (l)}$$

Moles of each reactant present:

$$12.0 \text{ g } H_2 \cdot \dfrac{1 \text{mol } H_2}{2.016 \text{ g } H_2} = 5.95 \text{ mol } H_2$$

$$74.5 \text{ g } CO \cdot \dfrac{1 \text{ mol } CO}{28.01 \text{ g } CO} = 2.66 \text{ mol } CO$$

moles-required ratio: $\dfrac{2 \text{ mol } H_2}{1 \text{ mol } CO}$

moles-available ratio: $\dfrac{5.95 \text{ mol H}_2}{2.66 \text{ mol CO}} = \dfrac{2.24 \text{ mol H}_2}{1 \text{ mol CO}}$

Since we require 2 moles of hydrogen per mole of CO, and we have available 2.24 moles of hydrogen per mole of CO, **CO is the limiting reagent** and **H$_2$ is present in excess**.

To determine the mass of excess reagent remaining after the reaction is complete, calculate the amount needed:

$2.66 \text{ mol CO} \cdot \dfrac{2 \text{ mol H}_2}{1 \text{ mol CO}} \cdot \dfrac{2.016 \text{ g H}_2}{1 \text{ mol H}_2} = 10.7 \text{ g H}_2 \text{ needed}$

Excess H$_2$ = 12.0 - 10.7 = 1.3 g

We calculate the theoretical yield of CH$_3$OH from the amount of limiting reagent:

$2.66 \text{ mol CO} \cdot \dfrac{1 \text{ mol CH}_3\text{OH}}{1 \text{ mol CO}} \cdot \dfrac{32.04 \text{ g CH}_3\text{OH}}{1 \text{ mol CH}_3\text{OH}} = 85.2 \text{ g CH}_3\text{OH}$

46.   The limiting reagent can be determined by calculating the mole-available and mole-needed ratios for the equation:

$$\text{CaO (s)} + 2 \text{ NH}_4\text{Cl (s)} \rightarrow 2 \text{ NH}_3 \text{ (g)} + \text{H}_2\text{O (g)} + \text{CaCl}_2 \text{ (s)}$$

Calculate the moles of CaO and of NH$_4$Cl :

$112 \text{ g CaO} \cdot \dfrac{1 \text{ mol CaO}}{56.08 \text{ g CaO}} = 2.00 \text{ mol CaO}$

$224 \text{ g NH}_4\text{Cl} \cdot \dfrac{1 \text{ mol NH}_4\text{Cl}}{53.49 \text{ g NH}_4\text{Cl}} = 4.19 \text{ mol NH}_4\text{Cl}$

moles-required ratio: $\dfrac{2 \text{ mol NH}_4\text{Cl}}{1 \text{ mol CaO}}$

moles-available ratio: $\dfrac{4.19 \text{ mol NH}_4\text{Cl}}{2.00 \text{ mol CaO}} = \dfrac{2.10 \text{ mol NH}_4\text{Cl}}{1.00 \text{ mol CaO}}$

CaO is the limiting reagent, and will determine the maximum amount of products obtainable:

$112 \text{ g CaO} \cdot \dfrac{1 \text{ mol CaO}}{56.08 \text{ g CaO}} \cdot \dfrac{2 \text{ mol NH}_3}{1 \text{ mol CaO}} \cdot \dfrac{17.03 \text{ g NH}_3}{1 \text{ mol NH}_3} = 68.0 \text{ g NH}_3$

The balanced equation shows that for each mole of CaO, 2 moles of NH_4Cl are required. So 2.00 mol of CaO would require 4.00 mol of NH_4Cl, leaving (4.19 - 4.00) 0.19 mol of NH_4Cl in excess. This number of moles would have a mass of:

$$0.19 \text{ mol } NH_4Cl \cdot \frac{53.49 \text{ g } NH_4Cl}{1 \text{ mol } NH_4Cl} = 10. \text{ g } NH_4Cl$$

48.   Calculate the moles-required and moles-available ratio for the two reactants for the equation given as:

$$SO_2 \text{ (g)} + 2 \text{ Cl}_2 \text{ (g)} \rightarrow OSCl_2 \text{ (g)} + Cl_2O \text{ (g)}$$

Moles of reactants present:

$$150 \text{ g } SO_2 \cdot \frac{1 \text{ mol } SO_2}{64.1 \text{ g } SO_2} = 2.3 \text{ mol } SO_2$$

$$150 \text{ g } Cl_2 \cdot \frac{1 \text{ mol } Cl_2}{70.9 \text{ g } Cl_2} = 2.1 \text{ mol } Cl_2$$

Calculating the appropriate ratios:

moles-required ratio: $\dfrac{2 \text{ mol } Cl_2}{1 \text{ mol } SO_2}$

moles-available ratio: $\dfrac{2.1 \text{ mol } Cl_2}{2.3 \text{ mol } SO_2} = \dfrac{0.91 \text{ mol } Cl_2}{1 \text{ mol } SO_2}$

Cl_2 is the limiting reagent, so some SO_2 will remain at the completion of the reaction. The mass of products available is determined by the limiting reagent, Cl_2.

$$150 \text{ g } Cl_2 \cdot \frac{1 \text{ mol } Cl_2}{70.9 \text{ g } Cl_2} \cdot \frac{1 \text{ mol } OSCl_2}{2 \text{ mol } Cl_2} \cdot \frac{119 \text{ g } OSCl_2}{1 \text{ mol } OSCl_2} = 130 \text{ g } OSCl_2$$

$$150 \text{ g } Cl_2 \cdot \frac{1 \text{ mol } Cl_2}{70.9 \text{ g } Cl_2} \cdot \frac{1 \text{ mol } Cl_2O}{2 \text{ mol } Cl_2} \cdot \frac{86.9 \text{ g } Cl_2O}{1 \text{ mol } Cl_2O} = 92 \text{ g } Cl_2O$$

50.   According to the balanced equation, 1.00 mol of Zn should produce 1.00 mol of $ZnCl_2$. The theoretical mass of $ZnCl_2$ would be:

$$1.00 \text{ mol } ZnCl_2 \cdot \frac{136.3 \text{ g } ZnCl_2}{1 \text{ mol } ZnCl_2} = 136 \text{ g } ZnCl_2$$

The actual yield--obtained from the statements of the question-- is 115 g $ZnCl_2$.

So the percent yield is:

$$\frac{actual}{theoretical} \times 100 = \frac{115 \text{ g } ZnCl_2}{136 \text{ g } ZnCl_2} \times 100 \text{ or } 84.6\% \text{ yield.}$$

52.   The theoretical yield of PBr_3 is:

$$12.7 \text{ g } Br_2 \cdot \frac{1 \text{ mol } Br_2}{159.8 \text{ g } Br_2} \cdot \frac{2 \text{ mol } PBr_3}{3 \text{ mol } Br_2} \cdot \frac{270.7 \text{ g } PBr_3}{1 \text{ mol } PBr_3} = 14.3 \text{ g } PBr_3$$

The actual yield of PBr_3 is 10.9 g, so the percent yield is:

$$\frac{actual}{theoretical} \times 100 = \frac{10.9 \text{ g } PBr_3}{14.3 \text{ g } PBr_3} \times 100 \text{ or } 76.0\% \text{ yield}$$

54.   To calculate the empirical formula of a hydrocarbon from combustion data requires that the mass of C and H present in the original sample be determined from the masses of CO_2 and H_2O which result upon combustion.

Mass of C present in the original sample:

$$1.760 \text{ g } CO_2 \cdot \frac{1 \text{ mol } CO_2}{44.010 \text{ g } CO_2} \cdot \frac{1 \text{ mol } C}{1 \text{ mol } CO_2} \cdot \frac{12.011 \text{ g } C}{1 \text{ mol } C} = 0.480 \text{ g } C$$

Since the compound originally contained only carbon and hydrogen, the balance of the original mass is attributable to hydrogen: (0.580 g - 0.480 g) or 0.100 g H.

From the data above: $0.480 \text{ g } C \cdot \dfrac{1 \text{ mol } C}{12.011 \text{ g } C} = 0.0400 \text{ mol C and}$

$$0.100 \text{ g } H \cdot \frac{1 \text{ mol } H}{1.008 \text{ g } H} = 0.0992 \text{ mol H}$$

Now we can express the ratio in which carbon and hydrogen combine as: $C_{0.0400}H_{0.0992}$. Since we know that atoms DO NOT COMBINE ON A FRACTIONAL BASIS, we divide the two subscripts by the smaller to obtain a ratio of:

$\dfrac{0.0400}{0.0400}$ or 1 for C ,   and $\dfrac{0.0992}{0.0400}$ or 2.5 for H

We could express this as $C_1H_{2.5}$, but we recognize that this ratio is more properly expressed as C_2H_5.

56. The carbon found in this cumene sample, upon combustion is located in the CO_2.

$$2.752 \text{ g } CO_2 \cdot \frac{1 \text{ mol } CO_2}{44.010 \text{ g } CO_2} \cdot \frac{1 \text{ mol C}}{1 \text{ mol } CO_2} = 0.06253 \text{ mol C}$$

Likewise, the hydrogen from the cumene sample is located in the H_2O.

$$0.751 \text{ g } H_2O \cdot \frac{1 \text{ mol } H_2O}{18.02 \text{ g } H_2O} \cdot \frac{2 \text{ mol H}}{1 \text{ mol } H_2O} = 0.0834 \text{ mol H}$$

Expressing the ratio of C and H atoms we obtain:

$$\frac{0.0834 \text{ mol H}}{0.06253 \text{ mol C}} = \frac{1.333 \text{ mol H}}{1 \text{ mol C}} \quad \text{or} \quad \frac{4 \text{ mol H}}{3 \text{ mol C}}$$

The empirical formula for cumene would then be C_3H_4, with a formula weight of approximately 40 g. The molar mass of the compound (120 g/mol) indicates that there are 3 formula weight units per mole or a molecular formula of C_9H_{12}.

58. Since we carry out the combustion analysis of a compound by placing the sample in an environment with excess oxygen, we cannot determine the amount of oxygen directly by examining the amount of water or carbon dioxide. We can, however, calculate the amount of carbon and hydrogen in the original sample and subtract their total masses from the original sample mass to obtain the mass of oxygen in the original sample.

Mass of C:

$$0.00600 \text{ g } CO_2 \cdot \frac{1 \text{ mol } CO_2}{44.01 \text{ g } CO_2} \cdot \frac{1 \text{ mol C}}{1 \text{ mol } CO_2} \cdot \frac{12.01 \text{ g C}}{1 \text{ mol C}} = 0.00164 \text{ g C}$$

Mass of H:

$$0.001632 \text{ g } H_2O \cdot \frac{1 \text{ mol } H_2O}{18.015 \text{ g } H_2O} \cdot \frac{2 \text{ mol H}}{1 \text{ mol } H_2O} \cdot \frac{1.00797 \text{ g H}}{1 \text{ mol H}} = 0.000183 \text{ g H}$$

Mass of O in original sample : $0.00400 - (0.00164 + 0.000183) = 0.00218 \text{ g O}$

Calculate the moles of each element in the original compound:

$$0.00164 \text{ g C} \cdot \frac{1 \text{ mol C}}{12.01 \text{ g C}} = 0.000136 \text{ mol C}$$

$$0.000183 \text{ g H} \cdot \frac{1 \text{ mol H}}{1.00797 \text{ g H}} = 0.000181 \text{ mol H}$$

$$0.00218 \text{ g O} \cdot \frac{1 \text{ mol O}}{15.9994 \text{ g O}} = 0.000136 \text{ mol O}$$

Expressing these amounts in a ratio gives $C_1H_{1.33}O_1$ or an empirical formula of $C_3H_4O_3$.

60.   Calculate the mass of Si in the silicon compound:

$$12.02 \text{ g SiO}_2 \cdot \frac{1 \text{ mol SiO}_2}{60.084 \text{ g SiO}_2} \cdot \frac{1 \text{ mol Si}}{1 \text{ mol SiO}_2} \cdot \frac{28.09 \text{ g Si}}{1 \text{ mol Si}} = 5.619 \text{ g Si or}$$

$$0.2000 \text{ mol Si}$$

and the mass of H:

$$5.40 \text{ g H}_2\text{O} \cdot \frac{1 \text{ mol H}_2\text{O}}{18.02 \text{ g H}_2\text{O}} \cdot \frac{2 \text{ mol H}}{1 \text{ mol H}_2\text{O}} \cdot \frac{1.008 \text{ g H}}{1 \text{mol H}} = 0.604 \text{ g H or}$$

$$0.600 \text{ mol H}$$

Calculating the ratio in which silicon and hydrogen have combined in this compound yields: SiH_3, with a formula mass of 31.1 grams. The experimental molar mass of 62.2 g/mol indicates that there are **two** empirical formula units in a molecule--or a molecular formula of Si_2H_6.

62.   Observing the stoichiometry of the equation:
$$Co_x(CO)_y \text{ (s)} + \text{some O}_2 \text{ (g)} \rightarrow \text{some Co}_2O_3 \text{ (s)} + y \text{ CO}_2 \text{ (g)}$$
we see that all the carbon dioxide arises from the carbon originally found in the cobalt complex. Determine the amount of CO_2 and you will know something about the stoichiometry of the original complex.

$$3.52 \text{ g CO}_2 \cdot \frac{1 \text{ mol CO}_2}{44.01 \text{ g CO}_2} = 0.0800 \text{ mol CO}_2 \text{ implying that there was that amount of}$$

CO in the original complex. The mass of CO corresponding to that number of moles is: 2.24 g CO. The mass of the original complex was 3.42 g, therefore the difference in these masses reflects the mass of cobalt contained therein.

>   3.42 g   mass of cobalt complex
>   2.24 g   mass of carbon monoxide
>   1.18 g   mass of cobalt

Calculating the moles of cobalt to which this mass corresponds:

$$1.18 \text{ g Co} \cdot \frac{1 \text{ mol Co}}{58.93 \text{ g Co}} = 0.0200 \text{ mol Co}$$

Now note that 0.0800 mol of CO combined with 0.0200 mol of Co implying that the formula for the cobalt complex is : $Co_{0.0200}(CO)_{0.0800}$ or more properly written: $Co(CO)_4$.

64.   The equation indicates that all the CO_2 originated in the metal carbonate. Determine the amount of CO_2 present:

$$0.395 \text{ g } CO_2 \cdot \frac{1 \text{ mol } CO_2}{44.01 \text{ g } CO_2} = 8.98 \times 10^{-3} \text{ mol } CO_2$$

The stoichiometry of the equation indicates that for each mole of the metal carbonate, one mole of CO_2 is formed. The calculation above tells us that the original sample (mass = 1.324 g) contained 8.98×10^{-3} mol of the metal carbonate. This allows us to determine a molar mass:

$$\frac{1.324 \text{ g}}{8.98 \times 10^{-3} \text{ mol}} = 147 \text{ g/mol}$$

Knowing that the molar mass is approximately 147, we subtract the masses of the other known substituents from this mass:

       147   g/mol of metal carbonate
   - 12   g/ mol of carbon
   - 48   g/ mol of oxygen (3 x 16)
        87   g/mol of metal

Examine a periodic table for a metal with an atomic mass of approximately 87. The closest metal is **strontium** with an atomic mass of 87.6.

66.   To determine the amount of $CaCO_3$ originally present in the sample, calculate the amount of CO_2 produced upon decomposition of the sample.

$$0.657 \text{ g } CO_2 \cdot \frac{1 \text{ mol } CO_2}{44.01 \text{ g } CO_2} = 0.0149 \text{ mol } CO_2$$

According to the stoichiometry of the equation, this also corresponds to the number of moles of $CaCO_3$ originally present. The mass of this amount is $CaCO_3$ is:

$$0.0149 \text{ mol } CaCO_3 \cdot \frac{100.1 \text{ g } CaCO_3}{1 \text{ mol } CaCO_3} = 1.49 \text{ g } CaCO_3$$

The weight percentage of $CaCO_3$ is then : $\dfrac{1.49 \text{ g } CaCO_3}{1.605 \text{ g limestone}} \times 100 = 93.1 \%$

68. One can determine the weight percent of TiO_2 in the ore sample by noting from the balanced equation that for **each mole of TiO_2, one mole of O_2 gas is formed**. Calculate the amount of O_2 formed. Note that there will be 1 mol of the titanium oxide for each mole of O_2, and convert the number of moles of TiO_2 to grams, using the molar mass:

$$36.1 \times 10^{-3} \text{ g } O_2 \cdot \frac{1 \text{ mol } O_2}{32.00 \text{ g } O_2} \cdot \frac{1 \text{ mol } TiO_2}{1 \text{ mol } O_2} \cdot \frac{79.90 \text{ g } TiO_2}{1 \text{ mol } TiO_2} = 0.0901 \text{ g } TiO_2$$

The weight percent of TiO_2 in the sample is then: $\dfrac{0.0901 \text{ g } TiO_2}{1.586 \text{ g ore}} \times 100 = 5.68 \text{ % } TiO_2$

The mass of Ti in the sample is: $0.0901 \text{ g } TiO_2 \cdot \dfrac{47.9 \text{ g Ti}}{79.90 \text{ g } TiO_2} = 0.0540 \text{ g Ti}$

The weight percent of Ti in the sample is then: $\dfrac{0.0540 \text{g Ti}}{1.586 \text{ g ore}} \times 100 = 3.41 \text{ % Ti}$

70. Balance the following chemical equations:

a. $6 \, UO_3 \text{ (s)} + 8 \, BrF_3 \text{ (l)} \rightarrow 6 \, UF_4 \text{ (s)} + 4 \, Br_2 \text{ (l)} + 9 \, O_2 \text{ (g)}$

As indicated in the beginning of this chapter, one should attempt to balance the elements involved in the more complex substances first. In this equation, the **fluorine** content of BrF_3 and UF_4 dictates that the (coefficients for these compounds x the number of fluorine atoms per molecule) be equal. Next the **uranium** balance in UO_3 is determined by UF_4. At this point a fractional coefficient for O_2 will appear. Doubling all coefficients will solve this dilemma. Oxygen and bromine can be balanced easily thereafter.

b. $3 \, IO_2F \text{ (s)} + 4 \, BrF_3 \text{ (l)} \rightarrow 3 \, IF_5 \text{ (l)} + 2 \, Br_2 \text{ (l)} + 3 \, O_2 \text{ (g)}$

Note that the coefficients for IO_2F, IF_5, and O_2 must be equal. The atom balance in IO_2F and IF_5 requires that BrF_3 provide an **even** number of fluorine atoms. If one assigns <u>variables as the coefficients</u> of IO_2F, BrF_3, and IF_5, the following equation results as a **fluorine** inventory.

$$x \, IO_2F \; + \quad y \, BrF_3 \; = \quad x \, IF_5$$

Counting the number of fluorine atoms in each substance the equation may be rewritten:

$$1x \quad + \quad 3y \quad = \quad 5x$$

Simplifying this equation yields $3y = 4x$, which values of $y = 4$ and $x = 3$ satisfy. The remaining species are balanced trivially.

c. $16\,S_2F_2\,(g) \rightarrow 3\,S_8\,(s) + 8\,SF_4\,(g)$

The **fluorine** balance in S_2F_2 and SF_4 require that the coefficient of S_2F_2 be twice that of SF_4. Following the methodology explained in part b, write an equation involving the sulfur balance.

$$xS_2F_2 \quad = \quad yS_8 + \quad zSF_4$$

Counting the number of sulfur atoms in each substance the equation may be rewritten:

$$2x \quad = \quad 8y + \quad 1z \quad \text{and since } x = 2z \text{ we can}$$

substitute and solve to obtain $2(2z) = 8y + 1z$ or $3z = 8y$. Values of $z = 8$ and $y = 3$ satisfy the equation and x equals 16.

d. $2\,Na_2S\,(aq) + 2\,O_2\,(g) + H_2O\,(l) \rightarrow Na_2S_2O_3\,(aq) + 2\,NaOH\,(aq)$

The **sulfur** content of $Na_2S_2O_3$ requires that the coefficient of Na_2S be two times its coefficient--try 2. Changing the coefficient of NaOH to 2 provides **sodium** balance. All other elements can then be balanced easily.

72.   Begin by calculating the amount of carbon and of hydrogen originally present in cyclooctatetraene:

$$0.940 \text{ g } CO_2 \cdot \frac{1 \text{ mol } CO_2}{44.01 \text{ g } CO_2} \cdot \frac{1 \text{ mol C}}{1 \text{ mol } CO_2} = 0.0214 \text{ mol C}$$

$$0.192 \text{ g } H_2O \cdot \frac{1 \text{ mol } H_2O}{18.02 \text{ g } H_2O} \cdot \frac{2 \text{ mol H}}{1 \text{ mol } H_2O} = 0.0213 \text{ mol H}$$

This molar ratio indicates that in cyclooctatetraene, carbon and hydrogen are present in a 1:1 ratio. So the empirical formula is CH, with a formula weight of about 13 g /mol. The experimental value for the molar mass is about 105 g/mol. Calculate the number of empirical formulas present in one molecular formula.

$$\frac{105 \text{ g/ molecular formula}}{13 \text{ g/ empirical formula}} = 8.0 \text{ empirical formulas/molecular formula, and we}$$

predict the molecular formula to be C_8H_8.

74.   The empirical formula of potassium superoxide is derived by first calculating the molar amounts of potassium and oxygen present in the compound:

          7.11 g potassium superoxide

      - <u>3.91 g potassium metal</u>

          3.20 g oxygen

Calculate the moles of each element:

$$3.91 \text{ g K} \cdot \frac{1 \text{ mol K}}{39.10 \text{ g K}} = 0.100 \text{ mol K and}$$

$$3.20 \text{ g O} \cdot \frac{1 \text{ mol O}}{16.00 \text{ g O}} = 0.200 \text{ mol O and the empirical formula is then } KO_2.$$

76.   From the mass of scandium, we can calculate the number of moles of Sc present:

$$0.929 \text{ g Sc} \cdot \frac{1 \text{ mol Sc}}{44.96 \text{ g Sc}} = 0.0207 \text{ mol Sc}$$

We can calculate the mass of oxygen by subtracting the mass of the metal from the mass of the oxide :   (1.423 g oxide - 0.929 g Sc) = 0.494 g O

The number of moles of oxygen is then: $0.494 \text{ g O} \cdot \dfrac{1 \text{ mol O}}{16.00 \text{ g O}} = 0.0309 \text{ mol O}$

and $\dfrac{0.0309 \text{ mol O}}{0.0207 \text{ mol Sc}} = \dfrac{1.49 \text{ mol O}}{1 \text{ mol Sc}}$ The oxide is most probably Sc_2O_3.

The combustion of this oxide may be represented as:

$$Sc_2O_3 \text{ (s)} + 3 H_2 \text{ (g)} \rightarrow 3 H_2O \text{ (g)} + 2 Sc \text{ (s)}$$

If 0.929 g Sc were produced, then we can calculate the amount of water produced.

$$0.929 \text{ g Sc} \cdot \frac{1 \text{ mol Sc}}{44.96 \text{ g Sc}} \cdot \frac{3 \text{ mol H}_2O}{2 \text{ mol Sc}} \cdot \frac{18.02 \text{ g H}_2O}{1 \text{ mol H}_2O} = 0.559 \text{ g H}_2O$$

78.   Determine the mass of $MgCO_3$ needed by calculating the mass of SO_2 in 20. million tons:

$$20. \times 10^6 \text{ tons} \cdot \frac{2000. \text{ lb}}{1 \text{ ton}} \cdot \frac{454 \text{ g}}{1 \text{ lb}} \cdot \frac{1 \text{ mol SO}_2}{64.06 \text{ g SO}_2} = 2.8 \times 10^{11} \text{ mol SO}_2$$

From this, we can calculate the mass of $MgCO_3$:

$$2.8 \times 10^{11} \text{ mol } SO_2 \cdot \frac{1 \text{ mol MgO}}{1 \text{ mol } SO_2} \cdot \frac{1 \text{ mol } MgCO_3}{1 \text{ mol MgO}} \cdot \frac{84.3 \text{ g } MgCO_3}{1 \text{ mol } MgCO_3} \cdot$$

$$\frac{1 \text{ lb}}{454 \text{ g}} \cdot \frac{1 \text{ ton}}{2000. \text{ lb}} = 2.6 \times 10^7 \text{ ton } MgCO_3$$

$MgSO_4$ is produced in a 1:1 mol ratio with SO_2 consumed. So if 2.8×10^{11} mol SO_2 must be removed per year, then 2.8×10^{11} mol $MgSO_4$ will be produced. The mass of that amount of $MgSO_4$ is:

$$2.8 \times 10^{11} \text{ mol } MgSO_4 \cdot \frac{120 \text{ g } MgSO_4}{1 \text{ mol } MgSO_4} \cdot \frac{1 \text{ lb}}{454 \text{ g}} \cdot \frac{1 \text{ ton}}{2000. \text{ lb}}$$

$$= 3.8 \times 10^7 \text{ ton } MgSO_4$$

80.   a. The mass of phosphoric acid obtainable from 1.00 ton of apatite:

$$9.08 \times 10^5 \text{ g apatite} \cdot \frac{1 \text{ mol apatite}}{504 \text{ g apatite}} \cdot \frac{3 \text{ mol } H_3PO_4}{1 \text{ mol apatite}} \cdot \frac{98.00 \text{ g } H_3PO_4}{1 \text{ mol } H_3PO_4}$$

$$= 5.30 \times 10^5 \text{ g } H_3PO_4$$

b. The percent yield of phosphoric acid is:

$$\frac{\text{actual}}{\text{theoretical}} \times 100 = \frac{2.50 \times 10^5 \text{ g } H_3PO_4}{5.30 \times 10^5 \text{ g } H_3PO_4} \times 100 = 47.2 \%$$

c. The number of moles of apatite present are:

$$9.08 \times 10^5 \text{ g apatite} \cdot \frac{1 \text{ mol apatite}}{504 \text{ g apatite}} = 1.80 \times 10^3 \text{ mol apatite}$$

In one ton of H_2SO_4 there are:

$$9.08 \times 10^5 \text{ g } H_2SO_4 \cdot \frac{1 \text{ mol } H_2SO_4}{98.07 \text{ g } H_2SO_4} = 9.26 \times 10^3 \text{ mol } H_2SO_4$$

Looking at the moles-available and moles-required ratios :

$$\text{moles-available:} \frac{9.26 \times 10^3 \text{ mol } H_2SO_4}{1.80 \times 10^3 \text{ mol apatite}} = \frac{5.14 \text{ mol } H_2SO_4}{1.00 \text{ mol apatite}}$$

$$\text{moles-required:} \frac{5 \text{ mol } H_2SO_4}{1 \text{ mol apatite}} \text{ so } \textbf{apatite is the limiting reagent}.$$

82.   The percent yield of SF_4 can be calculated if we know the actual yield. Calculate the mass of SF_4 obtainable from the SCl_2 given:

$$15.67 \text{ g } SCl_2 \cdot \frac{1 \text{ mol } SCl_2}{102.97 \text{ g } SCl_2} \cdot \frac{1 \text{ mol } SF_4}{3 \text{ mol } SCl_2} \cdot \frac{108.06 \text{ g } SF_4}{1 \text{ mol } SF_4} = 5.48 \text{ g } SF_4$$

The percent yield of SF_4 is then:

$$\frac{actual}{theoretical} \times 100 = \frac{3.56 \text{ g } SF_4}{5.48 \text{ g } SF_4} \times 100 = 64.9 \text{ \%}$$

Mass of S_2Cl_2 produced :

$$3.56 \text{ g } SF_4 \cdot \frac{1 \text{ mol } SF_4}{108.1 \text{ g } SF_4} \cdot \frac{1 \text{ mol } S_2Cl_2}{1 \text{ mol } SF_4} \cdot \frac{135.0 \text{ g } S_2Cl_2}{1 \text{ mol } S_2Cl_2} = 4.45 \text{ g } S_2Cl_2$$

84.   The following ratios in the production of nitric acid apply:

$$\frac{4 \text{ mol NO}}{4 \text{ mol } NH_3} \cdot \frac{2 \text{ mol } NO_2}{2 \text{ mol NO}} \cdot \frac{2 \text{ mol } HNO_3}{3 \text{ mol } NO_2}$$

Applying these ratios to the amount of NH_3 given we obtain:

$$545 \text{ g } NH_3 \cdot \frac{1 \text{ mol } NH_3}{17.03 \text{ g } NH_3} \cdot \frac{4 \text{ mol NO}}{4 \text{ mol } NH_3} \cdot \frac{2 \text{ mol } NO_2}{2 \text{ mol NO}} \cdot \frac{2 \text{ mol } HNO_3}{3 \text{ mol } NO_2}$$

$$\cdot \frac{63.01 \text{ g } HNO_3}{1 \text{ mol } HNO_3} = 1340 \text{ g } HNO_3$$

Percent yield of HNO_3 is:

$$\frac{actual}{theoretical} \times 100 = \frac{346 \text{ g } HNO_3}{1340 \text{ g } HNO_3} \times 100 = 25.8 \text{ \%}$$

86.   To perform any stoichiometric calculations, we should balance the equation:

$$2 \text{ } NH_3 \text{ (g)} + 3 \text{ CuO (s)} \rightarrow N_2 \text{ (g)} + 3 \text{ Cu (s)} + 3 \text{ } H_2O \text{ (g)}$$

As we have masses of both reactants, we first determine the limiting reagent:

$$5.589 \text{ g CuO} \cdot \frac{1 \text{ mol CuO}}{79.545 \text{ g CuO}} = 0.07026 \text{ mol CuO}$$

$$0.686 \text{ g } NH_3 \cdot \frac{1 \text{ mol } NH_3}{17.03 \text{ g } NH_3} = 0.0403 \text{ mol } NH_3$$

Determine the appropriate molar ratios:

$$\text{moles-required: } \frac{3 \text{ mol CuO}}{2 \text{ mol NH}_3} = \frac{1.5 \text{ mol CuO}}{1 \text{ mol NH}_3}$$

$$\text{moles-available: } \frac{0.07026 \text{ mol CuO}}{0.0403 \text{ mol NH}_3} = \frac{1.74 \text{ mol CuO}}{1 \text{ mol NH}_3}$$

Ammonia is the **limiting reagent**. Determine the number of grams of nitrogen using the amount of ammonia:

$$0.0403 \text{ mol NH}_3 \cdot \frac{1 \text{ mol N}_2}{2 \text{ mol NH}_3} \cdot \frac{28.01 \text{ g N}_2}{1 \text{ mol N}_2} = 0.564 \text{ g N}_2$$

The percent yield of nitrogen is then:

$$\frac{\text{actual}}{\text{theoretical}} \times 100 = \frac{0.345 \text{g N}_2}{0.564 \text{ g N}_2} \times 100 = 61.2\%$$

The mass of copper expected is:

$$0.345 \text{ g N}_2 \cdot \frac{1 \text{ mol N}_2}{28.01 \text{ g N}_2} \cdot \frac{3 \text{ mol Cu}}{1 \text{ mol N}_2} \cdot \frac{63.55 \text{ g Cu}}{1 \text{ mol Cu}} = 2.35 \text{ g Cu}$$

88. To calculate the empirical formula of this aluminum compound, determine the amount of aluminum present in the compound:

$$0.398 \text{ g Al}_2\text{O}_3 \cdot \frac{1 \text{ mol Al}_2\text{O}_3}{102.0 \text{ g Al}_2\text{O}_3} \cdot \frac{2 \text{ mol Al}}{1 \text{ mol Al}_2\text{O}_3} \cdot \frac{26.98 \text{ g Al}}{1 \text{ mol Al}} = 0.211 \text{ g Al}$$

The mass of carbon and hydrogen in the compound is then : $(0.563 \text{ g} - 0.211 \text{ g}) = 0.352 \text{ g}$

We know that the carbon and hydrogen is contained in the compound in units of CH_3. Calculate the number of moles of CH_3 units:

$$0.352 \text{ g CH}_3 \cdot \frac{1 \text{ mol CH}_3}{15.04 \text{ g CH}_3} = 0.0234 \text{ mol CH}_3$$

From the initial calculation above, there are 0.211 g Al (0.00782 mol Al).

Examining the ratio of moles of CH_3 to moles of Al, we find the empirical formula to be

$$\frac{0.0234 \text{ mol CH}_3}{0.00782 \text{ mol Al}} = \frac{2.99 \text{ mol CH}_3}{1 \text{ mol Al}} \quad \text{or } \text{Al(CH}_3)_3.$$

90.   HCl may be produced in the laboratory by the reaction:

$$2\ NaCl\ (s) + H_2SO_4\ (aq)\ \rightarrow\ 2\ HCl\ (g) + Na_2SO_4\ (aq)$$

a. The mass of H_2SO_4 required to completely react with 20. g of NaCl:

$$20.\ g\ NaCl\ \cdot\ \frac{1\ mol\ NaCl}{58.4\ g\ NaCl}\ \cdot\ \frac{1\ mol\ H_2SO_4}{2\ mol\ NaCl}\ \cdot\ \frac{98.1\ g\ H_2SO_4}{1\ mol\ H_2SO_4}\ =\ 17\ g\ H_2SO_4$$

b. Calculate the percent yield of HCl by first calculating the theoretical yield of HCl:

$$20.\ g\ NaCl\ \cdot\ \frac{1\ mol\ NaCl}{58.4\ g\ NaCl}\ \cdot\ \frac{2\ mol\ HCl}{2\ mol\ NaCl}\ \cdot\ \frac{36.5\ g\ HCl}{1\ mol\ HCl}\ =\ 13\ g\ HCl$$

$$\frac{actual}{theoretical}\ x\ 100\ =\ \frac{5.6\ g\ HCl}{13\ g\ HCl}\ x\ 100\ =\ 45\ \%$$

c. The mass of sodium sulfate obtainable with 5.6 g HCl:

$$5.6\ g\ HCl\ \cdot\ \frac{1\ mol\ HCl}{36.5\ g\ HCl}\ \cdot\ \frac{1\ mol\ Na_2SO_4}{2\ mol\ HCl}\ \cdot\ \frac{142g\ Na_2SO_4}{1\ mol\ Na_2SO_4}\ =\ 11\ g\ Na_2SO_4$$

92.   Use the masses of GeO_2 and H_2O to obtain the moles of Ge and H present in the compound Ge_xH_y:

$$0.523\ g\ GeO_2\ \cdot\ \frac{1\ mol\ GeO_2}{104.6\ g\ GeO_2}\ \cdot\ \frac{1\ mol\ Ge}{1\ mol\ GeO_2}\ =\ 0.005\ mol\ Ge$$

$$0.180\ g\ H_2O\ \cdot\ \frac{1\ mol\ H_2O}{18.02\ g\ H_2O}\ \cdot\ \frac{2\ mol\ H}{1\ mol\ H_2O}\ =\ 0.0200\ mol\ H$$

Since these were the amounts of Ge and H present in the original compound, the empirical formula for this compound is: $Ge_{0.005}H_{0.0200}$ or GeH_4.

94.   Assume that we begin with 100 g each of zinc and iodine.

The number of moles of each element obtained: $100\ g\ Zn\ \cdot\ \frac{1\ mol\ Zn}{65.39\ g\ Zn}\ =\ 1.53\ mol\ Zn$

and $100\ g\ I_2\ \cdot\ \frac{1\ mol\ I_2}{254\ g\ I_2}\ =\ 0.394\ mol\ I_2$

The equation may be written: $Zn + I_2 \rightarrow ZnI_2$

The 0.394 mol I_2 would require 0.394 mol Zn to form the salt. This amount of zinc would have a mass of:

$$0.394 \text{ mol Zn} \cdot \frac{65.39 \text{ g Zn}}{1 \text{ mol Zn}} = 25.8 \text{ g Zn}$$

Unreacted zinc = (100 - 25.8) = 74 g Zn and the fraction of zinc remaining is 0.74

96.   a. The balanced equation is: $2 Al (s) + 3 Br_2 (l) \rightarrow 2 AlBr_3 (s)$

b. To calculate the number of moles of each substance we use their densities and their volumes:

For Al: V = 0.06500 cm • 12.0 cm • 12.0 cm = 7.20 cm^3

Mass of Al   = Density • Volume

= 2.70 g/cm^3 • 7.20 cm^3 = 19.4 g Al and

Moles of Al   = $19.4 \text{ g Al} \cdot \frac{1 \text{ mol Al}}{26.98 \text{ g Al}}$ = 0.721 mol Al

For Br_2:

Mass of Br_2 = $3.10 \text{ g/cm}^3 \cdot \frac{1 \text{ cm}^3}{1 \text{ mL}} \cdot \frac{19.50 \text{ mL}}{1}$ = 60.5 g Br_2

Moles of Br_2 = $60.5 \text{ g Br}_2 \cdot \frac{1 \text{ mol Br}_2}{159.8 \text{ g Br}_2}$ = 0.378 mol Br_2

c. The theoretical yield of $AlBr_3$:

Moles-available ratio: $\frac{0.378 \text{ mol Br}_2}{0.721 \text{ mol Al}} = \frac{0.524 \text{ mol Br}_2}{1 \text{ mol Al}}$

Moles-required ratio: $\frac{3 \text{ mol Br}_2}{2 \text{ mol Al}} = \frac{1.5 \text{ mol Br}_2}{1 \text{ mol Al}}$

So Br_2 is the **limiting reagent**.

$$0.378 \text{ mol Br}_2 \cdot \frac{2 \text{ mol AlBr}_3}{3 \text{ mol Br}_2} \cdot \frac{266.7 \text{ g AlBr}_3}{1 \text{ mol AlBr}_3} = 67.2 \text{ g AlBr}_3$$

98.   a. Determine the empirical formula by first determining the number of moles of uranium and oxygen in the oxide.

Oxide U_aO_b is 83.22 % U (and 16.78 % O). The moles of U present in 100 g of the oxide:

$$83.22 \text{ g U} \cdot \frac{1 \text{ mol U}}{238.03 \text{ g U}} = 0.3496 \text{ mol U}$$

The moles of O present:

$$16.78 \text{ g O} \cdot \frac{1 \text{ mol O}}{15.999 \text{ g O}} = 1.049 \text{ mol O}$$

Calculating the ratio in which these combine gives an empirical formula of:

$$\frac{1.049 \text{ mol O}}{0.3496 \text{ mol U}} = \frac{3 \text{ mol O}}{1 \text{ mol U}} \quad \text{or } UO_3 \text{ -- uranium(VI) oxide.}$$

b. The second oxide, U_nO_m , is 84.8 % U (and 15.2 % O). The moles of each element present in 100 g of the oxide is then:

$$84.8 \text{ g U} \cdot \frac{1 \text{ mol U}}{238.0 \text{ g U}} = 0.356 \text{ mol U}$$

$$15.2 \text{ g O} \cdot \frac{1 \text{ mol O}}{16.00 \text{ g O}} = 0.950 \text{ mol O}$$

Dividing the smaller molar amount into the larger gives:

$$\frac{0.950 \text{ mol O}}{0.356 \text{ mol U}} = \frac{2.66 \text{ mol O}}{1 \text{ mol U}}$$

Recognizing 2.66 as another way of writing the fraction 8/3, we may also write the empirical formula of this second oxide as U_3O_8.

c. Schematically we can write the statement of this part of the problem as:

$$UO_x(NO_3)_y \cdot z\,H_2O \rightarrow \quad UO_x(NO_3)_y \rightarrow \quad U_nO_m$$
$$1.328 \text{ g} \qquad\qquad 1.042 \text{ g} \qquad\qquad 0.742 \text{ g}$$

The mass of water removed upon heating is then: 1.328 g - 1.042 g or 0.286 g

The number of moles of water is:

$$0.286 \text{ g H}_2\text{O} \cdot \frac{1 \text{ mol H}_2\text{O}}{18.02 \text{ g H}_2\text{O}} = 0.0159 \text{ mol H}_2\text{O}$$

In part b, we calculated the formula of the oxide, U_nO_m , to be U_3O_8. Using this formula we can calculate the amount of uranium in 0.742 g of the oxide.

$$0.742 \text{ g U}_3\text{O}_8 \cdot \frac{1 \text{ mol U}_3\text{O}_8}{842.1 \text{ g U}_3\text{O}_8} \cdot \frac{3 \text{ mol U}}{1 \text{ mol U}_3\text{O}_8} \cdot \frac{238.0 \text{ g U}}{1 \text{ mol U}} = 0.629 \text{ g U.}$$

Moles of U present are: $0.629 \text{ g U} \cdot \dfrac{1 \text{ mol U}}{238.0 \text{ g U}} = 2.64 \times 10^{-3} \text{ mol U}$

Comparing the moles of U and of H_2O :

$$\frac{1.59 \times 10^{-2} \text{ mol H}_2\text{O}}{2.64 \times 10^{-3} \text{ mol U}} = \frac{6 \text{ mol H}_2\text{O}}{1 \text{ mol U}} \quad \text{so the value for } \mathbf{z = 6.00} .$$

Noting the formula $UO_x(NO_3)_y$ contains one U atom tells us that the number of moles of the compound is also 2.64×10^{-3} mole. This number of moles has a mass of 1.328 g. From this we can calculate the molar mass of this compound:

$$2.64 \times 10^{-3} \text{ mol} = \frac{1.328 \text{ g}}{\text{Molar Mass}} \quad \text{or} \quad \text{Molar Mass} = \frac{1.328 \text{ g}}{2.64 \times 10^{-3} \text{ mol}} = 502 \text{ g/mol}$$

Since we now know the molar mass of the compound and some of the pieces of the formula, subtract the known masses of the formula from the molar mass:

$$UO_x(NO_3)_y \cdot 6 \text{ H}_2\text{O} = 502$$

Substituting the known atomic masses into the formula we can write:

$$
\begin{array}{llllll}
1 \text{ U} + & x \text{ O} + & y \text{ N} + & 3y \text{ O} + & 6 \text{ H}_2\text{O} & = 502 \\
1(238) + & x(16.0) + & y(14.0) + & 3y(16.0) + & 6(18.0) & = 502 \\
& x(16.0) + & y(14.0) + & 3y(16.0) & & = 502 - 238 - 6(18)
\end{array}
$$

Grouping the "16.0" terms:

$$16.0(x + 3y) + 14.0 \text{ y} = 156$$

and simplifying : $16.0 \text{ x} + 48.0 \text{ y} + 14.0 \text{ y} = 156$

or $16.0 \text{ x} + 62.0 \text{ y} = 156$

Now **x represents the number of oxygen atoms**, and **y represents the number of N atoms and equivalently, the number of "nitrate" groups.**

Inspection reveals that x = 2 and y = 2 satisfies this equation. Hence the formula for the original hydrated uranium compound is:

$$UO_2(NO_3)_2 \cdot 6 \text{ H}_2\text{O}$$

CHAPTER 5: Reactions in Aqueous Solution

16. a. $CuCl_2$ b. $AgNO_3$ c. KCl, K_2CO_3, $KMnO_4$

18. a. any alkali metal, e.g. $NaC_2H_3O_2$
    b. any cation except alkali, alkaline earths, or NH_4^+ e.g. CdS
    c. any alkali metal hydroxide, e.g. $NaOH$
    d. Ag^+, Hg^{2+}, or Pb^{2+} form insoluble chlorides although $PbCl_2$ is slightly soluble at temperatures above $\sim 30\,°C$.

20.
Compound	Cation	Anion
a. NaI	Na^+	I^-
b. K_2SO_4	K^+	SO_4^{2-}
c. $KHSO_4$	K^+	HSO_4^-
d. $NaCN$	Na^+	CN^-

22.
Compound	Water Soluble	Cation	Anion
a. $BaCl_2$	yes	Ba^{2+}	Cl^-
b. $Cr(NO_3)_2$	yes	Cr^{2+}	NO_3^-
c. $Pb(NO_3)_2$	yes	Pb^{2+}	NO_3^-
d. $BaSO_4$	no		

24. Anions to produce a soluble Cu^{2+} salt: SO_4^{2-}, NO_3^-, $C_2H_3O_2^-$, ClO_3^-, ClO_4^-, halides (F^-, Cl^-, Br^-, I^-),

    Anions to produce an insoluble Cu^{2+} salt: S^{2-}, O^{2-}, $C_2O_4^{2-}$, SO_3^{2-}, OH^-

26. $HNO_3\,(aq) + H_2O\,(l) \rightarrow H_3O^+\,(aq) + NO_3^-\,(aq)$
    alternatively $HNO_3\,(aq) \rightarrow H^+\,(aq) + NO_3^-\,(aq)$

28. a. $Zn\,(s) + 2\,HCl\,(aq) \rightarrow H_2\,(g) + ZnCl_2\,(aq)$

    1. Write species as they exist in aqueous solution.
       $Zn\,(s) + 2H^+\,(aq) + 2Cl^-\,(aq) \rightarrow H_2\,(g) + Zn^{2+}\,(aq) + 2\,Cl^-\,(aq)$

    2. Remove any species which appear <u>in exactly the same form</u> on both sides of the equation. For this reaction: $2\,Cl^-\,(aq)$

3. Examine the remaining species to see if a reduction of <u>every</u> coefficient is possible:

$$Zn \ (s) \ + \ 2 \ H^+ \ (aq) \ \rightarrow \ H_2 \ (g) \ + \ Zn^{2+} \ (aq)$$

b. $Mg(OH)_2 \ (s) + 2 \ HCl \ (aq) \ \rightarrow MgCl_2 \ (aq) + 2 \ H_2O \ (l)$

1. $Mg(OH)_2 \ (s) \ + 2 \ H^+ \ (aq) + 2 \ Cl^- \ (aq) \ \rightarrow \ Mg^{2+} \ (aq) + 2 \ Cl^- \ (aq) + 2 \ H_2O \ (l)$
2. Spectator ions: $2 \ Cl^- \ (aq)$
3. $Mg(OH)_2 \ (s) \ + \ 2 \ H^+ \ (aq) \ \rightarrow \ Mg^{2+} \ (aq) \ + \ 2 \ H_2O \ (l)$

c. $2 \ HNO_3 \ (aq) + CaCO_3 \ (s) \rightarrow Ca(NO_3)_2 \ (aq) + \ H_2O \ (l) \ + CO_2 \ (g)$

1. $2 \ H^+ \ (aq) + \ 2 \ NO_3^- \ (aq) \ + \ CaCO_3 \ (s) \rightarrow$
$$Ca^{2+} \ (aq) \ + \ 2 \ NO_3^- \ (aq) \ + \ H_2O \ (l) + CO_2 \ (g)$$
2. Spectator ions: $2 \ NO_3^- \ (aq)$
3. $2 \ H^+ \ (aq) \ + \ CaCO_3 \ (s) \rightarrow \ Ca^{2+} \ (aq) \ + \ H_2O \ (l) + CO_2 \ (g)$

d. $4 \ HCl \ (aq) \ + \ MnO_2 \ (s) \ \rightarrow \ MnCl_2 \ (aq) \ + \ Cl_2 \ (g) \ + \ 2 \ H_2O \ (l)$

1. $4 \ H^+ \ (aq) \ + \ 4 \ Cl^- \ (aq) \ + \ MnO_2 \ (s) \ \rightarrow$
$$Mn^{2+} \ (aq) \ + \ 2 \ Cl^- \ (aq) + \ Cl_2 \ (g) \ + \ 2 \ H_2O \ (l)$$
2. Spectator ions: $2 \ Cl^- \ (aq)$ <u>Note</u> only $2 \ Cl^- \ (aq)$ are spectators.
3. $4 \ H^+ \ (aq) \ + \ 2 \ Cl^- \ (aq) \ + \ MnO_2 \ (s) \ \rightarrow \ Mn^{2+} \ (aq) \ + \ Cl_2 \ (g) \ + \ 2H_2O \ (l)$

30.   Balance the following equations, and then write the net ionic equation:
a. $Ba(OH)_2 \ (s) + \ 2 \ HNO_3 \ (aq) \ \rightarrow \ Ba(NO_3)_2 \ (aq) \ + \ 2 \ H_2O \ (l)$
$$OH^- \ (aq) + \ H^+ \ (aq) \ \rightarrow \ H_2O(l)$$

b. $BaCl_2 \ (aq) \ + \ Na_2CO_3 \ (aq) \rightarrow BaCO_3 \ (s) \ + \ 2 \ NaCl \ (aq)$
$$Ba^{2+} \ (aq) \ + \ CO_3^{2-} \ (aq) \ \rightarrow \ BaCO_3 \ (s)$$

c. $Na_2S \ (aq) \ + \ Ni(NO_3)_2 \ (aq) \rightarrow \ NiS \ (s) \ + \ 2 \ NaNO_3 \ (aq)$
$$S^{2-} \ (aq) \ + \ Ni^{2+} \ (aq) \ \rightarrow \ NiS \ (s)$$

32.   Acid-base reactions (AB) usually produce a salt and water. Precipitation reactions (PR) always form a salt which is insoluble (usually in water).

A chart of solubility rules may come in handy if you haven't already learned the rules.

a. HCl (aq) + KOH (aq) $\rightarrow$ KCl (aq) + H_2O (l)   AB

b. $AgNO_3$ (aq) + KCl (aq) $\rightarrow$ $AgCl$ (s) + KNO_3 (aq)   PR

c. H_2SO_4 (aq) + 2 $NaOH$ (aq) $\rightarrow$ Na_2SO_4 (aq) + 2 H_2O (l)   AB

34.   Acid-base (AB), Precipitation reactions (PR), or Gas-forming (GF)

a. $SrCO_3$ (s) + 2 HNO_3 (aq) $\rightarrow$ $Sr(NO_3)_2$ (aq) + H_2O (l) + CO_2 (g) GF

b. $BaCl_2$ (aq) + H_2SO_4 (aq) $\rightarrow$ $BaSO_4$ (s) + 2 HCl (aq)   PR

c. $MnCl_2$ (aq) + $(NH_4)_2S$ (aq) $\rightarrow$ 2 NH_4Cl (aq) + MnS (s)   PR

36.   Balanced equations showing the preparation of a salt by an acid-base reaction:

a. $NaOH$ (aq) + HNO_3 (aq) $\rightarrow$ $NaNO_3$ (aq) + H_2O (l)

b. KOH (aq) + HCl (aq) $\rightarrow$ KCl (aq) + H_2O (l)

c. 3 KOH (aq) + H_3PO_4 (aq) $\rightarrow$ K_3PO_4 (aq) + 3 H_2O (l)

d. 2 $CsOH$ (aq) + H_2SO_4 (aq) $\rightarrow$ Cs_2SO_4 (aq) + 2 H_2O (l)

38.   Balanced equations showing the preparation of a salt by a precipitation reaction:

a. $Ni(NO_3)_2$ (aq) + 2 $NaOH$ (aq) $\rightarrow$ $Ni(OH)_2$ (s) + 2 $NaNO_3$ (aq)

b. $Ca(NO_3)_2$ (aq) + Na_2CO_3 (aq) $\rightarrow$ $CaCO_3$ (s) + 2 $NaNO_3$ (aq)

c. $NiCl_2$ (aq) + Na_2S (aq) $\rightarrow$ NiS (s) + 2 $NaCl$ (aq)

d. $BaCl_2$ (aq) + K_2SO_4 (aq) $\rightarrow$ $BaSO_4$ (s) + 2 KCl (aq)

40.   Balanced equations showing the formation of a salt by a gas-forming reaction:

a. K_2CO_3 (s) $+$ 2 HNO_3 (aq) $\rightarrow$ 2 KNO_3 (aq) $+$ CO_2 (g) $+$ H_2O (l)

b. $CaCO_3$ (s) $+$ 2 HCl (aq) $\rightarrow$ $CaCl_2$ (aq) $+$ CO_2 (g) $+$ H_2O (l)

c. $FeCO_3$ (s) $+$ 2 HNO_3 (aq) $\rightarrow$ $Fe(NO_3)_2$ (aq) $+$ CO_2 (g) $+$ H_2O (l)

42.   Balanced equations showing the formation of compounds by exchange reactions:

a. $NaOH$ (aq) $+$ HNO_3 (aq) $\rightarrow$ $NaNO_3$ (aq) $+$ H_2O (l)

b. $SrCl_2$ (aq) $+$ H_2SO_4 (aq) $\rightarrow$ $SrSO_4$ (s) $+$ 2 HCl (aq)

c. $ZnCl_2$ (aq) $+$ Na_2S (aq) $\rightarrow$ ZnS (s) $+$ 2 $NaCl$ (aq)

d. $Ba(OH)_2$ (aq) $+$ 2 HCl (aq) $\rightarrow$ $BaCl_2$ (aq) $+$ 2 H_2O (l)

e. 2 NH_3 (aq) $+$ H_2SO_4 (aq) $\rightarrow$ $(NH_4)_2SO_4$ (aq)

44.   Calculate the molarity by determining the number of moles, and expressing the volume in units of liters:

$$\frac{2.60 \text{ g NaBr}}{160. \text{ mL solution}} \cdot \frac{1000 \text{ mL solution}}{1 \text{ L solution}} \cdot \frac{1 \text{ mol NaBr}}{103.0 \text{ g NaBr}} = 0.158 \text{ molar NaBr}$$

The volume of 0.120 M NaBr which contains 2.60 g NaBr:

$$\frac{2.60 \text{ g NaBr}}{1} \cdot \frac{1 \text{ mol NaBr}}{103.0 \text{ g NaBr}} \cdot \frac{1000 \text{ mL solution}}{0.120 \text{ mol NaBr}} = 211 \text{ mL solution}$$

46.   The mass of solute needed to prepare a given volume of solution:

a. $$\frac{0.10 \text{ mol AgNO}_3}{1 \text{ L}} \cdot \frac{0.10 \text{ L}}{1} \cdot \frac{170. \text{ g AgNO}_3}{1 \text{ mol AgNO}_3} = 1.7 \text{ g AgNO}_3$$

b. $$\frac{0.05 \text{ mol NaCN}}{1 \text{ L}} \cdot \frac{5.0 \text{ mL}}{1} \cdot \frac{1 \text{ L}}{1000 \text{ mL}} \cdot \frac{49.0 \text{ g NaCN}}{1 \text{ mol NaCN}} = 1 \times 10^{-2} \text{ g NaCN}$$

c. $$\frac{0.10 \text{ mol BaCl}_2}{1 \text{ L}} \cdot \frac{0.10 \text{ L}}{1} \cdot \frac{208 \text{ g BaCl}_2}{1 \text{ mol BaCl}_2} = 2.1 \text{ g BaCl}_2$$

60

48.   The molar concentration of solute in each of the following solutions:

a. $\dfrac{5.6 \text{ g NaClO}_4}{0.50 \text{ L}}$ $\cdot$ $\dfrac{1 \text{ mol NaClO}_4}{122 \text{ g NaClO}_4}$ $=$ 0.092 M NaClO$_4$

b. $\dfrac{2.3 \text{ g KNO}_3}{0.10 \text{ L}}$ $\cdot$ $\dfrac{1 \text{ mol KNO}_3}{101 \text{ g KNO}_3}$ $=$ 0.23 M KNO$_3$

c. $\dfrac{1.5 \text{ g C}_4\text{H}_8\text{O}}{0.25 \text{ L}}$ $\cdot$ $\dfrac{1 \text{ mol C}_4\text{H}_8\text{O}}{72.1 \text{ g C}_4\text{H}_8\text{O}}$ $=$ 0.083 M C$_4$H$_8$O

50.   For soluble ionic compounds, placement in water results in the separation of the ion pairs which form the solid. The concentrations are determined easily by examination of the stoichiometry of the salt and the concentration of the "formula unit". These represent the ideal or maximum ion concentrations.

a. 1.0 M NaBr $\rightarrow$ 1.0 M Na$^+$ and 1.0 M Br$^-$

b. 0.50M Na$_3$PO$_4$ $\rightarrow$ 1.5M Na$^+$ and 0.50 M PO$_4^{3-}$

c. 0.10M ZnCl$_2$ $\rightarrow$ 0.10M Zn^{2+} and 0.20 M Cl$^-$

52.   To prepare 100. mL of 0.15 M CuSO$_4$ you need:

$\dfrac{0.15 \text{ mol CuSO}_4}{L}$ $\cdot$ 0.100 L $=$ 0.015 mol CuSO$_4$

To get this amount of CuSO$_4$ from a 0.50 M CuSO$_4$ solution you need:

    [Remember   # moles = Molarity $\cdot$ V]

0.015 mol CuSO$_4$ $=\dfrac{0.50 \text{ mol CuSO}_4}{L}$ $\cdot$ V

0.015 mol CuSO$_4$ $\cdot$ $\dfrac{L}{0.50 \text{ mol CuSO}_4}$ $=$ V

0.030 L or 30. mL $=$ V

54.   Dilution of a solution of known concentration with water affects the volume of the solution, but not the number of moles of solute, so we can write:

$$M_c \cdot V_c \;=\; M_d \cdot V_d$$

where c and d represent concentrated and dilute solutions, respectively.

Substituting our data yields:

$$0.300 \text{ M HCl} \cdot 0.0500 \text{ L} = M_d \cdot 0.300 \text{ L}$$

$$\frac{0.300 \text{ M HCl} \cdot 0.0500 \text{ L}}{0.300 \text{ L}} = M_d$$

$$0.0500 \text{ M HCl} = M_d$$

56.   The volume is changed from 2.00 mL to 10.00 mL (a factor of 5). The resulting solution will have a concentration which is one-fifth the original:

$$M_c \cdot V_c = M_d \cdot V_d$$
$$0.15 \text{ M} \cdot 2.00 \text{ mL} = M_d \cdot 10.00 \text{ mL}$$
$$0.030 = M_d$$

58.   To prepare 2.50 L of a 0.0200 M $KMnO_4$ solution using 0.500 M $KMnO_4$ requires:

$$\frac{0.0200 \text{ mol } KMnO_4}{L} \cdot 2.50 \text{ L} = 0.0500 \text{ mol } KMnO_4$$

In a 0.500 M $KMnO_4$ solution this amount of $KMnO_4$ is found in:

$$0.0500 \text{ mol } KMnO_4 = \frac{0.500 \text{ mol } KMnO_4}{L} \cdot V$$

$$0.0500 \text{ mol } KMnO_4 \cdot \frac{L}{0.500 \text{ mol } KMnO_4} = V$$

$$0.100 \text{ L} = V$$

60. $2 \text{ NaCl (aq)} + 2 \text{ H}_2\text{O} \xrightarrow{\text{electricity}} 2 \text{ NaOH (aq)} + \text{H}_2\text{(g)} + \text{Cl}_2 \text{ (g)}$

$$\frac{0.15 \text{ mol NaCl}}{1 \text{ L}} \cdot \frac{1.0 \text{ L}}{1} \cdot \frac{1 \text{ mol } H_2}{2 \text{ mol NaCl}} \cdot \frac{2.02 \text{ g } H_2}{1 \text{ mol } H_2} = 0.15 \text{ g } H_2$$

$$\frac{0.15 \text{ mol NaCl}}{1 \text{ L}} \cdot \frac{1.0 \text{ L}}{1} \cdot \frac{2 \text{ mol NaOH}}{2 \text{ mol NaCl}} \cdot \frac{40.0 \text{ g NaOH}}{1 \text{ mol NaOH}} = 6.0 \text{ g NaOH}$$

$$\frac{0.15 \text{ mol NaCl}}{1 \text{ L}} \cdot \frac{1.0 \text{ L}}{1} \cdot \frac{1 \text{ mol } Cl_2}{2 \text{ mol NaCl}} \cdot \frac{70.9 \text{ g } Cl_2}{1 \text{ mol } Cl_2} = 5.3 \text{ g } Cl_2$$

62.   The preparation of B_2H_6 may be written as:

$$2\ NaBH_4\ (aq)\ +\ H_2SO_4\ (aq)\ \rightarrow\ 2\ H_2\ (g)\ +\ Na_2SO_4\ (aq)\ +\ B_2H_6\ (g)$$

To prepare 4.14 g of B_2H_6 we need:

$$4.14\ g\ B_2H_6\ \cdot\ \frac{1\ mol\ B_2H_6}{27.67\ g\ B_2H_6}\ \cdot\ \frac{2\ mol\ NaBH_4}{1\ mol\ B_2H_6}\ \cdot\ \frac{37.83\ g\ NaBH_4}{1\ mol\ NaBH_4}\ =\ 11.3\ g\ NaBH_4$$

The volume of 0.875 M H_2SO_4 required is:

$$4.14\ g\ B_2H_6\ \cdot\ \frac{1\ mol\ B_2H_6}{27.67\ g\ B_2H_6}\ \cdot\ \frac{1\ mol\ H_2SO_4}{1\ mol\ B_2H_6}\ \cdot\ \frac{1\ L}{0.875\ mol\ H_2SO_4}\ =\ 0.171\ L$$

or 171 mL.

64.   Calculate the moles of AgBr:

$$0.125\ g\ AgBr\ \cdot\ \frac{1\ mol\ AgBr}{187.8\ g\ AgBr}\ \cdot\ \frac{2\ mol\ Na_2S_2O_3}{1\ mol\ AgBr}\ \cdot\ \frac{1\ L}{0.0256\ mol\ Na_2S_2O_3}$$

$$\cdot\ \frac{1000\ mL}{1\ L}\ =\ 52.0\ mL\ Na_2S_2O_3$$

66.   Calculate:
1. Amount of ammonia present

2. Moles (and mass) of AgCl that react with that amount of ammonia

$$0.0050\ L\ \cdot\ \frac{4.5\ mol\ NH_3}{1\ L}\ \cdot\ \frac{1\ mol\ AgCl}{2\ mol\ NH_3}\ \cdot\ \frac{143\ g\ AgCl}{1\ mol\ AgCl}\ =\ 1.6\ g\ AgCl$$

The calculation indicates that there is ample NH_3 to dissolve 1.274 g AgCl.

68.   To calculate the volume of HCl needed, we calculate the moles of NaOH in 2.50 g, then use the stoichiometry of the balanced equation:

$$HCl\ (aq)\ +\ NaOH\ (aq)\ \rightarrow\ NaCl\ (aq)\ +\ H_2O\ (l)$$

$$2.50\ g\ NaOH\ \cdot\ \frac{1\ mol\ NaOH}{40.00\ g\ NaOH}\ \cdot\ \frac{1\ mol\ HCl}{1\ mol\ NaOH}\ \cdot\ \frac{1\ L}{0.250\ mol\ HCl}\ \cdot\ \frac{1000\ mL}{1\ L}$$

$$=\ 250.\ mL\ HCl$$

70. Calculate:

1. moles of KOH corresponding to 1.56 g KOH

2. moles of H_2SO_4 that react with that number of moles (using the balanced equation)

3. the mass of H_2SO_4 corresponding to that number of moles

The balanced equation is:

$$2\ KOH\ (aq) + H_2SO_4\ (aq) \rightarrow K_2SO_4\ (aq) + 2\ H_2O\ (l)$$

$$1.56\ g\ KOH \cdot \frac{1\ mol\ KOH}{56.11\ g\ KOH} \cdot \frac{1\ mol\ H_2SO_4}{2\ mol\ KOH} \cdot \frac{1\ L}{0.0540\ mol\ H_2SO_4} \cdot \frac{1000\ mL}{1\ L}$$

$$= 258\ mL\ H_2SO_4$$

72. Calculate:

1. moles of Na_2CO_3 corresponding to 0.251 g Na_2CO_3

2. moles of HCl that react with that number of moles (using the balanced equation)

3. the molarity of HCl, knowing that 42.43 mL of the solution contained that number of moles

The balanced equation is:

$$Na_2CO_3\ (aq) + 2\ HCl\ (aq) \rightarrow 2\ NaCl\ (aq) + H_2O\ (l) + CO_2\ (g)$$

$$0.251\ g\ Na_2CO_3 \cdot \frac{1\ mol\ Na_2CO_3}{106.0\ g\ Na_2CO_3} \cdot \frac{2\ mol\ HCl}{1\ mol\ Na_2CO_3} \cdot \frac{1}{0.04243\ L}$$

$$= 0.112\ M\ HCl$$

74. Calculate:

1. moles of NaOH contained in 25.67 mL of the solution

2. moles of $HC_2H_3O_2$ which react with that amount of NaOH(balanced equation)

3. mass of $HC_2H_3O_2$ corresponding to that number of moles

The balanced equation is:

$$HC_2H_3O_2\ (aq) + NaOH\ (aq) \rightarrow NaC_2H_3O_2\ (aq) + H_2O\ (l)$$

$$25.67\ mL \cdot \frac{1\ L}{1000\ mL} \cdot \frac{0.674\ mol\ NaOH}{1\ L} \cdot \frac{1\ mol\ HC_2H_3O_2}{1\ mol\ NaOH}$$

$$\cdot \frac{60.05\ g\ HC_2H_3O_2}{1\ mol\ HC_2H_3O_2} = 1.04\ g\ HC_2H_3O_2$$

76.   Calculate:

1. moles of NaOH corresponding to 32.58 mL of 0.541 M NaOH

2. moles of oxalic acid that react with that number of moles (balanced equation)

3. mass of oxalic acid corresponding to that number of moles of oxalic acid

The balanced equation is:

$$H_2C_2O_4 \text{ (aq) } + 2 \text{ NaOH (aq)} \rightarrow Na_2C_2O_4 \text{ (aq) } + 2 H_2O \text{ (l)}$$

$$0.03258 \text{ L} \cdot \frac{0.541 \text{ mol NaOH}}{L} \cdot \frac{1 \text{ mol } H_2C_2O_4}{2 \text{ mol NaOH}} \cdot \frac{90.04 \text{ g } H_2C_2O_4}{1 \text{ mol } H_2C_2O_4}$$

$$= 0.794 \text{ g } H_2C_2O_4$$

Weight percent of oxalic acid $= \dfrac{0.794 \text{ g } H_2C_2O_4}{2.036 \text{ g mixture}}$ x 100 = 39.0 %

78.   To calculate the amount of barium in the ore:

$$\frac{8.23 \text{ g BaSO}_4}{20.0 \text{ g ore}} \cdot \frac{137.3 \text{ g Ba}}{233.4 \text{ g BaSO}_4} = \frac{0.242 \text{ g Ba}}{1 \text{ g ore}} \text{ or } 24.2\% \text{ Barium}$$

80.   Calculate:

1. moles of complex

2. using the stoichiometry of the complex, the number of moles (and mass) of aluminum

$$0.264 \text{ g Al}(C_9H_6NO)_3 \cdot \frac{1 \text{ mol Al}(C_9H_6NO)_3}{459.4 \text{ g Al}(C_9H_6NO)_3} \cdot \frac{1 \text{ mol Al}}{1 \text{ mol Al}(C_9H_6NO)_3}$$

$$\cdot \frac{26.98 \text{ g Al}}{1 \text{ mol Al}} = 0.0155 \text{ g Al}$$

Using the moles of Al calculated above, and the volume of the solution (100.0 mL), the molarity is:

$$\frac{5.75 \times 10^{-4} \text{ mol Al}}{0.1000 \text{ L}} = 5.75 \times 10^{-3} \text{ M Al}$$

82.   For questions on oxidation number, read the symbol (x) as "the oxidation number of x."

        a. BrO⁻   (Br) + (O) = -1

Since oxygen almost always has an oxidation number of -2, we can substitute this value and solve for the oxidation number of Br.

$$(Br) + (-2) = -1$$
$$(Br) = +1$$

b. $C_2O_4^{2-}$

$$2\,(C) + 4\,(O) = -2$$
$$2\,(C) + 4\,(-2) = -2$$
$$2\,(C) + -8 = -2$$
$$2\,(C) = +6$$
$$(C) = +3$$

c. I_2 The oxidation number for any free element is zero.

d. IO_3^-

$$(I) + 3\,(O) = -1$$
$$(I) + 3\,(-2) = -1$$
$$(I) = +5$$

e. $HClO_4$

$$(H) + (Cl) + 4(O) = 0$$
$$(+1) + (Cl) + 4(-2) = 0$$
$$(Cl) = +7$$

f. SO_4^{2-}

$$(S) + 4(O) = -2$$
$$(S) + 4(-2) = -2$$
$$(S) = +6$$

84.   Determine which reactant is oxidized and which is reduced:

a. $2\,Mg\,(s) + O_2\,(g) \rightarrow 2\,MgO\,(s)$

specie	ox. number before	after	has experienced
Mg	0	+2	oxidation
O	0	-2	reduction

b. $C_2H_4\,(g) + 3\,O_2\,(g) \rightarrow 2\,CO_2\,(g) + 2\,H_2O\,(g)$

specie	ox. number before	after	has experienced
C	-2	+4	oxidation
H	+1	+1	no change
O	0	-2	reduction

c. $Si\ (s)\ +\ 2\ Cl_2\ (g)\ \rightarrow\ SiCl_4\ (l)$

specie	ox. number before	after	has experienced
Si	0	+4	oxidation
Cl	0	-1	reduction

86. Balance the following:

	reactant is	overall process is
a. $Cr\ (s)\ \rightarrow\ Cr^{3+}\ (aq)\ +\ 3\ e^-$	reducing agent	oxidation
b. $Fe^{3+}\ (aq)\ +\ 1\ e^-\ \rightarrow\ Fe^{2+}\ (aq)$	oxidizing agent	reduction
c. $AsH_3\ (g)\ \rightarrow\ As\ (s)\ +\ 3\ H^+\ (aq)\ +\ 3\ e^-$	reducing agent	oxidation
d. $VO_3^-\ (aq)\ +\ 6\ H^+\ (aq)\ +\ 3\ e^-\ \rightarrow$ $V^{2+}\ (aq)\ +\ 3\ H_2O\ (l)$	oxidizing agent	reduction

Note: e^- are used to balance charge; H^+ balances only H atoms; H_2O balances both H and O atoms.

88. Balance the following (in acidic solution)

	reactant is	overall process is
a. $Cr_2O_7{}^{2-}\ (aq)\ +\ 14\ H^+\ (aq)\ +\ 6\ e^-$ $\rightarrow\ 2\ Cr^{3+}\ (aq)\ +\ 7\ H_2O\ (l)$	oxidizing agent	reduction
b. $N_2H_5^+\ (aq)\ \rightarrow\ N_2\ (g)\ +\ 5\ H^+\ (aq)\ +\ 4\ e^-$	reducing agent	oxidation
c. $CH_3CHO\ (aq)\ +\ H_2O\ (l)\ \rightarrow$ $CH_3COOH\ (aq)\ +\ 2\ H^+\ (aq)\ +\ 2\ e^-$	reducing agent	oxidation
d. $Bi^{3+}\ (aq)\ +\ 3\ H_2O\ (l)\ \rightarrow$ $HBiO_3\ (aq)\ +\ 5\ H^+\ (aq)\ +\ 2\ e^-$	reducing agent	oxidation

90. Balance the following (in basic solution)

	reactant is	overall process is
a. $Sn\ (s)\ +\ 4\ OH^-\ (aq) \rightarrow$ $Sn(OH)_4{}^{2-}\ (aq)\ +\ 2\ e^-$	reducing agent	oxidation
b. $MnO_4^-\ (aq)\ +\ 2\ H_2O\ (l)\ +\ 3\ e^-\ \rightarrow$ $MnO_2\ (s)\ +\ 4\ OH^-\ (aq)$	oxidizing agent	reduction
c. $ClO^-\ (aq)\ +\ H_2O\ (l)\ +\ 2\ e^-\ \rightarrow$ $Cl^-\ (aq)\ +\ 2\ OH^-\ (aq)$	oxidizing agent	reduction

92.   Balancing redox equations in neutral or acidic solutions may be accomplished in several steps. They are:

1. Separating the equation into two equations which represent reduction and oxidation
2. Balancing mass of elements (other than H or O)
3. Balancing mass of O by adding H_2O
4. Balancing mass of H by adding H^+
5. Balancing charge by adding electrons
6. Balancing electron gain (in the reduction half-equation) with electron loss (in the oxidation half-equation)
7. Combining the two half equations

For the parts (a-d) of this problem, each step will be identified with a number corresponding to the list above. In addition, the physical states of all species will be omitted in all but the final step. While this omission is <u>not generally recommended</u>, it should increase the clarity of the steps involved. In addition when a step leaves a half equation unchanged from the previous step, we have omitted the half equation.

a. $Cl_2 (aq) + Br^- (aq) \rightarrow Br_2 (aq) + Cl^- (aq)$

1.	$Cl_2 \rightarrow Cl^-$	$Br^- \rightarrow Br_2$
2.	$Cl_2 \rightarrow 2 Cl^-$	$2 Br^- \rightarrow Br_2$
5.	$Cl_2 + 2e^- \rightarrow 2 Cl^-$	$2 Br^- \rightarrow Br_2 + 2e^-$
7.	$Cl_2 (aq) + 2 Br^- (aq) \rightarrow 2 Cl^- (aq) + Br_2 (aq)$	

b. $Sn (s) + H_3O^+ (aq) \rightarrow Sn^{2+} (aq) + H_2 (g)$

1.	$Sn \rightarrow Sn^{2+}$	$H_3O^+ \rightarrow H_2$
3.		$H_3O^+ \rightarrow H_2 + H_2O$
4.		$2 H_3O^+ \rightarrow H_2 + 2 H_2O$
5.	$Sn \rightarrow Sn^{2+} + 2e^-$	$2 H_3O^+ + 2e^- \rightarrow H_2 + 2 H_2O$
7.	$Sn (s) + 2 H_3O^+ (aq) \rightarrow Sn^{2+} (aq) + H_2 (g) + 2 H_2O (l)$	

c. $Al\ (s) + Sn^{4+}\ (aq) \rightarrow Al^{3+}\ (aq) + Sn^{2+}\ (aq)$

1.	$Al \rightarrow Al^{3+}$	$Sn^{4+} \rightarrow Sn^{2+}$
5.	$Al \rightarrow Al^{3+} + 3\ e^-$	$Sn^{4+} + 2\ e^- \rightarrow Sn^{2+}$
6.	$2\ Al \rightarrow 2\ Al^{3+} + 6\ e^-$	$3\ Sn^{4+} + 6\ e^- \rightarrow 3\ Sn^{2+}$

7. $2\ Al\ (s) + 3\ Sn^{4+}\ (aq) \rightarrow 2\ Al^{3+}\ (aq) + 3\ Sn^{2+}\ (aq)$

d. $Zn\ (s) + VO^{2+}\ (aq) \rightarrow Zn^{2+}\ (aq) + V^{3+}\ (aq)$

1.	$Zn \rightarrow Zn^{2+}$	$VO^{2+} \rightarrow V^{3+}$
3.		$VO^{2+} \rightarrow V^{3+} + H_2O$
4.		$VO^{2+} + 2H^+ \rightarrow V^{3+} + H_2O$
5.	$Zn \rightarrow Zn^{2+} + 2\ e^-$	$VO^{2+} + 2\ H^+ + e^- \rightarrow V^{3+} + 2\ H_2O$
6.		$2\ VO^{2+} + 4\ H^+ + 2\ e^- \rightarrow 2\ V^{3+} + 2\ H_2O$

7. $Zn\ (s) + 2\ VO^{2+}\ (aq) + 4\ H^+\ (aq) \rightarrow Zn^{2+}\ (aq) + 2\ V^{3+}\ (aq) + 2\ H_2O\ (l)$

94.   See problem 92 for explanation of step numbers.

a. $Ag+\ (aq) + HCHO\ (aq) \rightarrow Ag\ (s) + HCOOH\ (aq)$

1.	$Ag^+ \rightarrow Ag$	$HCHO \rightarrow HCOOH$
3.		$H_2O + HCHO \rightarrow HCOOH$
4.		$H_2O + HCHO \rightarrow HCOOH + 2\ H^+$
5.	$Ag^+ + 1e^- \rightarrow Ag$	$H_2O + HCHO \rightarrow HCOOH + 2\ H^+ + 2\ e^-$
6.	$2\ Ag^+ + 2e^- \rightarrow 2\ Ag$	

7. $2\ Ag^+\ (aq) + H_2O\ (l) + HCHO\ (aq) \rightarrow 2\ Ag\ (s) + HCOOH\ (aq) + 2\ H^+\ (aq)$

b. $MnO_4^-\ (aq) + C_2H_5OH\ (aq) \rightarrow Mn^{2+}\ (aq) + C_2H_4O\ (aq)$

1.	$MnO_4^- \rightarrow Mn^{2+}$	$C_2H_5OH \rightarrow C_2H_4O$
3.	$MnO_4^- \rightarrow Mn^{2+} + 4\ H_2O$	
4.	$MnO_4^- + 8\ H^+ \rightarrow Mn^{2+} + 4\ H_2O$	$C_2H_5OH \rightarrow C_2H_4O + 2\ H^+$
5.	$MnO_4^- + 8\ H^+ + 5\ e^- \rightarrow Mn^{2+} + 4\ H_2O$	$C_2H_5OH \rightarrow C_2H_4O + 2\ H^+ + 2e^-$
6.	$2\ MnO_4^- + 16\ H^+ + 10e^- \rightarrow 2\ Mn^{2+} + 8\ H_2O$	$5\ C_2H_5OH \rightarrow 5\ C_2H_4O + 10\ H^+ + 10\ e^-$

7. $2\ MnO_4^-\ (aq) + 6\ H^+\ (aq) + 5\ C_2H_5OH\ (aq) \rightarrow 2\ Mn^{2+}\ (aq) + 8\ H_2O\ (l) + 5\ C_2H_4O\ (aq)$

Note the removal of 10 H^+ (aq) from both sides of the equation.

69

c. H_2S (aq) + $Cr_2O_7^{2-}$ (aq) →   S (s) + Cr^{3+} (aq)

1.	H_2S → S	$Cr_2O_7^{2-}$ → Cr^{3+}
2.		$Cr_2O_7^{2-}$ → 2 Cr^{3+}
3.		$Cr_2O_7^{2-}$ → 2 Cr^{3+} + 7 H_2O
4	H_2S → S + 2 H^+	14 H^+ + $Cr_2O_7^{2-}$ → 2 Cr^{3+} + 7 H_2O
5	H_2S → S + 2 H^+ + 2e^-	14 H^+ + $Cr_2O_7^{2-}$ + 6 e^- → 2 Cr^{3+} + 7 H_2O

6.   3 H_2S → 3 S + 6 H^+ + 6e^-

7   3 H_2S (aq) + 8 H^+ (aq) + $Cr_2O_7^{2-}$ (aq) → 3 S (s) + 2 Cr^{3+} (aq) + 7 H_2O (l)

Note the removal of 6 H^+ (aq) from both sides of the equation.

d.   Zn (s) + VO_3^- (aq) → V^{2+} (aq) + Zn^{2+} (aq)

1.	Zn → Zn^{2+}	VO_3^- → V^{2+}
3.		VO_3^- → V^{2+} + 3 H_2O
4.		6 H^+ + VO_3^- → V^{2+} + 3 H_2O
5.	Zn → Zn^{2+} + 2 e^-	6 H^+ + VO_3^- + 3 e^- → V^{2+} + 3 H_2O
6.	3 Zn → 3 Zn^{2+} + 6 e^-	12 H^+ + 2 VO_3^- + 6 e^- → 2 V^{2+} + 6 H_2O

7.   3 Zn (s) + 12 H^+ (aq) + 2 VO_3^- (aq) → 3 Zn^{2+} (aq) + 2 V^{2+} (aq) + 6 H_2O (l)

e. U^{4+} (aq) + MnO_4^- (aq) → Mn^{2+} (aq) + UO_2^+ (aq)

1.	U^{4+} → UO_2^+	MnO_4^- → Mn^{2+}
3.	U^{4+} + 2 H_2O → UO_2^+	MnO_4^- → Mn^{2+} + 4 H_2O
4.	U^{4+} + 2 H_2O → UO_2^+ + 4 H^+	MnO_4^- + 8 H^+ → Mn^{2+} + 4 H_2O
5.	U^{4+} + 2 H_2O → UO_2^+ + 4 H^+ + 1 e^-	MnO_4^- + 8 H^+ + 5 e^- → Mn^{2+} + 4 H_2O

6. 5 U^{4+} + 10 H_2O → 5 UO_2^+ + 20 H^+ + 5 e^-

7. 5 U^{4+} (aq) + 6 H_2O (l) + MnO_4^- (aq) → 5 UO_2^+ (aq) + Mn^{2+} (aq) + 12 H^+ (aq)

96.   One method for balancing redox reactions in basic solution is quite similar to the procedure in question 92 <u>with two exceptions:</u>

3.   Balance mass of O by adding OH^-. Use twice as many OH^- as you need oxygens.

4.   Balance H by adding H_2O.

Once again to clarify the steps, states of matter will be noted only in the final step. When no change is made from one step to the next, the equation will be written only for the first step.

a. $Zn \ (s) + ClO^- \ (aq) \rightarrow Zn(OH)_2 \ (s) + Cl^- \ (aq)$

1.	$Zn \rightarrow Zn(OH)_2$	$ClO^- \rightarrow Cl^-$
3.	$Zn + 2\,OH^- \rightarrow Zn(OH)_2$	$ClO^- \rightarrow Cl^- + 2\,OH^-$
4.		$H_2O + ClO^- \rightarrow Cl^- + 2\,OH^-$
5.	$Zn + 2\,OH^- \rightarrow Zn(OH)_2 + 2\,e^-$	$2\,e^- + H_2O + ClO^- \rightarrow Cl^- + 2\,OH^-$
7.	$Zn \ (s) + H_2O \ (l) + ClO^- \ (aq) \rightarrow Zn(OH)_2 \ (s) + Cl^- \ (aq)$	

b. $ClO^- \ (aq) + CrO_2^- \ (aq) \rightarrow Cl^- \ (aq) + CrO_4^{2-} \ (aq)$

1.	$ClO^- \rightarrow Cl^-$	$CrO_2^- \rightarrow CrO_4^{2-}$
3.	$ClO^- \rightarrow Cl^- + 2\,OH^-$	$4\,OH^- + CrO_2^- \rightarrow CrO_4^{2-}$
4.	$H_2O + ClO^- \rightarrow Cl^- + 2\,OH^-$	$4\,OH^- + CrO_2^- \rightarrow CrO_4^{2-} + 2\,H_2O$
5.	$H_2O + ClO^- + 2\,e^- \rightarrow Cl^- + 2\,OH^-$	$4\,OH^- + CrO_2^- \rightarrow CrO_4^{2-} + 2\,H_2O + 3\,e^-$
6.	$3\,H_2O + 3\,ClO^- + 6\,e^- \rightarrow 3\,Cl^- + 6\,OH^-$	$8\,OH^- + 2\,CrO_2^- \rightarrow 2\,CrO_4^{2-} + 4\,H_2O + 6\,e^-$
7.	$3\,ClO^- \ (aq) + 2\,OH^- \ (aq) + 2\,CrO_2^- \ (aq) \rightarrow 3\,Cl^- \ (aq) + 2\,CrO_4^{2-} \ (aq) + H_2O \ (l)$	

c. $Br_2 \ (aq) \rightarrow Br^- \ (aq) + BrO_3^- \ (aq)$

1.	$Br_2 \rightarrow Br^-$	$Br_2 \rightarrow BrO_3^-$
2	$Br_2 \rightarrow 2\,Br^-$	$Br_2 \rightarrow 2\,BrO_3^-$
3.		$12\,OH^- + Br_2 \rightarrow 2\,BrO_3^-$
4.		$12\,OH^- + Br_2 \rightarrow 2\,BrO_3^- + 6\,H_2O$
5.	$2e^- + Br_2 \rightarrow 2\,Br^-$	$12\,OH^- + Br_2 \rightarrow 2\,BrO_3^- + 6\,H_2O + 10\,e^-$
6.	$10\,e^- + 5\,Br_2 \rightarrow 10\,Br^-$	
7.	$6\,Br_2 \ (aq) + 12\,OH^- \ (aq) \rightarrow 10\,Br^- \ (aq) + 2\,BrO_3^- \ (aq) + 6\,H_2O \ (l)$	

Note that all coefficients in Step 7 are divisible by two. The overall balanced equation is then

$$3\,Br_2 \ (aq) + 6\,OH^- \ (aq) \rightarrow 5\,Br^- \ (aq) + BrO_3^- \ (aq) + 3\,H_2O \ (l)$$

98. MnO_4^- (aq) + 8 H$^+$ (aq) + 5 Fe^{2+} (aq) $\rightarrow$ Mn^{2+} (aq) + 5 Fe^{3+} (aq) + 4 H$_2$O (l)

$$1.050 \text{ g Fe(NH}_4)_2(SO_4)_2 \cdot 6 \text{ H}_2O \ \cdot \ \frac{1 \text{mol Fe(NH}_4)_2(SO_4)_2 \cdot 6 \text{ H}_2O}{392.13 \text{ g Fe(NH}_4)_2(SO_4)_2 \cdot 6 \text{ H}_2O}$$

$$\cdot \ \frac{1 \text{ mol Fe}^{2+}}{1 \text{ mol Fe(NH}_4)_2(SO_4)_2 \cdot 6 \text{ H}_2O} \ \cdot \ \frac{1 \text{ mol KMnO}_4}{5 \text{ mol Fe}^{2+}} \ \cdot \ \frac{1}{0.03245 \text{ L}}$$

$$= 0.01650 \text{ molar KMnO}_4$$

100.   2 MnO_4^- (aq) + 5 H$_2$C$_2$O$_4$ (aq) + 6 H$_3$O$^+$ (aq) $\rightarrow$ 2 Mn^{2+} (aq) + 10 CO$_2$ (g) + 14 H$_2$O (l)

$$0.892 \text{ g H}_2C_2O_4 \ \cdot \ \frac{1 \text{ mol H}_2C_2O_4}{90.04 \text{ g H}_2C_2O_4} \ \cdot \ \frac{2 \text{ mol MnO}_4^-}{5 \text{ mol H}_2C_2O_4} \ \cdot \ \frac{1 \text{ mol KMnO}_4}{1 \text{ mol MnO}_4^-}$$

$$\cdot \ \frac{1000 \text{ mL}}{0.0200 \text{ mol KMnO}_4} \ = \ 198 \text{ mL KMnO}_4 \text{ solution}$$

102.   Calculate:

1. moles of Ce^{4+} contained in 42.34 mL of the 0.133 M Ce(SO$_4$)$_2$ solution
2. moles of Fe^{2+} which react with that amount of Ce^{4+} (balanced equation)
3. mass of Fe^{2+} corresponding to that number of moles

The balanced equation is:

Fe^{2+} (aq) + Ce^{4+} (aq) $\rightarrow$ Fe^{3+} (aq) + Ce^{3+} (aq)

$$42.34 \text{ mL} \ \cdot \ \frac{1 \text{ L}}{1000 \text{ mL}} \ \cdot \ \frac{0.133 \text{ mol Ce(SO}_4)_2}{\text{L}} \ \cdot \ \frac{1 \text{ mol Ce}^{4+}}{1 \text{ mol Ce(SO}_4)_2} \ \cdot \ \frac{1 \text{ mol Fe}^{2+}}{1 \text{ mol Ce}^{4+}}$$

$$\cdot \ \frac{55.85 \text{ g Fe}}{1 \text{ mol Fe}} = 0.315 \text{ g Fe}$$

The weight percent of iron is then : $\dfrac{0.315 \text{ g Fe}}{15.45 \text{ g ore}}$ x 100 = 2.04 % Fe

104.   The balanced equation is:

PbO$_2$ (s) + 4 H$^+$ (aq) + 2 I$^-$ (aq) $\rightarrow$ Pb^{2+} (aq) + I$_2$ (aq) + 2 H$_2$O (l)

The reaction of I_2 with $Na_2S_2O_3$ is shown in question 99.

$$\frac{0.0500 \text{ mol } Na_2S_2O_3}{L} \cdot \frac{0.03523 \text{ L}}{1} \cdot \frac{1 \text{ mol } I_2}{2 \text{ mol } S_2O_3{}^{2-}} \cdot \frac{1 \text{ mol } Pb^{2+}}{1 \text{ mol } I_2}$$

$$\cdot \frac{1 \text{ mol Pb}}{1 \text{ mol } Pb^{2+}} \cdot \frac{207.2 \text{ g Pb}}{1 \text{ mol Pb}} = 0.182 \text{ g Pb}$$

Weight percent of lead in the ore $= \dfrac{0.182 \text{ g Pb}}{0.576 \text{ g ore}}$ x $100 = 31.7\%$ Pb

106.   Indicate if the substance is water-soluble, and if water-soluble-tell what ions are produced.

   a. $NiCO_3$ insoluble

   b. $Zn(CH_3COO)_2$ soluble Zn^{2+} and CH_3COO^-

   c. KCN   soluble K^+ and CN^-

   d. $Cu(ClO_4)_2$ soluble Cu^{2+} and ClO_4^-

   e. $K_2Cr_2O_7$ soluble K^+ and $Cr_2O_7{}^{2-}$

   f. FeS   insoluble

108.   Complete and balance the following equations:
   a. $CaCO_3$ (s) + 2 HNO_3 (aq) $\rightarrow$ $Ca(NO_3)_2$ (aq) + H_2O (l) + CO_2 (g)
   b. $SrCO_3$ (s) + heat $\rightarrow$ SrO (s) + CO_2 (g)
   c. K_2CO_3 (aq) + H_2SO_4 (aq) $\rightarrow$ K_2SO_4 (aq) + H_2O (l) + CO_2 (g)
   d. $CuCO_3$ (s) + 2 $HClO_4$ (aq) $\rightarrow$ $Cu(ClO_4)_2$ (aq) + H_2O (l) + CO_2 (g)

110. $\dfrac{3.4 \text{ g } C_{12}H_{22}O_{11}}{250 \text{ mL}} \cdot \dfrac{1 \text{ mol } C_{12}H_{22}O_{11}}{342 \text{g } C_{12}H_{22}O_{11}} \cdot \dfrac{1000 \text{ mL}}{1 \text{ L}} = 0.040 \text{ M } C_{12}H_{22}O_{11}$

112.   250 students $\cdot \dfrac{25 \text{ mL}}{1 \text{ student}} \cdot \dfrac{1 \text{ L}}{1000 \text{ mL}} = 6.3 \text{ L needed}$

To determine the mass of $CuSO_4 \cdot 5\, H_2O$ contained in 6.3 L:

6.3 L $\cdot \dfrac{0.15 \text{ mol } CuSO_4}{1 \text{ L}} \cdot \dfrac{249.7 \text{ g } CuSO_4 \cdot 5\, H_2O}{1 \text{ mol } CuSO_4 \cdot 5\, H_2O} = 230 \text{ g } CuSO_4 \cdot 5\, H_2O$

114. The concentration of Fe^{3+} in the final solution can be calculated by successive applications of the "dilution formula" -- $M_c \cdot V_c = M_d \cdot V_d$

The concentration of Fe^{3+} in the first prepared sample is:

$$0.025 \text{ M } Fe^{3+} \cdot 0.010 \text{ L} = M_d \cdot 0.100 \text{ L}$$

$$\frac{0.025 \text{ M } Fe^{3+} \cdot 0.010 \text{ L}}{0.100 \text{ L}} = M_d$$

$$2.5 \times 10^{-3} \text{ M } Fe^{3+} = M_d$$

Similarly in the second prepared sample:

$$2.5 \times 10^{-3} \text{ M } Fe^{3+} \cdot 0.025 \text{ L} = M_d \cdot 0.100 \text{ L}$$

$$\frac{2.5 \times 10^{-3} \text{ M } Fe^{3+} \cdot 0.025 \text{ L}}{0.100 \text{ L}} = M_d$$

$$6.3 \times 10^{-4} \text{ M } Fe^{3+} = M_d$$

116. $MnO_2 \text{ (s)} + 2 I^- \text{ (aq)} + 4 H_3O^+ \text{ (aq)} \rightarrow I_2 \text{ (aq)} + Mn^{2+} \text{ (aq)} + 2 H_2O \text{ (l)}$

a. Mass of MnO_2 to produce the I_2 which reacts with 25.67 mL of 0.100 M $S_2O_3^{2-}$

$$\frac{0.100 \text{ mol } S_2O_3^{2-}}{1 \text{ L}} \cdot \frac{0.02567 \text{ L}}{1} \cdot \frac{1 \text{ mol } I_2}{2 \text{ mol } S_2O_3^{2-}} \cdot \frac{1 \text{ mol } MnO_2}{1 \text{ mol } I_2} \cdot \frac{86.94 \text{ g } MnO_2}{1 \text{ mol } MnO_2}$$

$$= 0.112 \text{ g } MnO_2$$

b. To prepare 500. ml of 0.100 M $S_2O_3^{2-}$ one needs

$$\frac{0.100 \text{ mol } S_2O_3^{2-}}{1 \text{ L}} \cdot \frac{0.500 \text{ L}}{1} = 0.0500 \text{ mol } S_2O_3^{2-}$$

Our original source of $S_2O_3^{2-}$ is a solution that is 0.432 M in $S_2O_3^{2-}$. The volume of this solution which contains 0.0500 mol $S_2O_3^{2-}$ may be calculated as follows:

$$0.0500 \text{ mol } S_2O_3^{2-} = \frac{0.432 \text{ mol } S_2O_3^{2-}}{1 \text{ L}} \cdot V$$

$$0.0500 \text{ mol } S_2O_3^{2-} \cdot \frac{1 \text{ L}}{0.432 \text{ mol } S_2O_3^{2-}} = 0.116 \text{ L}$$

c. The mass of $Na_2S_2O_3$ required to prepare 500. mL of 0.432 M $Na_2S_2O_3$:

$$\frac{0.432 \text{ mol } Na_2S_2O_3}{1 \text{ L}} \cdot \frac{0.500 \text{ L}}{1} \cdot \frac{158.1 \text{ g } Na_2S_2O_3}{1 \text{ mol } Na_2S_2O_3} = 34.1 \text{ g } Na_2S_2O_3$$

74

118.   To identify the acid, one can calculate the molar mass of the sample by determining the number of moles of acid contained in the 1.021 g sample of the unknown acid. To determine the amount of acid, calculate the moles of NaOH in the 24.32 mL sample:

$$0.206 \text{ mol/L NaOH} \cdot 0.02432 \text{ L} \cdot \frac{1 \text{ mol } H_3O^+}{1 \text{ mol NaOH}} = 0.00501 \text{ mol } H_3O^+$$

The molar mass of this unknown acid is: $\frac{1.021 \text{ g acid}}{0.00501 \text{ mol}} = 204$

Using the atomic weights of each element, calculate the molar masses of the two possible acids: molar masses

$KC_8H_5O_4$,   potassium biphthalate   204

and $KC_4H_5O_6$,   potassium bitartrate   188

The unknown acid must be potassium biphthalate.

120.   Mixture mass: before heating: 1.292 g

                after heating: 1.181 g

                mass of water removed: 0.105 g

a. Weight percentage of water $= \frac{0.105 \text{ g}}{1.292 \text{ g}} \times 100 = 8.13\% \text{ } H_2O$

b. Weight percentage of $BaCl_2 \cdot 2 H_2O$:

        The formula indicates that each mole of hydrated salt contains two moles of water.

From a mass viewpoint:

        1 mol $BaCl_2 \cdot 2 H_2O$:   244.28 g

        1 mol $BaCl_2$:   208.25 g

        Mass of water: 36.03 g

The mass ratio of hydrated salt to water is: $\frac{244.28 \text{ g hydrated salt}}{36.03 \text{ g water}}$

The mass of hydrated salt releasing 0.105 g of water is:

        $0.105 \text{ g water} \cdot \frac{244.28 \text{ g hydrated salt}}{36.03 \text{ g water}} = 0.712 \text{ g hydrated salt}$

and the weight percentage of $BaCl_2 \cdot 2\,H_2O$ is:

$$\frac{0.712 \text{ g hydrated salt}}{1.292 \text{ g mixture}} \times 100 = 55.1\% \ BaCl_2 \cdot 2\,H_2O$$

c. Weight percentage of Ba in the mixture:

    1. Determine mass of anhydrous salt in the mixture:

        1.292 g mixture

        <u>0.712</u> g hydrated salt

        0.580 g anhydrous salt

    2. Determine mass of Ba in each salt:

$$0.580 \text{ g } BaCl_2 \cdot \frac{137.34 \text{ g Ba}}{208.25 \text{ g } BaCl_2} = 0.383 \text{ g Ba}$$

$$0.712 \text{ g } BaCl_2 \cdot 2\,H_2O \cdot \frac{137.34 \text{ g Ba}}{244.28 \text{ g } BaCl_2 \cdot 2\,H_2O} = 0.400 \text{ g Ba}$$

    3. Percentage of Ba:

$$\frac{(0.383 + 0.400) \text{ g Ba}}{1.292 \text{ g mixture}} \times 100 = 60.6\% \text{ Ba}$$

122. The balanced equation is:

$$6\,Fe^{2+} (aq) + 14\,H^+ (aq) + Cr_2O_7^{2-} (aq) \rightarrow 6\,Fe^{3+} (aq) + 2\,Cr^{3+} (aq) + 7\,H_2O (l)$$

$$\frac{0.200 \text{ mol } Fe^{2+}}{L} \cdot 0.0402 \text{ L} \cdot \frac{2 \text{ mol } Cr^{3+}}{6 \text{ mol } Fe^{2+}} \cdot \frac{52.00 \text{ g Cr}}{1 \text{ mol Cr}}$$

$$= 0.139 \text{ g Cr}$$

$$\text{weight percent of chromium} = \frac{0.139 \text{ g Cr}}{1.70 \text{ g ore}} \times 100 = 8.20\% \text{ Cr}$$

124. $5.10 \text{ g coin} \cdot \dfrac{75.0 \text{ g Cu}}{100 \text{ g coin}} \cdot \dfrac{1 \text{ mol Cu}}{63.54 \text{ g Cu}} \cdot \dfrac{1 \text{ mol } I_2}{2 \text{ mol Cu}} \cdot \dfrac{2 \text{ mol } S_2O_3^{2-}}{1 \text{ mol } I_2} \cdot$

$$\frac{1000 \text{ mL}}{0.500 \text{ mol } Na_2S_2O_3} = 120. \text{ mL solution}$$

126. To solve each of the four parts of this problem, first calculate moles of the reactant, platinum.

a. Mass of H_2PtCl_6 obtainable from 12.3 g platinum:

$$12.3 \text{ g Pt} \cdot \frac{1 \text{ mol Pt}}{195.1 \text{ g Pt}} \cdot \frac{3 \text{ mol } H_2PtCl_6}{3 \text{ mol Pt}} \cdot \frac{409.8 \text{ g } H_2PtCl_6}{1 \text{ mol } H_2PtCl_6}$$
$$= 25.8 \text{ g } H_2PtCl_6$$

b. Mass of NO produced from 12.3 g platinum:

$$12.3 \text{ g Pt} \cdot \frac{1 \text{ mol Pt}}{195.1 \text{ g Pt}} \cdot \frac{4 \text{ mol NO}}{3 \text{ mol Pt}} \cdot \frac{30.01 \text{ g NO}}{1 \text{ mol NO}} = 2.52 \text{ g NO}$$

c. Volume (in mL) in 10.0 M HNO_3 to react with 12.3 g Pt:

$$12.3 \text{ g Pt} \cdot \frac{1 \text{ mol Pt}}{195.1 \text{ g Pt}} \cdot \frac{4 \text{ mol } HNO_3}{3 \text{ mol Pt}} \cdot \frac{1 \text{ L}}{10.0 \text{ mol } HNO_3} \cdot \frac{1000 \text{ mL}}{1 \text{ L}}$$
$$= 8.41 \text{ mL } HNO_3$$

d. The limiting reagent, if 10.0 g Pt react with 180. mL of 5.00 M HCl :

$$10.0 \text{ g Pt} \cdot \frac{1 \text{ mol Pt}}{195.1 \text{ g Pt}} = 0.0513 \text{ mol Pt}$$

$$\frac{5.00 \text{ mol HCl}}{L} \cdot 0.180 \text{ L} = 0.900 \text{ mol HCl}$$

Moles-required ratio: $\dfrac{18 \text{ mol HCl}}{3 \text{ mol Pt}} = \dfrac{6 \text{ mol HCl}}{1 \text{ mol Pt}}$

Moles-available ratio: $\dfrac{0.900 \text{ mol HCl}}{0.0513 \text{ mol Pt}} = \dfrac{17.5 \text{ mol HCl}}{1 \text{ mol Pt}}$

Therefore **platinum** is the limiting reagent.

CHAPTER 6: Thermochemistry

9.   Express the following energies in either joules or Calories:

a. 1670 kJ in Calories:

$$1670 \text{ kJ} \cdot \frac{1000 \text{ J}}{1 \text{ kJ}} \cdot \frac{1 \text{ cal}}{4.184 \text{ J}} \cdot \frac{1 \text{ Calorie}}{1000 \text{ cal}} = 399 \text{ Calories}$$

b. 1200 Calories in joules:

$$1200 \text{ Calories} \cdot \frac{1000 \text{ cal}}{1 \text{ Calorie}} \cdot \frac{4.184 \text{ J}}{1 \text{ cal}} = 5.0 \times 10^6 \text{ J}$$

c. 110 Calories in joules:

$$110 \text{ Calories} \cdot \frac{1000 \text{ cal}}{1 \text{ Calorie}} \cdot \frac{4.184 \text{ J}}{1 \text{ cal}} = 4.6 \times 10^5 \text{ J}$$

11.   To melt 1.00 pound of lead:

$$1.00 \text{ lb lead} \cdot \frac{454 \text{ g Pb}}{1.00 \text{ lb Pb}} \cdot \frac{1 \text{ mol Pb}}{207.2 \text{ g Pb}} \cdot \frac{1224 \text{ calories}}{1 \text{ mol Pb}} \cdot \frac{4.184 \text{ J}}{1 \text{ cal}} = 1.12 \times 10^4 \text{ J}$$

13.   Using the specific heat for aluminum (0.902 J/g $\cdot$ K) and the change in temperature (250 °C - 25 °C), calculate the energy added:

$$q = c \cdot m \cdot \Delta T$$

$$= (0.902 \text{ J/g} \cdot \text{K}) \cdot 500. \text{ g Al} \cdot 225 \text{ K} = 101,000 \text{ J or } 101 \text{ kJ}$$

15.   The energy needed to raise the iron ball's temperature is:

$$q = c \cdot m \cdot \Delta T$$

$$= (0.451 \text{ J/g} \cdot \text{K}) \cdot 16 \text{ lb} \cdot \frac{454 \text{ g Fe}}{1 \text{ lb Fe}} \cdot 21 \text{ K} = 6.9 \times 10^4 \text{ joules}$$

Note the specific heat for iron (0.451 J/g $\cdot$ K) and the temperature change (from 16 °C to 37 °C or a ΔT of 21 °C or equivalently 21 K.

17.   To accomplish this process, one must:

1. heat the lead from 25 °C to 327 °C
2. melt the lead at 327 °C

Using the specific heat for lead, the energy for the first step is:

$$(0.159 \text{ J/g} \cdot \text{K}) \cdot 453.6 \text{ g Pb} \cdot 302 \text{ K} = 21,800 \text{ J or } 21.8 \text{ kJ}$$

To melt the lead at 327°C, we need 24.7 J/g :

$$24.7 \text{ J/g} \cdot 453.6 \text{ g Pb} = 11,200 \text{ J or } 11.2 \text{ kJ}$$

The total heat energy required is : $(21.8 + 11.2) = 33.0$ kJ or 3.30×10^4 J

19.   The heat energy added to the water is:

$$(4.18 \text{ J/g} \cdot \text{K}) \cdot 1.00 \times 10^3 \text{g} \cdot (32.8 - 20.0 \text{ °C}) = 53,500 \text{ J}$$

Note that we may use the temperature difference expressed in °C, since the magnitude of the temperature change in °C or K is identical. The energy was supplied by the iron bar, and we calculate the change in temperature, ΔT, for the bar:

$$53,500 \text{ J} = -(0.451 \text{ J/g} \cdot \text{K}) \cdot 400. \text{ g} \cdot \Delta T$$

[Note that the - sign on the right side of the equation is used since $q_{iron} = - q_{water}$.]
Rearranging the previous equation yields:

$$\frac{53500 \text{ J}}{-(0.451 \text{ J/g} \cdot \text{K}) \cdot 400.\text{g}} = \Delta T \quad \text{and solving gives a value for } \Delta T = -297 \text{ K}$$

Or a change of 297 °C.

$$\text{By convention } \Delta T = T_{final} - T_{initial}$$
$$-297 \text{ °C} = T_{final} - T_{initial} \quad \text{and } T_{final} \text{ for the iron bar is 32.8,}$$
$$-297 \text{ °C} = 32.8 - T_{initial}$$

and **$T_{initial}$ is 330. °C**

21.   To calculate the quantity of heat for the process described, think of the problem in three steps:

1.   melt ice at 0°C to liquid water at 0°C
2.   warm liquid water from 0°C to 100°C
3.   convert liquid water at 100 °C to gaseous water at 100 °C

1.   Example 6.5 in your textbook calculates the energy for the first step to be: 2.00×10^4 J
2.   The energy required to warm the liquid water from 0°C to 100 °C ($\Delta T = 100$ K) is:

$$(4.18 \text{ J/g} \cdot \text{K}) \cdot 60.1 \text{ g} \cdot 100 \text{ K} = 2.51 \times 10^4 \text{ J}$$

3. To convert liquid water at 100 °C to gaseous water at 100 °C:

$$2260 \text{ J/g} \cdot 60.1 \text{ g} = 13.6 \times 10^4 \text{ J}$$

The total energy required is: $[2.00 \times 10^4 + 2.51 \times 10^4 \text{ J} + 13.6 \times 10^4 \text{J}] = 1.81 \times 10^5 \text{ J}$

23.   Molar heat capacity of:

a. aluminum : $0.902 \text{ J/g} \cdot \text{K} \cdot \dfrac{26.98 \text{ g Al}}{1 \text{ mol Al}} = 24.3 \text{ J/mol} \cdot \text{K}$

b. iron : $0.451 \text{ J/g} \cdot \text{K} \cdot \dfrac{55.85 \text{ g Fe}}{1 \text{ mol Fe}} = 25.2 \text{ J/mol} \cdot \text{K}$

c. CCl_4 (l) : $0.861 \text{ J/g} \cdot \text{K} \cdot \dfrac{153.8 \text{ g } CCl_4}{1 \text{ mol } CCl_4} = 132 \text{ J/mol} \cdot \text{K}$

25.   The reaction: $ADP + H_3PO_4 + 38 \text{ kJ} \rightarrow ATP + H_2O$

requires energy, therefore it is **endothermic**.

27.   The molar enthalpy of combustion may be obtained if you calculate the number of moles of ethanol which correspond to 0.115 g of the substance:

$$0.115 \text{ g } C_2H_5OH \cdot \dfrac{1 \text{ mol } C_2H_5OH}{46.07 \text{ g } C_2H_5OH} = 2.50 \times 10^{-3} \text{ mol } C_2H_5OH$$

The molar $\Delta H_{combustion}$ is : $\dfrac{-3.42 \text{ kJ}}{2.50 \times 10^{-3} \text{ mol } C_2H_5OH} = -1370 \text{ kJ/mol}$

The negative sign of ΔH indicates that the combustion of ethanol is **exothermic**.

29.   The energy produced by 28.4 g of the solid:

$$28.4 \text{ g } (NH_4)_2Cr_2O_7 \cdot \dfrac{1 \text{ mol } (NH_4)_2Cr_2O_7}{252.1 \text{ g } (NH_4)_2Cr_2O_7} \cdot \dfrac{315 \text{ kJ}}{1 \text{ mol } (NH_4)_2Cr_2O_7} = 35.5 \text{ kJ}$$

31.   To calculate the enthalpy change for the formation of one mole of $SrCO_3$ given the three equations, requires that (according to Hess' Law) we provide a series of reactions which-- when added together --result in the overall equation.

Numbering the equations as follows:

1. $Sr(s) + 1/2 \ O_2 \ (g) \rightarrow SrO \ (s)$ $\Delta H = -592 \ kJ$
2. $SrO \ (s) + CO_2 \ (g) \rightarrow SrCO_3 \ (s)$ $\Delta H = -234 \ kJ$
3. $C \ (graphite) + O_2 \ (g) \rightarrow CO_2 \ (g)$ $\Delta H = -394 \ kJ$

Add equations 1 and 2 to get:

$$Sr(s) + 1/2 \ O_2 \ (g) + CO_2 \ (g) \rightarrow SrCO_3 \ (s) \ \ \Delta H = (-592 + -234) = -826 \ kJ$$

adding equation 3 to this yields:

$$Sr(s) + 3/2 \ O_2 \ (g) + C \ (graphite) \rightarrow SrCO_3 \ (s)$$
$$\Delta H = (-826 + -394) \ or \ -1220 \ kJ$$

which is the desired <u>overall</u> equation.

33.   The overall equation may be seen as the "addition" of the two steps:

                                                        <u>ΔH (in kJ)</u>

1. $PbS \ (s) + 3/2 \ O_2 \ (g) \rightarrow PbO \ (s) + SO_2 \ (g)$ -413.7
2. <u>$PbO \ (s) + C \ (s) \rightarrow Pb \ (s) + CO \ (g)$</u>   <u>$+106.8$</u>

     $PbS \ (s) + C \ (s) + 3/2 \ O_2 \ (g) \rightarrow Pb \ (s) + SO_2 \ (g) + CO \ (g)$ -306.9

So **306.9 kJ** of energy is **liberated** for each mole of PbS (s) which reacts.

Moles of galena in 1.00 kg $= 1.00 \times 10^3 \ g \ PbS \ \bullet \dfrac{1 \ mol \ PbS}{239 \ g \ PbS} = 4.18 \ mol \ PbS$

Total heat energy evolved $= \dfrac{-306.9 \ kJ}{1 mol \ PbS} \bullet 4.18 \ mol \ PbS = -1280 \ kJ$

35.   The desired net equation is $C = C \ (g) \rightarrow 2 \ C \ (g)$.

The data provided is:

                                                        <u>ΔH , kJ</u>

1. $H_2C = CH_2 \ (g) \rightarrow 4 \ H \ (g) + C = C \ (g)$ $+1656$
2. $C \ (graphite) \rightarrow C \ (g)$ $+716.7$
3. $H_2 \ (g) \rightarrow 2 \ H \ (g)$ $+436.0$
4. $2 \ C \ (graphite) + 2 \ H_2 \ (g) \rightarrow H_2C = CH_2 \ (g)$ $+52.3$

Note that in the net equation C=C is a reactant and C is a product. An equation which provides either of these is a good start.

[*For this and all other Hess' Law problems, we will indicate the reverse of equation x as x'*]

		ΔH (kJ)
1'	$4 H (g) + C \equiv C (g) \rightarrow H_2C=CH_2 (g)$	- 1656
3 (x 2)	$2 H_2 (g) \rightarrow 4 H (g)$	+ 436.0 x 2
4'	$H_2C=CH_2 (g) \rightarrow 2 C (graphite) + 2 H_2 (g)$	- 52.3
2 (x 2)	$2 C (graphite) \rightarrow 2 C (g)$	+ 716.7 x 2

Adding these four modifications gives the desired equation   + 597 kJ

with the overall ΔH being the algebraic sum.

37.   The formation of one mole of each of the compounds are shown as:

$\Delta H_f°$ (in kJ/ mol)

a. $2 Al (s) + 3/2 O_2 (g) \rightarrow Al_2O_3 (s)$ - 1675.7

b. $Ti (s) + 2 Cl_2 (g) \rightarrow TiCl_4 (l)$ - 804.2

c. $Mg (s) + H_2 (g) + O_2 (g) \rightarrow Mg(OH)_2 (s)$ - 924.5

d. $N_2 (g) + 2 H_2 (g) + 3/2 O_2 (g) \rightarrow NH_4NO_3 (s)$ - 365.6

e. $C (graphite) + 1/2 O_2 (g) + Cl_2 (g) \rightarrow COCl_2 (g)$ - 218.8

39.   For the reaction: $SO_2 (g) + 1/2 O_2 (g) \rightarrow SO_3 (g)$

ΔH_f (kJ/mol)   -296.8   0   -395.7

$\Delta H_{rxn} = \Sigma \Delta H_f°$ products $- \Sigma \Delta H_f°$ reactants

$\Delta H_{rxn} = [(1 \text{ mol})(- 395.7 \text{ kJ/mol})] - [(1 \text{ mol})(- 296.8 \text{ kJ/mol}) + 0]$

$= (- 395.7 \text{ kJ}) - (- 296.8 \text{ kJ})$

$= - 98.9 \text{ kJ}$

41.   For the reaction: $CO_2 (g) + 2 H_2O (l) \rightarrow O_2 (g) + CH_4 (g)$

ΔH_f (kJ/mol)   - 393.5   - 285.8   0   - 74.81

a. $\Delta H_{rxn} = \Sigma \Delta H_f°$ products $- \Sigma \Delta H_f°$ reactants

$\Delta H_{rxn} = [0 \text{ kJ} + (1 \text{ mol})(- 74.81 \text{ kJ/mol})] -$

$[(1 \text{ mol})(- 393.5 \text{ kJ/mol}) + (2 \text{ mol})(- 285.8 \text{ kJ/mol})]$

$= (- 74.81 \text{ kJ}) - (- 965.1 \text{ kJ}) = + 890.3 \text{ kJ}$

b. The reaction is endothermic ($\Delta H_{rxn} = +$)

43.   The enthalpy change for the reaction: $Ca(OH)_2$ (s) + CO_2 (g) $\rightarrow$ $CaCO_3$ (s) + H_2O (g)
                ΔH_f (kJ/mol)   - 986.1   - 393.5   - 1206.9   + 241.8

ΔH_{rxn} = [(1 mol)(- 1206.9 kJ/mol) + (1 mol)(- 241.8 kJ/mol)] -

                              [(1 mol)(- 986.1 kJ/mol) + (1 mol)(- 393.5 kJ/mol)]

                 = (- 1448.7 kJ) - (- 1379.6 kJ)

                 = - 69.1 kJ

45.   The molar enthalpy of OF_2 may be calculated using:

                              OF_2 (g) + H_2O (g) $\rightarrow$ 2 HF (g) + O_2 (g) + 318 kJ
        $\Delta H_f°$ (kJ/mol)   ?   - 241.8   - 271.1   0

        - 318 kJ = [(2 mol)(- 271.1 kJ/mol)] - [(1 mol)($\Delta H_f°$ OF_2(g)) + (1 mol)(- 241.8)]

        + 17.6 kJ/mol = $\Delta H_f°$ OF_2 (g)

47.   Calculate the heat evolved when 1.00 mole of CaO (s) reacts with CO_2 (g):
                              CaO (s) + CO_2 (g) $\rightarrow$ $CaCO_3$ (s)
        $\Delta H_f°$ (kJ/mol)   - 635.1   - 393.5   - 1206.9

        ΔH_{rxn} = [(1mol)(- 1206.9 kJ/mol)] - [(1 mol)(- 635.1 kJ/mol) + (1 mol)(- 393.5 kJ/mol)]
                 = - 178.3 kJ

49.   The enthalpy of formation of Al_2O_3 is -1675.7 kJ/ mol. Determine the overall heat energy
        evolved when 2.70 g Al is burned:

        2.70 g Al $\cdot$ $\dfrac{1 \text{ mol Al}}{26.98 \text{ g Al}}$ $\cdot$ $\dfrac{1 \text{ mol Al}_2O_3}{2 \text{ mol Al}}$ $\cdot$ $\dfrac{- 1675.7 \text{ kJ}}{1 \text{ mol Al}_2O_3}$ = - 83.8 kJ

51.   To calculate the heat energy necessary to decompose 10.0 g $BaCO_3$, calculate:

        1. overall ΔH_{rxn} ($\Sigma \Delta H_f°$ products - $\Sigma \Delta H_f°$ reactants)
        2. number of moles of $BaCO_3$ corresponding to 10.0 g

        ΔH_{rxn} = [$\Delta H_f°$ BaO (s) + $\Delta H_f°$ CO_2 (g)] - $\Delta H_f°$ $BaCO_3$ (s)
                 = [(-553.5 kJ/mol)(1 mol) + (-393.5 kJ/mol)(1 mol)] - [(- 1216.3 kJ/mol)(1 mol)]
                 = (- 947.0 kJ) - (- 1216.3 kJ)
                 = 269.3 kJ/mol $BaCO_3$(s)

To decompose 10.0 g $BaCO_3$:

$$269.3 \text{ kJ/mol } BaCO_3 \cdot \frac{1 \text{ mol } BaCO_3}{197.3 \text{ g } BaCO_3} \cdot \frac{10.0 \text{ g } BaCO_3}{1} = 13.6 \text{ kJ}$$

53. $N_2H_4 (l) + O_2 (g) \rightarrow N_2 (g) + 2 H_2O (g)$

$\Delta H_f° \; N_2H_4 (l) = +50.6 \text{ kJ/mol}$

$\Delta H_{rxn} \; N_2H_4 (l) = [2 \Delta H_f° \; H_2O(g) + \Delta H_f° \; N_2(g)] - [\Delta H_f° \; N_2H_4(l) + \Delta H_f° \; O_2(g)]$

$$= -534.2 \text{ kJ}$$

On a per gram basis this corresponds to:

$$\frac{-534.2 \text{ kJ}}{1 \text{ mol } N_2H_4} \cdot \frac{1 \text{ mol } N_2H_4}{32.05 \text{ g } N_2H_4} = -16.7 \text{ kJ/g}$$

Similarly for the dimethylhydrazine:

$N_2H_2(CH_3)_2 (l) + 4 O_2 (g) \rightarrow 2 CO_2 (g) + 4 H_2O (g) + N_2 (g)$

$\Delta H_f° \; N_2H_2(CH_3)_2 (l) = +42.0 \text{ kJ/mol}$

$\Delta H_{rxn} \; N_2H_2(CH_3)_2 (l) = [2 \Delta H_f° \; CO_2 (g) + 4 \Delta H_f° \; H_2O (g) + \Delta H_f° \; N_2 (g)]$

$$- [\Delta H_f° \; N_2H_2(CH_3)_2 (l) + 4\Delta H_f° \; O_2 (g)] = -1796 \text{ kJ}$$

On a per gram basis this corresponds to:

$$\frac{-1796 \text{ kJ}}{1 \text{ mol } N_2H_2(CH_3)_2(l)} \cdot \frac{1 \text{ mol } N_2H_2(CH_3)_2(l)}{60.10 \text{ g } N_2H_2(CH_3)_2(l)} = -29.9 \text{ kJ/g}$$

Dimethylhydrazine gives more heat per gram.

55.   Given the equations below, calculate the standard molar enthalpy of formation of PbO(s).

1. Rearrange the equations so that their "sum" will correspond to :

   $Pb (s) + 1/2 O_2 (g) \rightarrow PbO (s)$

2. Add: $\underline{\Delta H \text{ (kJ)}}$

   1' (x2) $2 Pb (s) + 2 CO (g) \rightarrow 2 PbO (s) + 2 C \text{ (graphite)}$ - 106.8 x 2
   2. $2 C \text{ (graphite)} + O_2 (g) \rightarrow 2 CO (g)$ - 221.0

   Net: $2 Pb (s) + O_2 (g) \rightarrow 2 PbO (s)$ - 434.6

Remember the standard molar enthalpy of formation of a compound refers to the formation of <u>one mole</u> of the substance from its elements. So we divide the net equation (and the associated ΔH) by 2 to obtain: $\Delta H_f PbO = -217.3$ kJ/mol

57.   Heat evolved = Heat absorbed by bomb and water

$$-q_{rxn} = q_{bomb} + q_{water}$$

$q_{bomb} = C \cdot \Delta t = (650 \text{ J/ }°C)(3.33 \text{ K}) = 2.16 \times 10^3 \text{ J}$

(where C = heat capacity of bomb)

$q_{water} = S \cdot m \cdot \Delta t = (4.184 \text{ J/g} \cdot \text{K})(320. \text{ g})(3.33 \text{ K}) = 4.46 \times 10^3 \text{ J}$

(where S = specific heat of water)

$-q_{rxn} = 2.16 \times 10^3 \text{ J} + 4.46 \times 10^3 \text{ J} = 6.62 \times 10^3 \text{ J}$

$q_{rxn} = -6.62 \times 10^3 \text{ J}$

59.   Heat evolved by reaction = - Heat absorbed by surroundings

= - (Heat absorbed by bomb and water)

= - (q_{bomb} + q_{water})

$\Delta t = (27.38°C - 25.00°C) = 2.38 °C$ or 2.38 K

$q_{bomb} = (893 \text{ J/k}) \cdot 2.38 \text{ K}$ = 2130 J

$q_{water} = (775 \text{ g})(4.18 \text{ J/g} \cdot \text{K}) \cdot 2.38 \text{ K}$ = 7710 J

Heat absorbed by bomb + water = 9840 J or 9.84 kJ

The heat evolved by reaction of 0.300 g C is then: -9.84 kJ

Expressing this on a molar basis:

$$\frac{-9.84 \text{ kJ}}{0.300 \text{ g C}} \cdot \frac{12.01 \text{ g C}}{1 \text{ mol C}} = \frac{-394 \text{ kJ}}{1 \text{ mol C}}$$

61.   100. mL of 0.200 M CsOH and 50.0 mL of 0.400 M HCl each supply 0.0200 moles of base and acid respectively. If we assume the specific heat capacities of the solutions are 4.20 J/g • K, the **heat evolved** for 0.200 moles of CsOH is:

$q = (4.20 \text{ J/g} \cdot \text{K})(150 \text{ g})(24.28 \text{ °C} - 22.50°C)$ [and since $1.78°C = 1.78$ K]

$q = (4.20 \text{ J/g} \cdot \text{K})(150 \text{ g})(1.78 \text{ K})$

$q = 1120 \text{ J}$

The molar enthalpy of neutralization is : $\dfrac{-1120 \text{ J}}{0.200 \text{ mol CsOH}} = -56100 \text{ J/mol}$

63.   Since **gold** has the lower specific heat, the temperature of the gold will exceed 100°C first.

65.   For the reaction : $Mg\,(s) + 2\,HCl\,(aq) \rightarrow MgCl_2\,(aq) + H_2\,(g)$

$\Delta H_{rxn}° = \Sigma\,\Delta H_f°$ products $- \Sigma\,\Delta H_f°$ reactants

$= [(-801.2\ kJ/mol)(1\ mol)] - [0 + (-167.2\ kJ/mol)(2\ mol)]$

$= -466.8\ kJ$

67.   a. Determine if the reaction is exo- or endothermic:

$\Delta H_{rxn}° = \Sigma\,\Delta H_f°$ products $- \Sigma\,\Delta H_f°$ reactants

$= [(-157.3\ kJ/mol)(1\ mol) + 0] - [0 + (-241.8\ kJ/mol)(1\ mol)]$

$= +84.5\ kJ$

The reaction is **endothermic**.

b. The energy absorbed when 2.00 g of Cu react:

$$2.00\ g\ Cu \cdot \frac{1\ mol\ Cu}{63.55\ g\ Cu} \cdot \frac{84.5\ kJ}{1\ mol\ Cu} = 2.66\ kJ$$

69.   The heat energy evolved when 9.08×10^5 g P_4 is oxidized:

$$\frac{2984\ kJ}{1\ mol\ P_4O_{10}} \cdot \frac{1\ mol\ P_4O_{10}}{1\ mol\ P_4} \cdot \frac{1\ mol\ P_4}{123.9\ g\ P_4} \cdot 9.08 \times 10^5\ g\ P_4$$
$$= 2.19 \times 10^7\ kJ$$

71.   Hess' Law is useful in calculating the molar enthalpy of formation. Adding the two equations as written, we obtain:

	$\Delta H°,\ kJ$
$2\,C\,(graphite) + 2\,H_2\,(g) \rightarrow C_2H_4\,(g)$	$+52.3$
$Cl_2\,(g) + C_2H_4\,(g) \rightarrow C_2H_4Cl_2\,(g)$	-116
$2\,C\,(graphite) + 2\,H_2\,(g) + Cl_2\,(g) \rightarrow C_2H_4Cl_2\,(g)$	$-64\ kJ/mol$

73.   The molar enthalpy of hydrazine, N_2H_4 (l) may be calculated from:

$$\underline{\Delta H°, kJ}$$

$$N_2 (g) + 2 H_2O (g) \rightarrow N_2H_4 (l) + O_2 (g) \qquad + 534 \text{ kJ}$$

$$2[H_2 (g) + 1/2 O_2 (g) \rightarrow H_2O (g)] \qquad - 242 \text{ kJ x 2}$$

$$N_2 (g) + 2 H_2 (g) \rightarrow N_2H_4 (l) \qquad + 50. \text{ kJ/mol}$$

75.   Heat energy evolved when 8.00 kJ of ammonium nitrate react with excess aluminum:

$$2 Al (s) + 3 NH_4NO_3 (s) \rightarrow 3 N_2 (g) + 6 H_2O (g) + Al_2O_3 (s)$$

$$\Delta H°_{rxn} = [6 \Delta H_f° H_2O (g) + \Delta H_f° Al_2O_3 (s)] - [3 \Delta H_f° NH_4NO_3 (s)]$$

$$= -2030 \text{ kJ}$$

$$8.00 \times 10^3 \text{ g } NH_4NO_3 \cdot \frac{1 \text{ mol } NH_4NO_3}{80.04 \text{ g } NH_4NO_3} = 100. \text{ mol } NH_4NO_3$$

$$100. \text{ mol } NH_4NO_3 \cdot \frac{-2030 \text{ kJ}}{3 \text{ mol } NH_4NO_3} = - 6.76 \times 10^4 \text{ kJ/ mol}$$

77.   Use the following equations, added appropriately, to provide the desired overall equation:

$$UO_2 (s) + 4 HF (g) \rightarrow UF_4 (s) + 2 H_2O (g)$$

$$UF_4 (s) + F_2 (g) \rightarrow UF_6 (s)$$

$$UO_2 (s) + 4 HF (g) + F_2 (g) \rightarrow UF_6 (s) + 2 H_2O (g)$$

We can then calculate the $\Delta H°_{rxn}$:

$$\Delta H°_{rxn} = [2 \Delta H_f° H_2O (g) + \Delta H_f° UF_6 (s)] - [\Delta H_f° UO_2 (s) + 4 \Delta H_f° HF (g) + \Delta H_f° F_2 (g)]$$

$$= [2 \cdot (- 241.8) + (- 2147.4)] - [(- 1084.9) + 4 \cdot (- 271.1) + 0] = - 461.7 \text{ kJ}$$

$$225 \text{ tons } UF_6 \cdot \frac{9.08 \times 10^5}{1 \text{ ton}} \cdot \frac{1 \text{ mol } UF_6}{352.0 \text{ g } UF_6} = 5.80 \times 10^5 \text{ mol } UF_6$$

$$5.80 \times 10^5 \text{ mol } UF_6 \cdot \frac{- 461.7 \text{ kJ}}{1 \text{ mol } UF_6} = - 2.68 \times 10^8 \text{ kJ}$$

79.   To achieve the overall reaction, add the three given equations. As a step in the "right" direction, note that only equation **c** has FeO in it, so begin with this equation (reversed).

Note also that in the overall equation the coefficients of FeO and Fe are equal. In equation **b**, note the coefficient of iron (2) as opposed to the coefficient for FeO in equation **c** (3). Equalize these coefficients by multiplying equation **c by 2** and equation **b by 3**.

		ΔH
c' x 2	$6\ FeO\ (s)\ +\ 2\ CO_2\ (g) \rightarrow 2\ Fe_3O_4\ (s)\ +\ 2\ CO\ (g)$	- 19 kJ x 2
b x 3	$3\ Fe_2O_3\ (s)\ +\ 9\ CO\ (g) \rightarrow 6\ Fe\ (s)\ +\ 9\ CO_2\ (g)$	- 25 kJ x 3

To eliminate Fe_2O_3 and Fe_3O_4 add equation a':

$$2\ Fe_3O_4\ (s)\ +\ CO_2\ (g)\ \rightarrow\ 3\ Fe_2O_3\ (s)\ +\ CO\ (g) \qquad + 47\ kJ$$

Adding the three equations yields:

$$6\ FeO\ (s)\ +\ 6\ CO\ (g)\ \rightarrow\ 6\ Fe\ (s)\ +\ 6\ CO_2\ (g) \qquad - 66\ kJ$$

Dividing all coefficients (and the enthalpy change) by six (6) provides the desired equation with an enthalpy change of -11 kJ.

81.   Moles of air present in the room:

$$1800\ ft^2\ \bullet\ 8.0\ ft\ \bullet\ \left(\frac{12\ in}{1\ ft}\right)^3 \bullet \left(\frac{0.0254\ m}{1\ in}\right)^3 \bullet \frac{1\ L}{1\ x\ 10^3\ m^3} \bullet \frac{1.22\ g}{1\ L}$$

$$\bullet\ \frac{1\ mol\ air}{28.9\ g\ air}\ =\ 1.7\ x\ 10^4\ mol\ air$$

To heat that amount of air by 7 K requires:

$$(29.1\ J/mol\ \bullet\ K)\ \bullet\ 1.7\ x\ 10^4\ mol\ \bullet\ 7\ K\ =\ 3.5\ x\ 10^6\ J$$

The combustion of 1 mole of CH_4 liberates 802.3 kJ/mol :

$$CH_4\ (g)\ +\ 2\ O_2\ (g)\ \rightarrow\ 2\ H_2O\ (g)\ +\ CO_2\ (g)$$

$\Delta H°_{rxn}\ =\ [(2\ mol)(-\ 241.8\ kJ/mol)\ +\ (-\ 393.5\ kJ)]\ -\ [(-\ 74.8\ kJ/mol)(1\ mol)\ +\ 0]$
$\qquad =\ -\ 802.3\ kJ$

The amount of CH_4 needed is:

$$3.5\ x\ 10^6\ J\ \bullet\ \frac{1\ kJ}{1\ x\ 10^3\ J}\ \bullet\ \frac{1\ mol\ CH_4}{802.3\ kJ}\ \bullet\ \frac{16.0\ g\ CH_4}{1\ mol\ CH_4}\ =\ 70.\ g\ CH_4$$

83.   From Example 6.12 in the text, the combustion of octane liberates 48.1 kJ/g.

i.e. $\Delta H°_{rxn} = -48.1$ kJ/g

For the combustion of CH_3OH:

$$2\ CH_3OH\ (l)\ +\ 3\ O_2\ (g)\ \rightarrow\ 4\ H_2O\ (g)\ +\ 2\ CO_2\ (g)$$

$\Delta H°_{rxn}$ = [(- 393.5 kJ/mol)(2 mol) + (- 241.8 kJ/mol)(4 mol)] -

[(- 238.7 kJ/mol)(2 mol) + 0]

= [(- 787.0) + (- 967.2)] + 477.4 kJ

= - 1276.8 kJ

Express this on a per mol and per gram basis:

$$\frac{-\ 1276.8\ kJ}{2\ mol\ CH_3OH} \cdot \frac{1\ mol\ CH_3OH}{32.04\ g\ CH_3OH} = -19.9\ kJ/g$$

On a per gram basis, octane liberates the greater amount of heat energy.

84.   a. Mass of SO_2 contained in 440 million gallons of wine that is 100. ppm in SO_2:

$$440 \times 10^6\ gal \cdot \frac{4\ qt}{1\ gal} \cdot \frac{0.9463\ L}{1\ qt} \cdot \frac{1 \times 10^3\ cm^3}{1\ L} \cdot \frac{1.00\ g\ wine}{1\ cm^3}$$

$$\cdot \frac{100.\ g\ SO_2}{1 \times 10^6\ g\ wine} = 1.67 \times 10^8\ g\ SO_2$$

The number of moles to which this corresponds is:

$$1.67 \times 10^8\ g\ SO_2 \cdot \frac{1\ mol\ SO_2}{64.06\ g\ SO_2} = 1.07 \times 10^{10}\ mol\ SO_2$$

b. For the consumption of 20. million tons of SO_2 by MgO:

$$20. \times 10^6\ T\ SO_2 \cdot \frac{1\ mol\ SO_2}{64.06\ g\ SO_2} \cdot \frac{1\ mol\ MgO}{1\ mol\ SO_2} \cdot \frac{40.30\ g\ MgO}{1\ mol\ MgO}$$

$$= 1.3 \times 10^7\ T\ MgO$$

c. The heat energy change per mole of $MgSO_4$ by the reaction:

$$MgO \ (s) \ + \ SO_2 \ (g) \ + \frac{1}{2} \ O_2 \ (g) \rightarrow \ MgSO_4 \ (s)$$

$\Delta H_f°$ - 601.70   - 296.8   0   - 1284.9

$\Delta H_{rxn}° \ = \ \Sigma \ \Delta H_f°$ products - $\Sigma \ \Delta H_f°$ reactants

$= \ (- \ 1284.9 \ kJ/mol)(1 \ mol)$

$- \ [(-601.70 \ kJ/mol)(1 \ mol) + (- \ 296.8 \ kJ/mol)(1 \ mol) + 0]$

$= \ - \ 386.4 \ kJ$

d. Heat evolved by production of 750. T H_2SO_4 per day.

The overall ΔH_{rxn} is the sum of the three reactions given:

$\Delta H_{overall} \ = \ (- \ 296.8 \ kJ) + (- \ 98.9 \ kJ) \ + \ (-130.0 \ kJ) \ = \ - \ 525.7 \ kJ/mol \ H_2SO_4$

The number of moles of H_2SO_4 corresponding to 750. T is:

$$750 \ T \ H_2SO_4 \ \cdot \frac{9.08 \ x \ 10^5 \ g \ H_2SO_4}{1 \ T \ H_2SO_4} \cdot \frac{1 \ mol \ H_2SO_4}{98.08 \ g \ H_2SO_4} = 6.94 \ x \ 10^6 \ mol$$

The overall energy produced per day is then:

$$6.94 \ x \ 10^6 \ mol \ H_2SO_4 \ \cdot \frac{- \ 525.7 \ kJ}{mol \ H_2SO_4} = \ 3.65 \ x \ 10^9 \ kJ$$

CHAPTER 7: Nuclear Chemistry

12. Balance the following nuclear equations, supplying the missing particle.
   [The missing particle is emboldened]

   a. $^{54}_{26}\text{Fe} + {}^{4}_{2}\text{He} \rightarrow 2\,^{1}_{1}\text{H} + \mathbf{^{56}_{26}\text{Fe}}$

   b. $^{27}_{13}\text{Al} + {}^{4}_{2}\text{He} \rightarrow {}^{30}_{15}\text{P} + \mathbf{^{1}_{0}\text{n}}$

   c. $^{32}_{16}\text{S} + {}^{1}_{0}\text{n} \rightarrow {}^{1}_{1}\text{H} + \mathbf{^{32}_{15}\text{P}}$

   d. $^{96}_{42}\text{Mo} + {}^{2}_{1}\text{H} \rightarrow {}^{1}_{0}\text{n} + \mathbf{^{97}_{43}\text{Tc}}$

   e. $^{98}_{42}\text{Mo} + {}^{1}_{0}\text{n} \rightarrow {}^{99}_{43}\text{Tc} + \mathbf{^{0}_{-1}\text{e}}$

14. Balance the following nuclear equations, supplying the missing particle.
   [The missing particle is emboldened]

   a. $^{104}_{47}\text{Ag} \rightarrow {}^{104}_{48}\text{Cd} + \mathbf{^{0}_{-1}\text{e}}$

   b. $^{87}_{36}\text{Kr} \rightarrow {}^{0}_{-1}\text{e} + \mathbf{^{87}_{37}\text{Rb}}$

   c. $^{231}_{91}\text{Pa} \rightarrow {}^{227}_{89}\text{Ac} + \mathbf{^{4}_{2}\text{He}}$

   d. $^{230}_{90}\text{Th} \rightarrow {}^{4}_{2}\text{He} + \mathbf{^{226}_{88}\text{Ra}}$

   e. $^{82}_{35}\text{Br} \rightarrow {}^{82}_{36}\text{Kr} + \mathbf{^{0}_{-1}\text{e}}$

   f. $\mathbf{^{24}_{11}\text{Na}} \rightarrow {}^{24}_{12}\text{Mg} + {}^{0}_{-1}\text{e}$

16. $^{235}_{92}\text{U} \rightarrow {}^{231}_{90}\text{Th} \rightarrow {}^{231}_{91}\text{Pa} \rightarrow {}^{227}_{89}\text{Ac} \rightarrow {}^{227}_{90}\text{Th} \rightarrow {}^{223}_{88}\text{Ra}$

   $\qquad +\qquad\quad +\qquad\quad +\qquad\quad +\qquad\quad +$

   $\qquad {}^{4}_{2}\text{He}\qquad {}^{0}_{-1}\text{e}\qquad {}^{4}_{2}\text{He}\qquad {}^{0}_{-1}\text{e}\qquad {}^{4}_{2}\text{He}$

18.   The change in mass (Δm) for ^{10}B is:

$$\Delta m \quad = 10.01294 - [5(1.00783) + 5(1.00867)]$$
$$= 10.01294 - 10.0825$$
$$= -0.06956 \text{ g/mol}$$

while that for ^{11}B is:

$$\Delta m \quad = 11.00931 - [5(1.00783) + 6(1.00867)]$$
$$= 11.00931 - 11.09117$$
$$= -0.08186 \text{ g/mol}$$

The binding energies for the two isotopes are:

$$^{10}B : \quad \Delta E = (-6.956 \times 10^{-5} \text{ kg/mol})(2.998 \times 10^8 \text{ m/s})^2 \left(\frac{1 \text{ J}}{1 \text{ kg} \cdot \text{m}^2 \cdot \text{s}^{-2}} \right)$$
$$= -6.252 \times 10^{12} \text{ J/mol}$$

$$^{11}B : \quad \Delta E = (-8.186 \times 10^{-5} \text{ kg/mol})(2.998 \times 10^8 \text{ m/s})^2 \left(\frac{1 \text{ J}}{1 \text{ kg} \cdot \text{m}^2 \cdot \text{s}^{-2}} \right)$$
$$= -7.358 \times 10^{12} \text{ J/mol}$$

We see that the binding energy of ^{11}B is greater. Compare their relative abundances!

20.   For ^{64}Cu, $t_{1/2} = 128$ hr

The fraction remaining as ^{64}Cu following n half-lives is equal to $\left(\frac{1}{2}\right)^n$. Note that 2 days and 16 hrs (64 hrs) corresponds to exactly **five** half-lives.

The <u>fraction</u> remaining as ^{64}Cu is the $\left(\frac{1}{2}\right)^5$ or $\frac{1}{32}$ or 0.03125.

The mass remaining is:
$$(0.03125)(15.0 \text{ mg}) = 0.469 \text{ mg}.$$

22.   a. The equation for β–decay of ^{131}I is:

$$^{131}_{53}I \rightarrow ^{\quad 0}_{-1}e + ^{131}_{54}Xe$$

b. The amount of ^{131}I remaining after 32.2 days:

For ^{131}I, $t_{1/2}$ is 8.05 days--so 32.2 days is exactly **four** half-lives:

The fraction of ^{131}I remaining is $(\frac{1}{2})^4$ or $\frac{1}{16}$ or 0.0625

The amount of the original 25.0 mg remaining will be :

$$(0.0625)(25.0 \text{ mg}) = 1.56 \text{ mg}$$

24.   Mass of Cobalt-60 ($t_{1/2}$ = 5.3 yrs) remaining after 21.2 yr:

$$k = \frac{0.693}{t_{1/2}} = \frac{0.693}{5.3 \text{ yr}} = 0.131 \text{ yr}^{-1}$$

$$\ln\left(\frac{N}{N_o}\right) = -kt \qquad \text{where} \quad N = \text{amt. remaining after time t and}$$
$$N_o = \text{initial amount}$$

$$\ln\left(\frac{N}{N_o}\right) = -(0.131 \text{ yr}^{-1})(21.2 \text{ yr})$$

$$\ln\left(\frac{N}{10.0 \text{ mg}}\right) = -2.77$$

$$\frac{N}{10.0 \text{ mg}} = 0.0625$$

$$N = 0.625 \text{ mg}$$

Mass remaining after 1 century:

$$\ln\left(\frac{N}{10.0 \text{ mg}}\right) = -(0.131 \text{ yr}^{-1})(100 \text{ yr})$$

and $N = 2.10 \times 10^{-5} \text{ mg}$

26.   For a sample of NaI containing ^{131}I, the time required for activity to fall to 5.0 % of original activity:

This isotope has a half-life of 8.05 days. We begin by calculating the rate constant, k.

$$k = \frac{0.693}{t_{1/2}} = \frac{0.693}{8.05 \text{ days}} = 0.0861 \text{ days}^{-1}$$

The time required for the sample to have only 5% of the original activity is then:

$$\ln\left(\frac{N}{N_o}\right) = -(0.0861 \text{ days}^{-1})\, t$$

$$\ln\left(\frac{5.00}{100.}\right) = -(0.0861 \text{ days}^{-1})\, t$$

$$34.8 \text{ days} = t$$

28. Substituting the activities into the rate equation gives:

$$\ln\left(\frac{1600 \text{ cpm}}{6400 \text{ cpm}}\right) = -k \cdot 6.00 \text{ hr}$$

$$0.231 \text{ hr}^{-1} = k$$

$$\text{and } t_{1/2} = \frac{0.693}{0.231 \text{ hr}^{-1}} = 3.00 \text{ hr}$$

Here is another problem which can be quickly solved by a "common sense" approach. The cpm dropped to $\frac{1}{4}$ the original value in 6.00 hours. We would expect this behavior in **two** half-lives--i. e. 6.00 hours and therefore $t_{1/2} = 3.00$ hrs.

30. The alpha-particle decay of 226-Ra may be written: $^{226}_{88}\text{Ra} \rightarrow {}^{4}_{2}\text{He} + {}^{222}_{86}\text{Rn}$

For $t_{1/2} = 1620$ yrs or 8.52×10^8 min and $k = \dfrac{0.693}{8.52 \times 10^8 \text{ min}} = 8.13 \times 10^{-10} \text{ min}^{-1}$

In a 0.0010 g sample of ^{226}Ra there are:

$$0.0010 \text{ g Ra} \cdot \frac{1 \text{ mol Ra}}{226 \text{ g Ra}} \cdot \frac{6.02 \times 10^{23} \text{ atoms}}{1 \text{ mol Ra}} = 2.67 \times 10^{18} \text{ atoms}$$

and $A = (8.13 \times 10^{-10} \text{ min}^{-1})(2.67 \times 10^{18} \text{ atoms}) = 2.17 \times 10^9 \text{ atoms/min}$

32. Using the activity in disintegrations per minute per gram compared to the activity of ^{14}C in living material, we can calculate the elapsed time since the death of the tree, if we first calculate the rate constant, k:

$$k = \frac{0.693}{t_{1/2}} = \frac{0.693}{5.73 \times 10^3 \text{ yr}} = 1.21 \times 10^{-4} \text{ yr}^{-1}$$

$$\ln\left(\frac{11.2 \text{ dis/min} \cdot g}{14.0 \text{ dis/min} \cdot g}\right) = -1.21 \times 10^{-4} \text{ yr}^{-1} \cdot t$$

Solving for t yields approximately **1850 years**.

Subtracting the current year from 1850 gives an approximate date of 140 AD.

34.   The formation of ^{241}Am from ^{239}Pu may be written:

$$^{239}_{94}\text{Pu} + 2\,^{1}_{0}\text{n} \rightarrow \,^{241}_{95}\text{Am} + \,^{0}_{-1}\text{e}$$

36.   A proposed method for producing ^{246}Cf:

$$^{238}_{92}\text{U} + \,^{12}_{6}\text{C} \rightarrow \,^{246}_{98}\text{Cf} + 4\,^{1}_{0}\text{n}$$

38.   The energy liberated by one pound of ^{235}U:

$$\frac{2.1 \times 10^{10} \text{ kJ}}{1 \text{ mol } ^{235}\text{U}} \cdot \frac{1 \text{ mol } ^{235}\text{U}}{235 \text{ g U}} \cdot \frac{454 \text{ g U}}{1 \text{ pound U}} = \frac{4.1 \times 10^{10} \text{ kJ}}{1 \text{ pound U}}$$

comparing this amount of energy to the energy per ton of coal yields:

$$\frac{4.1 \times 10^{10} \text{ kJ}}{1 \text{ pound U}} \cdot \frac{1 \text{ ton coal}}{2.6 \times 10^7 \text{ kJ}} = \frac{1600 \text{ tons coal}}{1 \text{ pound U}}$$

40.   This "dilution" problem may be solved using the equation which is useful for solutions:

$$M_c \times V_c = M_d \times V_d$$

where c and d represent the concentrated and diluted states, respectively. We'll use the number of disintegrations/second as our "molarity."

$$(2.0 \times 10^6 \text{ dps} \cdot 1.0 \text{ mL} = 1.5 \times 10^4 \text{ dps} \cdot V_d$$

$$130 \text{ mL} = V_d$$

The approximate volume of the circulatory system is 130 mL.

42.   Balance the following nuclear reactions, supplying the missing particle.

[The missing particle is emboldened]

a. $^{13}_{6}C + ^{1}_{0}\mathbf{n} \rightarrow ^{14}_{6}C$

b. $^{40}_{18}Ar + ^{4}_{2}\mathbf{He} \rightarrow ^{43}_{19}K + ^{1}_{1}H$

c. $^{250}_{98}Cf + ^{11}_{5}B \rightarrow 4 ^{1}_{0}n + ^{257}_{103}\mathbf{Lr}$

d. $^{53}_{24}Cr + ^{4}_{2}He \rightarrow ^{1}_{0}\mathbf{n} + ^{56}_{26}Fe$

e. $^{212}_{84}Po \rightarrow ^{208}_{82}Pb + ^{4}_{2}\mathbf{He}$

f. $^{122}_{53}I \rightarrow ^{122}_{52}\mathbf{Te} + ^{0}_{1}e$

g. $^{23}_{10}\mathbf{Ne} \rightarrow ^{23}_{11}Na + ^{0}_{-1}e$

h. $^{137}_{53}I \rightarrow ^{1}_{0}n + ^{136}_{53}\mathbf{I}$

i. $^{22}_{11}Na \rightarrow ^{1}_{1}H + ^{21}_{10}\mathbf{Ne}$

44.   Balance the following nuclear reactions:

[The missing particle is emboldened]

a. $^{238}_{92}U + ^{14}_{7}N \rightarrow ^{247}_{99}\mathbf{Es} + 5 ^{1}_{0}n$

b. $^{238}_{92}U + ^{16}_{8}\mathbf{O} \rightarrow ^{249}_{100}Fm + 5 ^{1}_{0}n$

c. $^{253}_{99}Es + ^{4}_{2}\mathbf{He} \rightarrow ^{256}_{101}Md + ^{1}_{0}n$

d. $^{246}_{96}Cm + ^{12}_{6}\mathbf{C} \rightarrow ^{254}_{102}No + 4 ^{1}_{0}n$

e. $^{252}_{98}Cf + ^{10}_{5}\mathbf{B} \rightarrow ^{257}_{103}Lr + 5 ^{1}_{0}n$

46.   The number of Pm atoms in the sample may be determined by calculating the decay
        constant, k:

$$A = k \cdot N$$

$$1.0 \times 10^7 \text{ atom/s} = k \cdot N$$

Since $t_{1/2} = 2.64$ yr (or equivalently 8.33×10^7 s), $k = \dfrac{0.693}{8.33 \times 10^7 \text{ s}}$

$$1.0 \times 10^7 \text{ atom/s} = \dfrac{0.693}{8.33 \times 10^7 \text{ s}} \cdot N$$

$$1.2 \times 10^{15} \text{ atoms} = N$$

The number of grams of promethium can be calculated by using the molar mass of the
element (147).

$$1.2 \times 10^{15} \text{ atoms} \cdot \dfrac{1 \text{ mol Pm}}{6.022 \times 10^{23} \text{ atoms Pm}} \cdot \dfrac{147 \text{ g Pm}}{1 \text{ mol Pm}} = 2.9 \times 10^{-7} \text{ g Pm}$$

22. Observing Figure 8.3 we see:

    a. Orange and red light involve less energy than yellow light.

    b. Violet light has longer wavelengths therefore photons of greater energy.

    c. Blue light is of a shorter wavelength than green light, and therefore has a greater frequency.

24. Since frequency • wavelength = speed of light (c)

$$76 \text{ s}^{-1} \cdot \lambda = 3.0 \times 10^8 \text{ m/s}$$

$$\lambda = 3.9 \times 10^6 \text{ m}$$

wavelength in miles $= 3.9 \times 10^6 \text{ m} \cdot \dfrac{1 \text{ km}}{1.0 \times 10^3 \text{ m}} \cdot \dfrac{1 \text{ mi}}{1.61 \text{ km}} = 2.5 \times 10^3 \text{ mi}$

26. frequency $= \dfrac{\text{speed of light}}{\text{wavelength}} = \dfrac{3.00 \times 10^8 \text{ m/s}}{5.00 \times 10^2 \text{ nm}} \cdot \dfrac{1.00 \times 10^9 \text{ nm}}{1.00 \text{ m}}$

$$= 6.00 \times 10^{14} \text{ s}^{-1}$$

The energy of a photon of this light may be determined :

$$E = h\upsilon$$

Planck's constant, h, has a value of 6.626×10^{-34} J • s • photons^{-1}

$$E = (6.626 \times 10^{-34} \text{ J} \cdot \text{s} \cdot \text{photons}^{-1})(6.00 \times 10^{14} \text{ s}^{-1})$$

$$= 3.98 \times 10^{-19} \text{ J} \cdot \text{photon}^{-1}$$

Energy of 1.00 mol of photons $= 3.98 \times 10^{-19}$ J • photon^{-1} • $\dfrac{6.02 \times 10^{23} \text{ photons}}{1.00 \text{ mol photons}}$

$$= 2.39 \times 10^5 \text{ J/mol photon}$$

28. The region of the electromagnetic spectrum in which 200. nm is located is the **near ultraviolet** region.

The frequency of this light is: $\dfrac{3.00 \times 10^8 \text{ m/s}}{2.00 \times 10^2 \text{ nm}} \cdot \dfrac{1.00 \times 10^9 \text{ nm}}{1.00 \text{ m}} = 1.50 \times 10^{15} \text{ s}^{-1}$

The energy of a photon of this light :

$$E = h\upsilon = 6.626 \times 10^{-34} \text{ J} \cdot \text{s} \cdot \text{photons}^{-1})(1.50 \times 10^{15} \text{ s}^{-1}) = 9.94 \times 10^{-19} \text{ J} \cdot \text{photon}^{-1}$$

30.   Express the 589.0 nm wavelength in meters:

$$589.0 \text{ nm} \cdot \frac{1 \text{ m}}{1 \times 10^9 \text{ nm}} = 5.890 \times 10^{-7} \text{ m}$$

The frequency of the light is then: $\upsilon = \dfrac{c}{\lambda} = \dfrac{2.998 \times 10^8 \text{ m} \cdot \text{s}^{-1}}{5.890 \times 10^{-7} \text{ m}} = 5.090 \times 10^{14} \text{ s}^{-1}$

The energy is:

$$E = h\upsilon = (6.626 \times 10^{-34} \text{ J} \cdot \text{s} \cdot \text{photon}^{-1})(5.090 \times 10^{14} \text{ s}^{-1})(6.022 \times 10^{23} \text{ photons/mol})$$

and for one mole: $2.031 \times 10^5 \text{ J/mol} \cdot 1.00 \text{ mol} = 2.03 \times 10^5 \text{ J}$

32.   Since energy is proportional to frequency ($E = h\upsilon$), we can arrange the radiation in order of increasing energy per photon by listing the types of radiation in increasing frequency (or decreasing wavelength).

$\rightarrow$ Energy increasing $\rightarrow$

| radar | microwave | red neon | UV-light | gamma rays |
| signals | oven | sign | sun lamp | from nuclear reaction |

$\rightarrow$ Frequency (υ) increasing $\rightarrow$
$\leftarrow$ Wavelength (λ) increasing $\leftarrow$

34. $\text{Energy} = 220. \text{ kJ/mol} \cdot \dfrac{1 \text{ mol}}{6.02 \times 10^{23} \text{ photons}} \cdot \dfrac{1.00 \times 10^3 \text{ J}}{1.00 \text{ kJ}}$

$= 3.65 \times 10^{-19} \text{ J} \cdot \text{photons}^{-1}$

$E = h\upsilon = \dfrac{hc}{\lambda}$ or $\lambda = \dfrac{hc}{E} = \dfrac{(6.626 \times 10^{-34} \text{ J} \cdot \text{s} \cdot \text{photons}^{-1})(3.00 \times 10^8 \text{ m} \cdot \text{s}^{-1})}{3.65 \times 10^{-19} \text{ J} \cdot \text{photons}^{-1}}$

$= 5.44 \times 10^{-7} \text{ m or } 544 \text{ nm}$

Radiation of this wavelength--in the **visible** region of the electromagnetic spectrum-- would appear yellow.

36.   a.   <u>Transitions from</u>   <u>to</u>

            n = 4   n = 3, 2, or 1
            n = 3   n = 2 or 1
            n = 2   n = 1

      Six transitions are possible from these four quantum levels.

   b. Photons of the highest energy will be emitted in a transition from the level
      with **n = 4** to the level **n = 1**. This is easily seen with the aid of the equation

      $$\Delta E = Rhc\left(\frac{1}{n^2_f} - \frac{1}{n^2_i}\right).$$

      Since R, h, and c are constant for any transition, inspection shows that the largest
      change in energy results if $n_f = 1$ and $n_i = 4$.

   c. The emission line having the **longest wavelength** also has the **lowest frequency**
      and the **lowest energy**. A transition from $n_i = 4$ to $n_f = 3$ would provide the longest
      wavelength line.

38.   One can calculate the wavelength of any transition by applying the relationship:

         $$\frac{1}{\lambda} = R\left(\frac{1}{n^2_f} - \frac{1}{n^2_i}\right).$$

      For any line with a wavelength greater than 97.3 nm, the left side of the equation ($\frac{1}{\lambda}$)

      would be smaller than that for the 97.3 nm line. This will be true if the term $\left(\frac{1}{n^2_f} - \frac{1}{n^2_i}\right)$ is

      smaller than 15/16.

              For $n_f = 1$ and $n_i = 4$: $\frac{1}{1^2} - \frac{1}{4^2}$ = 15/16 or 0.94

      Substituting the suggested transitions into the term $\left(\frac{1}{n^2_f} - \frac{1}{n^2_i}\right)$ yields:

	transition between	$\left(\frac{1}{n^2_f} - \frac{1}{n^2_i}\right)$
a.	2 and 4	0.19
b.	1 and 3	0.89
c.	1 and 5	0.96
d.	3 and 5	0.07

      The transition between 1 and 5 would produce a line of shorter wavelength than 97.3 nm
      (higher energy). The remaining transitions would produce lines of longer wavelength (and
      lower energy) .

40. The wavelength of emitted light for the transition n = 4 to n = 1.

$$\frac{1}{\lambda} = R\left(\frac{1}{1^2} - \frac{1}{4^2} \right) = (1.0974 \times 10^7 \text{ m}^{-1})(15/16) = 1.0288 \times 10^7 \text{ m}^{-1}$$

$\lambda = 9.7199 \times 10^{-8}$ m or approximately 97 nm (far ultraviolet)

42. The energy needed to move an electron from n = 1 to n = 5 will be the same as the amount emitted as the electron relaxed from n = 5 to n = 1, i.e. 2.093×10^{-18} J.

44. Mass of an electron: 9.11×10^{-31} kg

Planck's constant: 6.626×10^{-34} J • s • photon^{-1}

Velocity of the electron: 3.0×10^8 cm • s^{-1} or 3.0×10^6 m • s^{-1}

$$\lambda = \frac{h}{m \bullet v} = \frac{6.626 \times 10^{-34} \text{ J} \bullet \text{s}}{(9.11 \times 10^{-31} \text{ kg} \bullet 3.0 \times 10^6 \text{ m} \bullet \text{s}^{-1})}$$

$= 2.4 \times 10^{-10}$ m = 2.4 Angstroms = 0.24 nm

46. Velocity of neutrons = 0.40 m • s^{-1}

mass of neutron = 1.67492×10^{-27} kg

Planck's constant: 6.626×10^{-34} J • s • photon^{-1}

$$\lambda = \frac{h}{m \bullet v} = \frac{6.626 \times 10^{-34} \text{ J} \bullet \text{s}}{(1.67492 \times 10^{-27} \text{ kg} \bullet 0.40 \text{ m} \bullet \text{s}^{-1})}$$

$= 9.9 \times 10^{-7}$ m = 990 nm

48. Complete the following table: (Answers are emboldened.)

QUANTUM NUMBER	ATOMIC PROPERTY DETERMINED
n	orbital size
m$_\ell$	relative orbital orientation
ℓ	orbital shape

50.   a. n = 3   possible ℓ values = 0, 1, 2   (ℓ = 0,1,2,...(n-1))

      b. ℓ = 3   possible m_ℓ values = -3, -2, -1, 0, +1, +2, +3   (-ℓ....0....+ℓ)

      c. orbital = 4s   n = 4; ℓ = 0; m_ℓ = 0

      d. orbital = 5f   n = 5; ℓ = 3; m_ℓ = -3, -2, -1, 0, +1, +2, +3

52.   An electron in a 4p orbital must have n = 4 and ℓ = 1. The possible m_ℓ values give rise to the
      following sets of n, ℓ, and m_ℓ

n	ℓ	m_ℓ
4	1	-1
4	1	0
4	1	+1

      Note that the **three values** of m describe **three orbital orientations**.

54.   The number of subshells within any given shell is always equal to the principal quantum
      number. The shell with principal quantum number **n = 4** would therefore have four
      subshells: s, p, d and f.

56.   Explain why each of the following is not a possible set of quantum numbers for an electron
      in an atom.

      a. n = 2, ℓ = 2, m_ℓ = 0   For n = 2, maximum value of ℓ is one (1).

      b. n = 3, ℓ = 0, m_ℓ = -2   For ℓ = 0, possible values of m_ℓ are ± ℓ and 0.

      c. n = 6, ℓ = 0, m_ℓ = 1   For ℓ = 0, possible values of m_ℓ are ± ℓ and 0.

58.   <u>quantum number designation</u>   <u>maximum number of orbitals</u>

      a.   n = 4; ℓ = 2   5 ("**d**" orbitals)

      b.   n = 5; ℓ = 0   1 ("**s**" orbital)
                      ℓ = 1   3 ("**p**" orbitals)
                      ℓ = 2   5 ("**d**" orbitals)
                      ℓ = 3   7 ("**f**" orbitals)
                      ℓ = 4   <u>9 ("**g**" orbitals)</u>
                                                  25 orbitals

      c.   n = 2; ℓ = 2   invalid set; for n = 2
                                                  the max number of ℓ = 1

      d.   n = 3; ℓ = 1; m_ℓ = -1   1 (a 3 "**p**" orbital)

60.   The number of planar nodes possessed by each of the following:

orbital	number of planar nodes
a. 2s	0
b. 5d	2
c. 5f	3

62.   Which of the following orbitals cannot exist and why:

2s exists	$n = 2$ permits ℓ values as large as 1
2d cannot exist	$\ell = 2$ is not permitted for $n < 3$
3p exists	$n = 3$ permits ℓ values as large as 2
3f cannot exist	$\ell = 3$ is not permitted for $n < 4$
4f exists	$\ell = 4$ permits ℓ values as large as 3
5s exists	$n = 5$ permits ℓ values as large as 4

64.   The complete set of quantum numbers for :

		$\underline{n}$	$\underline{\ell}$	$\underline{m_\ell}$	
a.	2p	2	1	-1, 0, +1	(3 orbitals)
b.	3d	3	2	-2, -1, 0, +1, +2	(5 orbitals)
c.	4f	4	3	-3, -2, -1, 0, +1, +2, +3	(7 orbitals)

66.   The general shapes of the electrons clouds for the orbitals requested are show below:

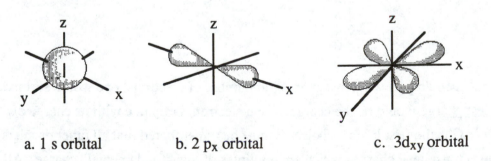

a. 1 s orbital   b. 2 p_x orbital   c. 3d_{xy} orbital

68.   All orbitals except **the s orbital** may have magnetic quantum numbers, $m_\ell = -1$.
        The s orbital has as its only value for the magnetic quantum number, $m_\ell = 0$

70.   The frequency 6.00×10^5 s^{-1} has a wavelength of:

$$\lambda = \frac{c}{\upsilon} = \frac{2.998 \times 10^8 \text{ m} \cdot s^{-1}}{6.00 \times 10^5 \text{ } s^{-1}} = 5.00 \times 10^2 \text{ m}$$

72.   Calculate the frequency of light with a wavelength of 560 nm:

$$\upsilon = \frac{c}{\lambda} = \frac{2.998 \times 10^8 \text{ m} \cdot s^{-1}}{5.6 \times 10^{-7} \text{ m}} = 5.4 \times 10^{14} \text{ } s^{-1}$$

Calculate the energy of this radiation:

$$E = h\upsilon = (6.626 \times 10^{-34} \text{ J} \cdot \text{s} \cdot \text{photon}^{-1})(5.4 \times 10^{14} \text{ } s^{-1}) = 3.6 \times 10^{-19} \text{ J} \cdot \text{photon}^{-1}$$

Five percent of the radiated energy is:

$$0.05 \cdot 100 \text{ W} \cdot \frac{1 \text{ J} \cdot s^{-1}}{\text{W}} = 5 \text{ J} \cdot s^{-1}$$

Since each photon has energy equal to 3.6×10^{-19} J, we calculate the number of photons:

$$\frac{5 \text{ J}}{1 \text{ s}} \cdot \frac{1 \text{ photon}}{3.6 \times 10^{-19} \text{ J}} = 1 \times 10^{19} \text{ photons per second.}$$

[Note that the calculation is restricted to one significant figure by the 5% value.]

74.

Drop	Charge on drop (in units of 10^{-19} C)
a	-3.2
b	-4.8
c	-6.4
d	-8.0
e	-1.6

Noting that each droplet will have an integral number of electrons, one would surmise that the smallest charge would be the charge of one electron. Droplet **e** with its charge of -1.6×10^{-19} C is the first logical choice. It must be remembered that all other droplets (**a-d** in this case) must have charges which are multiples of this "fundamental" charge. All charges can be arrived at with multiples of this charge (-1.6×10^{-19} C), so this data indicates that the charge of one electron is -1.6×10^{-19} coulombs.

76.   The energy of the electron in the n = 4 orbital, using Bohr's energy equation:

$$E = \frac{-R \cdot h \cdot c}{n^2}$$

$$= \frac{(-1.0974 \times 10^7 \text{ m}^{-1})(6.6261 \times 10^{-34} \text{ J} \cdot \text{s})(2.9979 \times 10^8 \text{ m} \cdot \text{s}^{-1})}{4^2}$$

$$= -1.3624 \times 10^{-19} \text{ J} \cdot \text{atom}^{-1}$$

Expressing this in kJ/mol we obtain:

$$-1.3624 \times 10^{-19} \text{ J} \cdot \text{atom}^{-1} \cdot \frac{1 \text{kJ}}{1000 \text{ J}} \cdot \frac{6.0221367 \times 10^{23} \text{ atom}}{1 \text{ mol}} = -82.049 \text{ kJ/mol}$$

78.   Increasing energy order: 1s < 2s = 2p < 3s = 3p = 3d < 4s

80.   a. The quantum number n describes the **size or diameter** of an atomic orbital and the quantum number ℓ describes its **shape or type**.

b. When n = 3, the possible values of ℓ are 0,1, and 2 [1.....(n-1)].

c. The **f** orbital corresponds to $\ell = 3$.

d. For a 4d orbital, the value of n is **4**, the value of ℓ is **2**, and a possible value of m_ℓ is **0, ± 1, or ± 2**.

e. The orbitals pictured have the following characteristics:

letter =	p	s	d
ℓ value =	1	0	2
nodal planes =	1	0	2

f. An atomic orbital with 3 nodal planes is **f**.

g. According to modern quantum theory, the **1d** and **4g** orbitals cannot exist.
        Maximum ℓ value for n = 1 is 0 (s) and for n = 4 (f).

h. The set **n = 2, ℓ = 1, and m_ℓ = 2** is not a valid set of quantum numbers:

i. The maximum number of orbitals associated with the following sets of quantum

numbers is:	orbitals
i. n = 2 and ℓ = 1	3 (the p orbitals)
ii. n = 3	1 s, 3 p's, and 5 d's = 9 total
iii n = 3 and ℓ = 3	none (max value of ℓ = (n - 1)
iv. n = 2, ℓ = 1, and m_ℓ = 0	1 (of the three 2p orbitals)

105

CHAPTER 9: Atomic Electron Configurations and Periodicity

20.   The electron configurations for Mg and Cl using both the orbital box and spectroscopic notations:

Orbital box notation	Spectroscopic notation

Mg   [↑↓]   [↑↓]   [↑↓|↑↓|↑↓]   [↑↓] $1s^2 2s^2 2p^6 3s^2$

Cl   [↑↓]   [↑↓]   [↑↓|↑↓|↑↓]   [↑↓]   [↑↓|↑↓|↑] $1s^2 2s^2 2p^6 3s^2 3p^5$

22.   Vanadium's electron configuration: $[Ar] 3d^3 4s^2$

24.   Germanium's electron configuration: $[Ar] 3d^{10} 4s^2 4p^2$

26.   Electron configurations, using the spectroscopic notation:

a. Sr: $[Kr] 5s^2$ b. Zr: $[Kr] 4d^2 5s^2$
c. Rh: $[Kr] 4d^8 5s^1$ d. Sn: $[Kr] 4d^{10} 5s^2 5p^2$

28.   Electron configurations for the following "rare earths", using the spectroscopic notation:

a. Eu: $[Xe] 4f^7 6s^2$ b. Yb: $[Xe] 4f^{14} 6s^2$

30.   Electron configurations for the following actinides, using the spectroscopic notation:

a. Pu: $[Rn] 5f^6 7s^2$ b. Es: $[Rn] 5f^{11} 7s^2$

32.   The orbital box representations for the following ions:

Orbital box notation	Spectroscopic notation

a. Na^+ [↑↓]   [↑↓]   [↑↓|↑↓|↑↓] $1s^2 2s^2 2p^6$

b. Al^{3+} [↑↓]   [↑↓]   [↑↓|↑↓|↑↓] $1s^2 2s^2 2p^6$

c. Cl^- [↑↓]   [↑↓]   [↑↓|↑↓|↑↓]   [↑↓]   [↑↓|↑↓|↑↓] $1s^2 2s^2 2p^6 3s^2 3p^6$

34.   Electron configurations of:

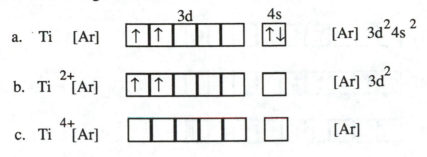

a. Ti   [Ar]   [Ar] $3d^2 4s^2$

b. Ti $^{2+}$ [Ar]   [Ar] $3d^2$

c. Ti $^{4+}$ [Ar]   [Ar]

Note that the Ti^{2+} ion contains two unpaired electrons, and is therefore paramagnetic.

36.   Element 25 is manganese.

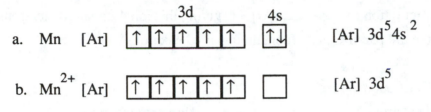

a. Mn   [Ar]   [Ar] $3d^5 4s^2$

b. Mn $^{2+}$ [Ar]   [Ar] $3d^5$

c. Having five (5) unpaired electrons, Mn $^{2+}$ is **paramagnetic**.

38.   For ruthenium:

a. The predicted and actual electron configurations, using the box notation are:

Predicted: Ru [Kr]   [Kr] $4d^6 5s^2$

Actual: Ru [Kr]   [Kr] $4d^7 5s^1$

b. The formation of the Ru^{3+} ion can be arrived at (from the predicted configuration) by
    loss of the two 5s electrons and one 4d electron. From the actual configuration, we see
    the loss of two 4d electrons (note the increased number of unpaired 4d electrons) and the
    5s electron.

Ru $^{3+}$ [Kr]   [Kr] $4d^5$

40.   The electron configuration for Sm and Ho and their ions using the orbital box method:

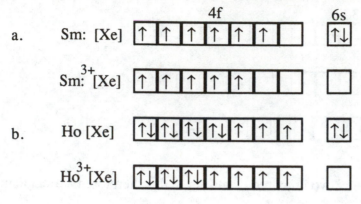

a.   Sm: [Xe]

        Sm³⁺ [Xe]

b.   Ho [Xe]

        Ho³⁺ [Xe]

42.   The electron configuration for Ti²⁺ is : [Ar] 3d² hence this ion has 2 unpaired electrons.
        The electron configuration for Co³⁺ is : [Ar] 3d⁶ giving this ion 4 unpaired electrons.
        Both ions therefore are paramagnetic (contain unpaired electrons).

44.   The 2+ ions for Ti through Zn are:

Ion	Unpaired electrons		Ion	Unpaired electrons
Ti²⁺	2		Co²⁺	3
V²⁺	3		Ni²⁺	2
Cr²⁺	4		Cu²⁺	1
Mn²⁺	5		Zn²⁺	0 **diamagnetic**
Fe²⁺	4			

46.   The electron configuration for Be using the orbital box method:

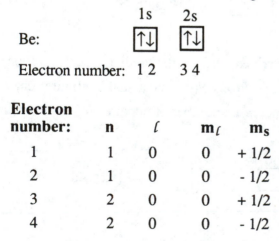

Be:

Electron number: 1 2   3 4

Electron number:	n	ℓ	m_ℓ	m_s
1	1	0	0	+ 1/2
2	1	0	0	- 1/2
3	2	0	0	+ 1/2
4	2	0	0	- 1/2

48.   The electron configuration for Ti using the orbital box method:

$$\text{Ti} \quad [\text{Ar}]$$

3d   4s

| ↑ | ↑ | | | | | ↑↓ |

Electron number   3   4   1 2

Electron number:	n	ℓ	m_ℓ	m_s
1	4	0	0	+ 1/2
2	4	0	0	- 1/2
3	3	2	-2	+ 1/2
4	3	2	-1	+ 1/2

50.   Maximum number of electrons associated with the following sets of quantum numbers:

	Characterized as	Maximum number of electrons
a. $n = 2$ and $\ell = 1$	2p electrons	6
b. $n = 5$	5s, 5p, 5d, 5f, 5g electrons	50
c. $n = 4$ and $\ell = 4$	[Maximum $\ell = n-1$]	none
d. $n = 3$, $\ell = 1$, $m_\ell = -1$, and $m_s = -1/2$	3p orbital	1 (4 q. n. completely describe 1 electron)
e. $n = 3$, $\ell = 0$, $m_\ell = +1$	[Maximum $m_\ell = +2$ or -2]	none

52.   Explain why the following sets of quantum numbers are not valid:

a. $n = 2$, $\ell = 2$, $m_\ell = 0$, $m_s = +1/2$:

       The maximum value of ℓ for any given n value is (n - 1).

b. $n = 2$, $\ell = 1$, $m_\ell = 1$, $m_s = 0$:

       The spin quantum number possesses only two possible values: +1/2 or -1/2.

54.   Possible sets of quantum numbers for an electron in a 4p orbital:

n	ℓ	m_ℓ	m_s
4	1	-1	+ 1/2
4	1	-1	- 1/2
4	1	0	+ 1/2
4	1	0	- 1/2
4	1	+1	+ 1/2
4	1	+1	- 1/2

56.   Estimate E-Cl bond distances: [The atomic radius for Cl = 100 pm.]

Element	Atomic Radius (pm)	E-Cl Bond Distance (pm)
C	77	177
Si	118	218
Ge	122	222
Sn	140	240
Pb	146	246

58.   Elements arranged in order of increasing size: C < B < Al < Na < K

60.   The specie in each pair with the larger radius:

   a. Cl⁻ is larger than Cl -- The ion has more electrons/proton than the atom.

   b. Al is larger than N -- Al is in period 3, while N is in period 2.

   c. In is larger than Sn -- Atomic radii decrease, in general, across a period.

62.   The group of elements with correctly ordered increasing ionization energy (IE):

                         c. Li < Si < C < Ne.

   Neon would have the greatest IE. Silicon, being slightly larger in atomic radius than
   carbon, has a lesser IE. Lithium, the largest atom of this group, would have the smallest
   IE.

64.   For the elements Li, K, C, and N:
       Increasing ionization energy: K < Li < C < N

66.   (e) 1st IE of K < (b) 1st IE of Li < (a) 1st IE of Be < (c) 2nd IE of Be < (d) 2nd IE of Na
       The first ionization energies of K, Li, and Be increase as the atomic sizes decrease. In
       general the second ionization energies are larger than the first ionization energies. The
       relative orders for the second IE of beryllium and sodium are anticipated if one recalls that
       beryllium is located in group 2A and sodium in group 1A. One should expect that the
       difference between the first and second ionization energies of sodium to be much greater
       than that of beryllium.

68.   For the elements Li, K, C, and N:
       a. The largest atomic radius: K

b. The largest electron affinity: C (most negative value for EA)

c. Increasing ionization energy: K < Li < C < N

70.   a. Increasing ionization energy: S < O < F

Ionization energy is inversely proportional to atomic size.

b. Largest ionization energy of O, S, or Se: O

Oxygen is the smallest of these Group 6A elements.

c. Largest electron affinity of Se, Cl, or Br: Cl

Chlorine is the smallest of these three elements. EA tends to increase (become more negative) on a diagonal from the lower left of the periodic table to the upper right.

d. Largest radius of O^{2-}, F^-, F: O^{2-}

The oxide ion has the largest electron : proton ratio.

72.   Group 2 A should form dipositive ions (as well as Sn and Pb in Group 4A).

74.   The electronic configuration for element 109 is: $[Rn] 5f^{14} 6d^7 7s^2$

Iridium would be another element in the same group.

76.   The set of quantum numbers which is incorrect is b. The maximum value of ℓ is (n - 1).

With an $\ell = 1$, this value could not exceed 0.

78.   The number of complete electron shells in element 71 is four (4).

Element 71 is Lu, which has all subshells of the fourth shell filled.

80.   The number of p orbital pairs in Se is: 7

2p (3 pairs), 3p (3 pairs), and 4p (1 pair)

82.   The order of first IE's for the elements given--Si < S < P < Cl-- is explained by noting that IE's generally increase across a period due to increasing effective nuclear charge. Phosphorus has a half-filled p subshell, and the addition of the "next" electron-- for S results in an increased electron-electron repulsion, with the resulting order of IE being S < P.

84.   Elements containing four unpaired electrons in the 3+ ion: Tc and Rh.

86.   For element A = ...$3p^64s^2$ and B = ...$3p^63d^{10}4s^24p^5$

   a. Element A is a metal (two electrons in the s sublevel).

   b. Element B is a nonmetal (seven electrons in the outer shell).

   c. Since B is "nonmetallic", it should have the larger ionization energy.

   d. B would be smaller as a result of the larger effective nuclear charge for B's outer
        electrons when compared to A's outer electrons.

88.   These 5 species are isoelectronic (18 electrons).

   Radius (pm):   170   167   152   114

$$S^{2-} > Cl^- > K^+ > Ca^{2+}$$

90.   An examination of the electron configurations for Group 4A and 5A shows that the gain of
        one electron by a Group 4A element would result in a half-filled p subshell. The 5A
        elements have a half-filled p subshell naturally, and the addition of an electron would result
        in increased electron-electron repulsions--resulting in lower electron affinities for Group
        5A.

92.   Since Mg is smaller than Na, we would expect the first IE of Mg to be greater than that of
        Na. Once Na has lost its "first" electron, it possesses a [Ne] core. Removal of an
        additional electron (the second) requires much energy. Mg, on the other hand, can lose a
        "second" electron to obtain the stable noble gas configuration with a much smaller amount
        of energy (smaller IE).

94.   For the elements Na, B, Al, and C:

   a. The largest atomic radius: Na (186 pm)

   b. The largest electron affinity: C (most negative value for EA)

   c. Increasing ionization energy: Na < Al < B < C

96. a. Sulfur is located in period 3.

   b. Orbital box notation for sulfur:

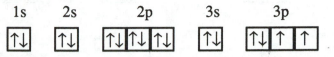

   c. Quantum numbers for the "last electron": $n = 3$, $\ell = 1$, $m_\ell = +1$, $m_s = -1/2$

   d. Element with the smallest ionization energy : S

      Element with the smallest radius: O

   e. S: Negative ions are always larger than the element from which they are derived.

   f. Grams of Cl_2 to make 675 g of $OSCl_2$:

   $$675 \text{ g } OSCl_2 \cdot \frac{1 \text{ mol } OSCl_2}{119.0 \text{ g } OSCl_2} \cdot \frac{2 \text{ mol } Cl_2}{1 \text{ mol } OSCl_2} \cdot \frac{70.91 \text{ g } Cl_2}{1 \text{ mol } Cl_2} = 804 \text{ g } Cl_2$$

   g. The theoretical yield of $OSCl_2$ if 10.0 g of SO_2 and 20.0 g of Cl_2 are used:

   $$\text{Moles of } SO_2 = 10.0 \text{ g } SO_2 \cdot \frac{1 \text{ mol } SO_2}{64.06 \text{ g } SO_2} = 0.156 \text{ moles } SO_2$$

   $$\text{Moles of } Cl_2 = 20.0 \text{ g } Cl_2 \cdot \frac{1 \text{ mol } Cl_2}{70.91 \text{ g } Cl_2} = 0.282 \text{ moles } Cl_2$$

   $$\text{Moles-available ratio: } \frac{0.156 \text{ moles } SO_2}{0.282 \text{ moles } Cl_2} = \frac{0.553 \text{ moles } SO_2}{1 \text{ mol } Cl_2}$$

   $$\text{Moles-required ratio: } \frac{1 \text{ mol } SO_2}{2 \text{ mol } Cl_2} = \frac{0.5 \text{ mol } SO_2}{1 \text{ mol } Cl_2}$$

   Cl_2 is the limiting reagent.

   $$0.282 \text{ moles } Cl_2 \cdot \frac{1 \text{ mol } OSCl_2}{2 \text{ moles } Cl_2} \cdot \frac{119.0 \text{ g } OSCl_2}{1 \text{ mol } OSCl_2} = 16.8 \text{ g } OSCl_2$$

   h. For the reaction:

   $$SO_2 \text{ (g)} + 2 Cl_2 \text{ (g)} \rightarrow OSCl_2 \text{ (g)} + Cl_2O \text{ (g)} \qquad \Delta H = +164.6 \text{ kJ}$$

   $\Delta H_{rxn} = [\Delta H_f \, OSCl_2 \text{ (g)} + \Delta H_f \, Cl_2O \text{ (g)}] - [\Delta H_f \, SO_2\text{(g)} + 2 \cdot \Delta H_f \, Cl_2 \text{ (g)}]$

   $+ 164.6 \text{ kJ} = [\Delta H_f \, OSCl_2 \text{ (g)} + 80.3 \text{ kJ/mol}] - [-296.8 \text{ kJ/mol} + 2 \cdot (0)]$

   $+ 164.6 \text{ kJ} = \Delta H_f \, OSCl_2 \text{ (g)} + 377.1 \text{ kJ}$

   $- 212.5 \text{ kJ} = \Delta H_f \, OSCl_2 \text{ (g)}$

CHAPTER 10: Basic Concepts of Chemical Bonding

27.

	Element	Number of Valence Electrons	Group Number
a.	N	5	5A
b.	B	3	3A
c.	S	6	6A
d.	Na	1	1A
e.	Mg	2	2A

29. The compound with the larger enthalpy of dissociation :

Since the $\Delta H_{dissociation}$ is inversely proportional to the distance between ion centers in the crystal lattice, the $\Delta H_{dissociation}$ increases as the distance decreases.

a. LiI > NaI since Li$^+$ < Na$^+$ b. LiF > KF since Li$^+$ < K$^+$

c. MgO > CaO since Mg^{2+} < Ca^{2+}

31. Assuming that the solubility in water increases with a decreasing $\Delta H_{dissociation}$, the solubility of the following compounds should be:

a. NaI more soluble than NaCl   b. CsF more soluble than KF

c. BaO more soluble than MgO

33. Assuming that the melting point of a compound is in direct proportion to the enthalpy of dissociation, we predict:

a. RbCl melts lower than NaCl (Rb$^+$ > Na$^+$)

b. BaO melts lower than MgO (Ba^{2+} > Mg^{2+})

c. NaCl melts lower than MgCl$_2$ (charge on Mg^{2+} > charge on Na$^+$)

35.

Group Number	Number of Bond Pairs
1A	1
2A	2
3A	3
4A	4
5A	3 (or 4 in species such as NH$_4$$^+$)
6A	2 (or 3 as in H$_3$O$^+$)
7A	1
8A	0

37.   a. NF_3 : $[1(5) + 3(7)]$ = 26 valence electrons

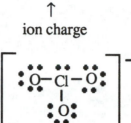

b. ClO_3^- : $[1(7) + 3(6) + 1]$ = 26 valence electrons

↑
ion charge

$$\left[\,:\!\ddot{O}\!-\!\overset{\cdot\cdot}{\underset{\overset{|}{:\ddot{O}:}}{Cl}}\!-\!\ddot{O}\!: \, \right]^{-}$$

c. HOCl : $[1(1) + 1(6) + 1(7)]$ = 14 valence electrons

$$H-\overset{\cdot\cdot}{\underset{\cdot\cdot}{O}}-\overset{\cdot\cdot}{\underset{\cdot\cdot}{Cl}}:$$

d. SO_3^{2-} : $[1(6) + 3(6) + 2]$ = 26 valence electrons

↑
ion charge

$$\left[\,:\!\ddot{O}\!-\!\overset{\cdot\cdot}{\underset{\overset{|}{:\ddot{O}:}}{S}}\!-\!\ddot{O}\!: \, \right]^{2-}$$

39.   a. $CHClF_2$: $[1(4) + 1(1) + 1(7) + 2(7)]$ = 26 valence electrons

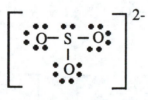

b. HCOOH: $[1(1) + 1(4) + 2(6) + 1(1)]$ = 18 valence electrons

c. H₃CCN: [3(1) + 2(4) + 1(5)] = 16 valence electrons

$$H-\underset{\underset{H}{|}}{\overset{\overset{H}{|}}{C}}-C\equiv N$$

d. H₃COH: [3(1) + 1(4) + 1(6) + 1(1)] = 14 valence electrons

$$H-\underset{\underset{H}{|}}{\overset{\overset{H}{|}}{C}}-\ddot{O}-H$$

41.   Resonance structures for:
       a. SO₂:

$$\ddot{O}=\dot{S}-\ddot{O}:\qquad\qquad:\ddot{O}-\dot{S}=\ddot{O}$$

       b. SO₃:

$$:\ddot{O}-\underset{}{\overset{\overset{:\overset{..}{O}:}{\|}}{S}}-\ddot{O}:\qquad:\ddot{O}=\underset{}{\overset{\overset{:\overset{..}{O}:}{|}}{S}}-\ddot{O}:\qquad:\ddot{O}-\underset{}{\overset{\overset{:\overset{..}{O}:}{|}}{S}}=O$$

       c. SCN⁻:

$$\left[:\ddot{N}-C\equiv S:\right]^{-}\quad\left[\ddot{N}=C=\ddot{S}\right]^{-}\quad\left[:N\equiv C-\ddot{S}:\right]^{-}$$

43. a. BrF₃ : [1(7) + 3(7)] = 28 valence electrons

$$:\ddot{F}-\underset{\underset{:\ddot{F}:}{|}}{Br}-\ddot{F}:$$

     b. I₃⁻ : [3(7) + 1] = 22 valence electrons

$$\left[:\ddot{I}-\ddot{I}-\ddot{I}:\right]$$

c. XeO_2F_2 : [1(8) + 2(6) + 2(7)] = 34 valence electrons

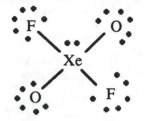

45.

Specie	σ	π	Bond Order : Bonded Atoms
a. HCOOH	4	1	1 : CH;C-O;O-H 2: C = O
b. SO_3^{2-}	3	0	1 : SO
c. NO_2^+	2	2	2 : NO

47.   In each case the shorter bond length should be between the atoms with smaller radii--if we assume that the bond orders are equal.

a. B-Cl   B is smaller than Ga   b. C-O   C is smaller than Sn

c. P-O   O is smaller than S   d. C=O   O is smaller than C

49.   The CO bond in carbon monoxide is shorter. The CO bond in carbon monoxide is a triple bond, thus it requires more energy to break than the CO double bond in H_2CO.

51.   The bond order for NO_2^+ is 2, while the bond order for NO_3^- is 4/3. The Lewis dot structure for the NO_2^+ ion indicates that both NO bonds are double, while in the nitrate ion, any resonance structure (there are three) shows one π bond and three σ bonds. Hence the NO bonds in nitrate will be longer than those in the NO_2^+ ion.

53.   Formation of :

a. NH_3 (g) : N_2 (g) + 3 H_2 (g) $\rightarrow$ 2 NH_3 (g)

	Bonds broken	1 mol N≡N	946 kJ/mol =	946 kJ
		3 mol H-H	436 kJ/mol =	1308 kJ
			Energy input:	2254 kJ

Bonds formed   6 mol N-H   389 kJ/mol = <u>2334 kJ</u>

Energy released   2334 kJ

Energy change = 2254 kJ - 2334 kJ = -80. kJ

For the formation of 1 mol of NH_3(g) : $\dfrac{-80.\ kJ}{2\ mol\ NH_3}$ = - 40. kJ/ mol NH_3

b. H_2O (g) : $2\ H_2$ (g) + O_2 (g) → $2\ H_2O$ (g)

Bonds broken   2 mol H-H = 2 mol • 436 kJ/mol = 872 kJ

1 mol O-O = 1 mol • 498 kJ/mol = <u>498 kJ</u>

Energy input   1370 kJ

Bonds formed   4 mol H-O = 4 mol • 464 kJ/mol = <u>1856 kJ</u>

Energy released   1856 kJ

For the formation of 1 mol of H_2O (g) = $\dfrac{-486\ kJ}{2\ mol\ H_2O}$ = - 243 kJ/mol

55.   Heat of formation of propane:

The formation of propane from its elements may be written:

$3\ C$ (graphite) + $4\ H_2$ (g) → C_3H_8 (g)

Energy input: 3 C (graphite) → 3 C (g)   = 3 mol • 717 kJ/mol = 2151 kJ

4 mol H-H   = 4 mol • 436 kJ/mol = <u>1744 kJ</u>

Total input   = 3895 kJ

Energy released: 2 mol C-C   = 2 mol • 347 kJ/mol = 694 kJ

8 mol C-H   = 8 mol • 414 kJ/mol = <u>3312 kJ</u>

Total release   = 4006 kJ

Energy change = 3895 kJ - 4006 kJ = - 111 kJ

57.   Heat of formation of glycine:

$2\ C$ (graphite) + $5/2\ H_2$ (g) + $1/2\ N_2$ (g) + O_2 (g) → $C_2H_5O_2N$ (s)

Energy input: 2 C (graphite) → 2 C (g) = 2 mol • 717 kJ/mol   = 1434 kJ

5/2 mol H_2 (g)   = 5/2 mol • 436 kJ/mol = 1090 kJ

1/2 mol N-N   = 1/2 mol • 946 kJ/mol = 473 kJ

1 mol O=O   = 1 mol • 498 kJ/mol = <u>498 kJ</u>

Total input   = 3495 kJ

Energy released: 2 mol C-H   = 2 mol • 414 kJ/mol   = 828 kJ

                       1 mol C-N   = 1 mol • 293 kJ/mol   = 293 kJ

                       2 mol N-H   = 2 mol • 389 kJ/mol   = 778 kJ

                       1 mol C-C   = 1 mol • 347 kJ/mol   = 347 kJ

                       1 mol C=O   = 1 mol • 745 kJ/mol   = 745 kJ

                       1 mol C-O   = 1 mol • 351 kJ/mol   = 351 kJ

                       1 mol O-H   = 1 mol • 464 kJ/mol   = 464 kJ

                                            Total release   = 3806 kJ

Energy change = 3495 kJ - 3806 kJ = - 311 kJ

59.   For the reaction H_2 (g) + I_2 (g) → 2 HI (g) ΔH = 26.48 kJ/mol HI

Energy input: H-H bond breaking = 1 mol • 436 kJ/mol   = 436 kJ

                I-I bond breaking = 1 mol • 151 kJ/mol   = 151 kJ

                I_2 (s) → I_2 (g)   = 1 mol • 62.44 kJ/mol = 62.44 kJ

                                        Total input   = 649. kJ

Energy released: 2 mol • H-I bond formation.

We may write:

       Energy change = Energy input - Energy release

(2 mol HI)(26.48 kJ/mol HI) = 649 kJ - (2 mol • H-I)

       52.96 kJ - 649 kJ   =   -(2 mol • H-I)

              -596 kJ   =   -2 mol • H-I

       298 kJ/mol   = H-I bond energy

From Table 10.4, the bond energy of H-I is 297 kJ/mol. An inspection of bond energies reveals that the H-X bond energy decreases as the atomic weight (and size) of X increases.

61. OF_2 (g) + H_2O (g) → O_2 (g) + 2 HF (g) ΔH = - 318 kJ

Energy input :   2 mol O-F = 2 x (where x = O-F bond energy)

                     2 mol O-H = 2 mol • 464 kJ/mol = 928 kJ

                                    Total input   = (928 + 2x) kJ

Energy release: 1 mol O=O = 1 mol • 498 kJ/mol =   498 kJ

                     2 mol H-F = 2 mol • 569 kJ/mol = 1138 kJ

                                    Total release   = 1636 kJ

$$- 318 \text{ kJ} = 928 \text{ kJ} + 2x - 1636 \text{ kJ}$$
$$390 \text{ kJ} = 2x$$
$$195 \text{ kJ/mol} = \text{O-F bond energy}$$

63.   Heat of reaction for : $CO \text{ (g)} + Cl_2 \text{ (g)} \rightarrow Cl_2CO \text{ (g)}$

Energy input: 1 mol C≡O = 1 mol • 1075 kJ/mol = 1075 kJ
                       1 mol Cl-Cl = 1 mol • 243 kJ/mol = 243 kJ
                                           Total input   = 1318 kJ

Energy release: 2 mol C-Cl = 2 mol • 330 kJ/mol = 660 kJ
                       1 mol C=O = 1 mol • 745 kJ/mol = 745 kJ
                                           Total released = 1405 kJ

Energy change: 1318 kJ - 1405 kJ = - 87 kJ

65.   Estimate the enthalpy of the hydrogenation reaction for propane:

$$\begin{array}{ccc}
& \text{H} \quad \text{H} & \qquad\qquad \text{H} \quad \text{H} \\
& | \quad\; | & \qquad\qquad | \quad\; | \\
H_3C - C=C-H \;+\; H_2 & \longrightarrow & H_3C - C - C - H \\
& & \qquad\qquad | \quad\; | \\
& & \qquad\qquad \text{H} \quad \text{H}
\end{array}$$

Energy input: 1 mol C=C = 1 mol • 611 kJ/mol =   611 kJ
                       1 mol H-H = 1 mol • 436 kJ/mol =   436 kJ
                                           Total input   = 1047 kJ

Energy release: 1 mol C-C = 1 mol • 347 kJ/mol = 347 kJ
                       2 mol C-H = 2 mol • 414 kJ/mol = 828 kJ
                                           Total released = 1175 kJ

Energy change: 1047 kJ - 1175 kJ = - 128 kJ

67.   Indicate the more polar bond (Arrow points toward the more negative atom in the dipole).

a. C-O >   C-N   b. P-O >   P-S
   →   →   →   →

c. P-H <   P-N   d. B-H < B-I
   not polar   →   →   →

69. For the bonds in acrolein the polarities are as follows:

	H-C	C-C	C=O
$\frac{\Delta\chi}{\Sigma\chi}$	0.09	0	0.16

a. The C-C bonds are nonpolar, the C-H bonds are only slightly polar, and the C=O bond is polar.

b. The most polar bond in the molecule is the C=O bond, with the oxygen atom being the negative end of the dipole.

71. Using the Lewis structures, the oxidation numbers of the atoms in the following:

a. water   b. hydrogen peroxide   c. sulfur dioxide

H–Ö–H   H–Ö–Ö–H   Ö=S̈–Ö:

H = +1; O = -2   H = +1; O = -1   S = +4 ; O = -2

d. nitrogen (I) oxide   e. hypochlorite ion

N̈=N=Ö   [:Ö–C̈l:]⁻

N = +1 ; O = -2   Cl = +1 ; O = -2

73.

	Atom	Formal Charge	Oxidation Number
a. H_2O	H	$1 - 1/2(2) = 0$	+1
	O	$6 - 4 - 1/2(4) = 0$	-2
b. CH_4	C	$4 - 1/2(8) = 0$	-4
	H	$1 - 1/2(2) = 0$	+1
c. NO_2^+	N	$5 - 1/2(8) = +1$	+5
	O	$6 - 4 - 1/2(4) = 0$	-2
d. HOF	H	$1 - 1/2(2) = 0$	+1
	O	$6 - 4 - 1/2(4) = 0$	0
	F	$7 - 6 - 1/2(2) = 0$	-1

75. Resonance structures for NO_2^- :

$$\left[\ddot{O}=\ddot{N}-\ddot{O}:\right]^- \qquad \left[:\ddot{O}-\ddot{N}=\ddot{O}\right]^-$$

   1   2   1   2

Formal charges:

O_1 6 - 4 - 1/2(4) = 0 O_1 6 - 6 - 1/2(2) = -1

N   5 - 2 - 1/2(6) = 0   N   5 - 2 - 1/2(6) = 0

O_2 6 - 6 - 1/2(2) = -1 O_2 6 - 4 - 1/2(4) = 0

77.   a. Resonance structures of N_2O :

$$:N\equiv N-\ddot{O}: \longleftrightarrow \ddot{N}=N=\ddot{O} \longleftrightarrow :\ddot{N}-N\equiv O:$$

   1   2   1   2   1   2

b. Formal Charges:

N_1	5 - 2 - 1/2(6) = 0	5 - 4 - 1/2(4) = -1	5 - 6 - 1/2(2) = -2
N_2	5 - 0 - 1/2(8) = +1	5 - 0 - 1/2(8) = +1	5 - 0 - 1/2(8) = +1
O	6 - 6 - 1/2(2) = -1	6 - 4 - 1/2(4) = 0	6 - 2 - 1/2(6) = +1

c. Of these three structures, the first is the most reasonable in that the most electronegative atom, O, bears a formal charge of -1.

79.   Using the Lewis structure describe the structural-pair and molecular geometry:

a.

$$H-\ddot{N}-\ddot{Cl}:$$
$$\;\;\;\;|$$
$$\;\;\;\;H$$

Structural-pair: tetrahedral
Molecular : pyramidal

b.

$$:\ddot{Cl}-\ddot{O}-\ddot{Cl}:$$

Structural-pair: tetrahedral
Molecular: bent

c.

$$\left[:N\equiv C-\ddot{S}:\right]^-$$

Structural-pair: linear
Molecular: linear

d.

H—Ö—F̈:   Structural-pair: tetrahedral
           Molecular: bent

81. Using the Lewis structure describe the structural-pair and molecular geometry:

a.

Ö=C=Ö   Structural-pair: linear
          Molecular : linear

b.

[Ö=N—Ö]⁻   Structural-pair: trigonal planar
           Molecular: bent

c.

Ö=S—Ö: Structural-pair: trigonal planar
           Molecular: bent

d.

:Ö—Ö=Ö   Structural-pair: trigonal planar
           Molecular: bent

e.

[:Ö—Cl—Ö:]⁻   Structural-pair: tetrahedral
              Molecular: bent

83.   Using the Lewis structure describe the structural-pair and molecular geometry:
       [Lone pairs on F have been omitted for clarity.]

a.

[F—Cl—F]⁻   Structural-pair: trigonal bipyramidal
             Molecular : linear

b.

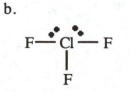   Structural-pair: trigonal bipyramidal
         Molecular: T-shaped

c.

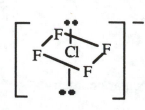

Structural-pair: octahedral
Molecular: square planar

d.

F

Structural-pair: octahedral
Molecular: pyramidal

85.   a. O-S-O angle in SO_2 :   Slightly less than 120°; The lone pair of S should
                                                     reduce the predicted 120° angle slightly.

       b. F-B-F angle in BF_3 :   120°

       c. (1) H-O-N angle in HNO_3 :   Slightly less than 109°; the lone pair on O should
                                                     reduce the predicted angle to approximately 105°.

          (2) O-N=O angle in HNO_3 :   120°

       d. (1) H-C-H angle in C_2H_3CN :   120°
          (2) C-C ≡N angle in C_2H_3CN :   180°

87.   For the molecule acetylacetone:
       a. The number of σ and π bonds:
              Counting the number of atom-to-atom links gives a total of 14 σ bonds.
              There are 2 π bonds.

       b. The estimated bond angles:
              Angle 1:   Slightly less than 109° ; the lone pairs on O should
                                                     reduce the angle to about 105 °.
              Angle 2:   120°
              Angle 3:   120°.

89.   Approximate values for the angles: (A = axial ; E= equatorial)

	Bond	Compound	Angles	
a.	F-Se-F	SeF_4	F(A)-Se-F(A)	180°
			F(A)-Se-F(E)	90°
			F(E)-Se-F(E)	120°
b.	O-S-F	OSF_4	O-S-F(A)	90°
			O-S-F(E)	120°
c.	F-Br-F	BrF_5	F(E)-Br-F(E)	90°
			F(A)-Br-F(E)	90°

91. NO_2^+ has two structural pairs around the N atom. [See study question 73] We predict that the O-N-O bond angle would be approximately 180°. NO_2^- has three stuctural pairs (one lone pair). The geometry around this central atom would be trigonal planar with a bond angle of approximately 120°.

93.   For the molecules:

H_2O	NH_3	CO_2	ClF	CCl_4

a. Using the electronegativities to determine bond polarity:

$\frac{\Delta\chi}{\Sigma\chi}$ $\frac{1.4}{5.6}$ $\frac{0.9}{5.1}$ $\frac{1.0}{6.0}$ $\frac{10}{7.0}$ $\frac{0.5}{5.5}$

Reducing these fractions to a decimal form indicates that the H-O bonds in water are the most polar of these bonds.

b. The nonpolar compounds are:

CO_2 The O-C-O bond angle is 180°, thereby cancelling the C-O dipoles.

CCl_4 The Cl-C-Cl bond angles are approximately 109°, with the Cl atoms directed at the corners of a tetrahedron. Such an arrangement results in a net dipole moment of zero.

c. The F atom in ClF is more negatively charged.(Electronegativity of F = 4.0, Cl = 3.0)

95.   Molecular polarity of the following: CO_2 , HBF_2, CH_3Cl, SO_3

CO_2 and SO_3 are nonpolar. For HBF_2, the hydrogen and fluorine atoms are arranged at the corners of a triangle. The "negative end" of the molecule lies on the plane between the fluorine atoms, and the H atom is the "positive end." For CH_3Cl, the chlorine atom is the negative end and the H atoms form the positive end.

97.

Group Number	Number of Valence Electrons
1A	1
3A	3
3B	3
4A	4

99. Lewis structures for the following molecules and ions:

a.

$$\ddot{O}=C=\ddot{O}$$

Structural-pair: linear
Molecular: linear

b.

$$[:N\equiv N-\ddot{N}:]^- \quad [\ddot{N}=N=\ddot{N}]^- \quad [:\ddot{N}-N\equiv N:]^-$$

Structural-pair: linear
Molecular: linear

c.

$$[:\ddot{O}-C\equiv N:]^- \quad [:\ddot{O}-C\equiv N:]^- \quad [:\ddot{O}-C\equiv N:]^-$$

Structural-pair: linear
Molecular: linear

Aside from the obvious similarities of the structural and molecular geometries, these species have identical numbers of valence electrons.

101.   a. Using bond energies, calculate the enthalpy of combustion for methane:

$$CH_4 (g) + 2 O_2 (g) \rightarrow CO_2 (g) + 2 H_2O (g)$$

Bonds broken:

4 mol C-H bond	= 4 mol • 414 kJ/mol =	1656 kJ
2 mol O=O bonds	= 2 mol • 498 kJ/mol =	996 kJ
	Energy input =	2652 kJ

Bonds formed:

4 mol H-O bond	= 4 mol • 464 kJ/mol =	1856 kJ
2 mol C=O bonds	= 2 mol • 803 kJ/mol =	1606 kJ
	Energy released =	3462 kJ

$\Delta H_{reaction}$ = 2652 - 3462 = - 810. kJ for 1 mole of CH_4 (16.04 g)

or - 50.5 kJ/ g CH_4

b. $C_3H_8 (g) + 5 O_2 (g) \rightarrow 3 CO_2 (g) + 4 H_2O (g)$

Bonds broken:

8 mol C-H bond	= 8 mol • 414 kJ/mol =	3312 kJ
2 mol C-C bond	= 2 mol • 347 kJ/mol =	694 kJ
5 mol O=O bonds	= 5 mol • 498 kJ/mol =	2490 kJ
	Energy input =	6496 kJ

Bonds formed:

8 mol H-O bond	= 8 mol • 464 kJ/mol =	3712 kJ
6 mol C=O bonds	= 6 mol • 803 kJ/mol =	4818 kJ
	Energy released =	8530 kJ

$\Delta H_{reaction}$ = 6496 - 8530 = - 2034 kJ for 1 mole of C_3H_8 (44.10 g)

or -46.1 kJ/ g C_3H_8

Methane provides more heat per gram.

103.   Using bonds energies calculate the enthalpy change for the reaction:

$$2 NH_3 (g) + CO (g) \rightarrow H_2N-C(O)-NH_2 (g) + H_2 (g)$$

Bonds broken:

6 mol N-H bond	= 6 mol • 389 kJ/mol =	2334 kJ
1 mol C ≡ O bond	= 1 mol • 1075 kJ/mol =	1075 kJ
	Energy input =	3409 kJ

Bonds formed:

4 mol N-H bond	= 4 mol • 389 kJ/mol =	1556 kJ
1 mol H-H bond	= 1 mol • 436 kJ/mol =	436 kJ
2 mol C-N bond	= 2 mol • 293 kJ/mol =	586 kJ
1 mol C=O bonds	= 1 mol • 745 kJ/mol =	745 kJ
	Energy released =	3323 kJ

$$\Delta H^{\circ}_{rxn} = 3409 - 3323 = 86 \text{ kJ}$$

105.   a. Bond angles in vanillin:

        Angle-1   120°

        Angle-2   109°

        Angle-3   120° (slightly smaller due to lone pairs on oxygen)

b. The C=O bond (from carbon-1) is the shortest CO bond in the vanillin molecule.

c. The most polar bond in the molecule is the O (oxygen-1)-H bond.

107.   Using the Lewis structure describe the structural-pair and molecular geometry:

a.

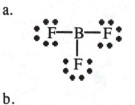

Structural-pair: trigonal planar

Molecular : trigonal planar

b.

Structural-pair: tetrahedral

Molecular: tetrahedral

c.

Structural-pair: tetrahedral

Molecular: pyramidal

d.

$$:\!\overset{\displaystyle ..}{\underset{\displaystyle ..}{F}}\!-\!\overset{\displaystyle ..}{\underset{\displaystyle ..}{O}}\!-\!\overset{\displaystyle ..}{\underset{\displaystyle ..}{F}}\!:$$

Structural-pair: tetrahedral

Molecular: bent

e.

$$H\!-\!\overset{\displaystyle ..}{\underset{\displaystyle ..}{F}}\!:$$

Structural-pair: tetrahedral

Molecular: linear

109. a. In XeF_2 the bonding pairs occupy the axial positions with the lone pairs located in the (preferred) equatorial plane.

   b. In ClF_3 two of the preferred equatorial positions are occupied by the lone pairs of electrons on the Cl atom.

111. Peroxyacetyl nitrate

   a. Number of σ bonds: 10   b. Number of π bonds : 2

   c. | Bond | 1 | 2 | 3 | 4 |
       |------|---|---|---|---|
       | Angle (degrees) | 109 | 120 | 109 | 120 |

113. For histidine:

   a. There are 17 σ bonds and 3 π bonds.

   b. For the bond angles: Angle 1 = 109°; Angle 2 = 120°; Angle 3 = 109°; Angle 4 = 120°; Angle 5 = 109°

   c. The C=C double bond is the shortest of the three carbon-carbon bonds.

   d. The C=C double bond should be the strongest carbon-carbon bond and have the largest bond energy.

115. a. The production of ClF_3 from the elements may be represented:

$$Cl_2 + 3 F_2 \rightarrow 2 ClF_3$$

   b. Mass of ClF_3 expected:

Moles Cl_2: $0.71 \text{ g } Cl_2 \cdot \dfrac{1 \text{ mol } Cl_2}{70.9 \text{ g } Cl_2} = 0.010 \text{ mol } Cl_2$

Moles F_2: $1.00 \text{ g } F_2 \cdot \dfrac{1 \text{ mol } F_2}{38.00 \text{ g } F_2} = 0.0263 \text{ mol } F_2$

Ratios: Moles required: $\dfrac{3 \text{ mol } F_2}{1 \text{ mol } Cl_2}$

Moles available: $\dfrac{0.0263 \text{ mol } F_2}{0.010 \text{ mol } Cl_2}$

F_2 is the **limiting reagent.**

$0.0263 \text{ mol } F_2 \cdot \dfrac{2 \text{ mol } ClF_3}{3 \text{ mol } F_2} \cdot \dfrac{92.45 \text{ g } ClF_3}{1 \text{ mol } ClF_3} = 1.62 \text{ g } ClF_3$

c. The electron dot structure for ClF_3 is:

   [The three lone pairs on F's have
   been omitted for clarity.]

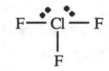

d. The structural pair geometry for the molecule is trigonal bipyramidal.

e. The molecular geometry is T-shape. The polarity of this molecule implies that there are polar bonds that are asymmetrically arranged. This **does not** unambiguously establish a geometry. Variations in the location of F atoms in axial or equatorial positions are possible.

f. The enthalpy of formation of ClF_3 from bond energies:

Bonds broken:

3 mol F-F bond	= 3 mol • 159 kJ/mol =	477 kJ
1 mol Cl-Cl bond	= 1 mol • 243 kJ/mol =	243 kJ
	Energy input =	720 kJ

Bonds formed:

6 mol Cl-F bond	= 6 mol • 255 kJ/mol =	1530 kJ
	Energy released =	1530 kJ

$\Delta H^\circ_{rxn} = 720 - 1530 = -810 \text{ kJ}$

The enthalpy change per mole is then : $\dfrac{-810 \text{ kJ}}{2 \text{ mol } ClF_3}$ or - 405 kJ/mol

CHAPTER 11: Further Concepts of Chemical Bonding

15. The molecule H_2S has :

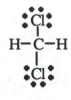

Structural pair geometry = tetrahedral

Molecular geometry = bent

The H-S bonds are formed as a result of overlap of hydrogen **s** orbitals with **sp³** hybrid orbitals of sulfur.

17. The Lewis electron dot structure of CH_2Cl_2 :

Structural pair geometry = tetrahedral

Molecular geometry = tetrahedral

The H-C bonds are a result of the overlap of the hydrogen **s** orbital with **sp³** hybrid orbitals on carbon. The Cl-C bonds are formed by the overlap of the **sp³** hybrid orbitals on carbon with the **p** orbitals on chlorine.

19. Orbital sets used by the underlined atoms:

a. $\underline{B}Cl_3$: sp²   b. $\underline{N}O_2^+$: sp   c. $H\underline{C}Cl_3$: sp³   d. $H_2\underline{C}O$: sp²

21. Hybrid orbital sets used by the underlined atoms:

a. $\underline{C}$: sp³ ; $\underline{O}$: sp³   b. $\underline{C}H_3$: sp³ ; $\underline{C}H$ and $\underline{C}H_2$: sp²   c. $\underline{C}H_2$: sp³, $\underline{C}$=O: sp²

23. For the molecule XeF_2 :

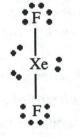

Structural pair geometry = trigonal bipyramidal

Molecular geometry = linear

Bonding is a result of overlap of **dsp³** orbitals on xenon with **p** orbitals on fluorine.

25. Orbital sets used by the underlined atoms:

a. $\underline{S}F_6$: sp³d²   b. $\underline{S}F_4$: sp³d   c. $\underline{I}Cl_2^-$: sp³d

27.   For the molecule CO_2:   Hybridization of $C = sp$

$$\left[\ddot{O} = C = \ddot{O} \right]$$

Bonding: 1 sigma bond between each oxygen and carbon
1 pi bond between each oxygen and carbon

29.   For the molecule $COCl_2$:   Hybridization of $C = sp^2$

$$:\ddot{O}:$$
$$\|$$
$$:\ddot{Cl} - C - \ddot{Cl}:$$

Bonding: 1 sigma bond between each chlorine and carbon
1 sigma bond between carbon and oxygen
1 pi bond between carbon and oxygen

31.   Configuration for H_2^+: $(\sigma_{1s})^1$

Bond order for H_2^+: 1/2 (no. bonding e^- - no. antibonding e^-) = 1/2

The bond order for molecular hydrogen is <u>one</u> (1), and the H-H bond is stronger in the H_2 molecule than in the H_2^+ ion.

33.   The molecular orbital diagram for C_2^{2-} , the acetylide ion:

σ^*_{2p}	____	
π^*_{2p}	____	____
σ_{2p}	↑↓	
π_{2p}	↑↓	↑↓
σ^*_{2s}	↑↓	
σ_{2s}	↑↓	

There are 2 net pi bonds and 1 net sigma bond in the ion, giving a bond order of 3. On adding two electrons to C_2 (added to σ_{2p})to obtain C_2^{2-}, the bond order increases by one.

35.   Using Figure 11.17 as a model for heteronuclear diatomic molecules, the electron configuration (showing only the outer level electrons) for CO is:

σ^*_{2p}	____	
π^*_{2p}	____	____
σ_{2p}	↑↓	
π_{2p}	↑↓	↑↓
σ^*_{2s}	↑↓	
σ_{2s}	↑↓	

There are no unpaired electrons, hence CO is diamagnetic. There is one net sigma bond, and two net pi bonds for an overall bond order of 3.

37.   The hybrid orbitals used by sulfur in: a. SO_2: sp^2 b. SO_3: sp^2 c. SO_4^{2-}: sp^3

39.   For the ion NO_3^- :

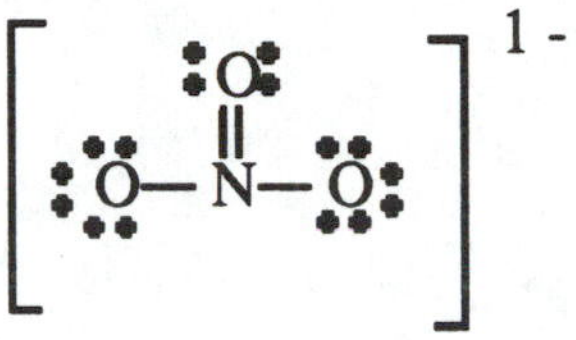

Hybridization of N: sp^2

The N=O bond is formed by overlap of an **sp^2** hybrid orbital on both N and O to form a σ bond, while the unhybridized **p** orbitals on N and O overlap side-to-side to form a π bond.

41.   For the oxime shown in the problem:

a. The methyl carbon (H_3C-) is sp^3 hybridized while the other carbon is sp^2 hybridized.

b. The approximate C-N-O bond angle is 120° (sp^2 hybridization on N).

43.   For acetylsalicylic acid:

a. Number of π bonds: 5   Number of σ bonds: 2l (don't forget to count those associated with the double bonds.)

b. angle A: 120°; angle B: 109°; angle C: 109°

c. Hybridization used by: C_1: sp^2; C_2: sp^2; O_3: sp^3

45.   For histamine:

a. Number of σ bonds: 17   Number of π bonds: 2

b. Hybridization of atom: N_1: sp^3 C_2: sp^3 C_3: sp^2 N_4: sp^2

c. Bond angles: A: 109°;   B: 120°;   C: 109°

47.   The valence bond picture for the peroxide ion includes 14 electrons [2(6) + 2]:

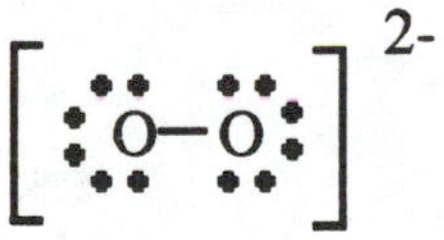

with a bond order of one. The MO picture would include (for the outer shell) $(\sigma_{2s})^2(\sigma^*_{2s})^2 (\pi_{2p})^4 (\sigma_{2p})^2 (\pi^*_{2p})^4$ with a bond order of: (4 bonding pairs) - (3 antibonding pairs) = 1. There are no unpaired electrons in either scheme, hence both theories predict the peroxide ion to be diamagnetic.

133

49. $H^+ + NH_3 \rightarrow NH_4^+$

                  sp^3 sp^3

The H-N-H bond angle in NH_3 is slightly less than 109° (owing to the large lone pair on the N atom), while it is expected to be exactly 109° in NH_4^+.

51.   The Lewis dot structures for IO_4^- and IO_5^{3-}:

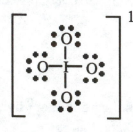

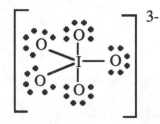

Structural pair geometry: tetrahedral   Structural pair geometry: trigonal bipyramidal
Molecular geometry: tetrahedral   Molecular geometry: trigonal bipyramidal
Hybridization of I: sp^3 Hybridization of I: sp^3d

53.

Molecule	MO electron configuration	Bond order
Li_2	$(\sigma_{2s})^2$	1
Be_2	$(\sigma_{2s})^2(\sigma^*_{2s})^2$	0
B_2	$(\sigma_{2s})^2(\sigma^*_{2s})^2 (\pi_{2p})^2$	1 †
C_2	$(\sigma_{2s})^2(\sigma^*_{2s})^2 (\pi_{2p})^4$	2
N_2	$(\sigma_{2s})^2(\sigma^*_{2s})^2 (\pi_{2p})^4 (\sigma_{2p})^2$	3
O_2	$(\sigma_{2s})^2(\sigma^*_{2s})^2 (\pi_{2p})^4 (\sigma_{2p})^2 (\pi^*_{2p})^2$	2 †
F_2	$(\sigma_{2s})^2(\sigma^*_{2s})^2 (\pi_{2p})^4 (\sigma_{2p})^2 (\pi^*_{2p})^4$	1
Ne_2	$(\sigma_{2s})^2(\sigma^*_{2s})^2 (\pi_{2p})^4 (\sigma_{2p})^2 (\pi^*_{2p})^4 (\sigma^*_{2p})^2$	0

† Note that for B_2 and O_2, filling of the two π MO's occurs in accordance with the Pauli exclusion principle. i. e. the electrons do not form pairs-- giving two unpaired electrons in each of these species. Nitrogen, with a bond order of three, has the greatest bond order.

55.   Possible molecular structures: [Lone pair electrons on F have been omitted for clarity.]

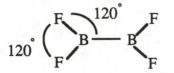 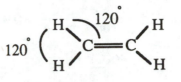

Hybridization of B= sp^2 Hybridization of C = sp^2
F-B-B bond angle: approximately 120°   H-C-C bond angle: approximately 120°

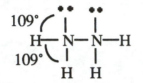 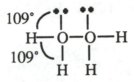

Hybridization of N = sp^3 Hybridization of O = sp^3
H-N-N bond angle: approximately 109°   H-O-O bond angle: approximately 109°

For both hydrazine (N_2H_4) and hydrogen peroxide (H_2O_2), the lone pairs on N and O respectively will reduce the bond angles to slightly less than 109 °.

Only the carbon compound contains a double bond.

57. | Atom | Hybridization in Reactant | Hybridization in Product |
|---|---|---|
| a. C | sp^2 | sp^3 |
| b. P | sp^3 | sp^3d |
| c. Xe | sp^3d | sp^3d^2 |
| d. Sn | sp^3 | sp^3d^2 |

59.   For the molecule: $CH_3-N=C=S$:

       a. Orbital sets used by:

              C_1 : sp^3 (bonds formed by C_1 to four atoms)

              C_2 : sp(bonds formed by C_2 to two atoms)

              N : sp^2 (bonds formed by N to two atoms)

       b. Bond angles:
              C-N=C angle: approximately 120°   N=C=S angle: approximately 180°

       c. For the C_2 atom, the s and p_x orbitals would form the two sp hybrid orbitals, leaving
          the p_y and p_z orbitals to form the pi bonds to N and S.

135

11.　a.　$740 \text{ mmHg} \cdot \dfrac{1 \text{ atm}}{760 \text{ mmHg}} = 0.97 \text{ atm}$

　　　b.　$1.25 \text{ atm} \cdot \dfrac{760. \text{ mmHg}}{1 \text{ atm}} = 950. \text{ mmHg}$

　　　c.　$740 \text{ mmHg} \cdot \dfrac{101.325 \text{ kPa}}{760 \text{ mmHg}} = 99 \text{ kPa}$

　　　d.　$0.50 \text{ atm} \cdot \dfrac{101.3 \text{ kPa}}{1.0 \text{ atm}} = 51 \text{ kPa}$

　　　e.　$542 \text{ mmHg} \cdot \dfrac{1.00 \text{ torr}}{1.00 \text{ mmHg}} = 542 \text{ torr}$

13.　The levels in the U-tube manometer indicate a gas pressure of 56.3 mmHg.

$56.3 \text{ mmHg} \cdot \dfrac{1 \text{ atm}}{760 \text{ mmHg}} = 0.0741 \text{ atm}$

$56.3 \text{ mmHg} \cdot \dfrac{1 \text{ torr}}{1 \text{ mmHg}} = 56.3 \text{ torr}$

$56.3 \text{ mmHg} \cdot \dfrac{101.3 \text{ kPa}}{760.0 \text{ mmHg}} = 7.50 \text{ kPa}$

15.　**Boyle's law** states that the pressure a gas exerts is inversely proportional to the volume it occupies, or for a given amount of gas--PV = constant . We can write this as:

$$P_1 V_1 = P_2 V_2$$

So　$(67.5 \text{ mmHg})(256 \text{ mL}) = (P_2)(135 \text{ mL})$

and　$\dfrac{(67.5 \text{ mmHg})(256 \text{ mL})}{135 \text{ mL}} = 128 \text{ mmHg}$

17.　**Charles' law** states that V α T (in Kelvin)　or　$\dfrac{V_1}{T_1} = \dfrac{V_2}{T_2}$

$$\dfrac{25.0 \text{ mL}}{295 \text{ K}} = \dfrac{V_2}{273 \text{ K}}$$

$$V_2 = \dfrac{(25.0 \text{ mL})(273 \text{ K})}{(295 \text{ K})} = 23.1 \text{ mL}$$

19.   Using the general gas law we can write: $\dfrac{P_1V_1}{T_1} = \dfrac{P_2V_2}{T_2}$

and for a fixed volume $\dfrac{P_1}{T_1} = \dfrac{P_2}{T_2}$ or $P_2 = P_1 \cdot \dfrac{T_2}{T_1}$

$$P_2 = 135 \text{ mmHg} \cdot \dfrac{291.8 \text{ K}}{298.7 \text{ K}} = 132 \text{ mmHg}$$

21.   The reaction can be written: $2 H_2 + O_2 \rightarrow 2 H_2O$. Avogadro's law states that 1.0 L of H_2 would require 0.50 L of O_2, if both volumes are measured at the same temperature and pressure.

23.   Using the general gas law: $\dfrac{P_1V_1}{T_1} = \dfrac{P_2V_2}{T_2}$ or $P_2 = \dfrac{P_1V_1T_2}{T_1V_2}$

so $P_2 = 3.50 \text{ atm} \cdot \left(\dfrac{250 \text{ cm}^3}{500 \text{ cm}^3} \right) \cdot \left(\dfrac{398 \text{ K}}{523 \text{ K}} \right)$ or (d)

25.   Since the amount of gas is constant in the two vessels, we can use the general gas law to solve for the pressure in the smaller flask (P_2):

Rearranging $\dfrac{P_1V_1}{T_1} = \dfrac{P_2V_2}{T_2}$, we can write

$$P_2 = \dfrac{P_1V_1T_2}{T_1V_2} = \dfrac{48.5 \text{ mmHg} \cdot 2560 \text{ mL} \cdot 295.8 \text{ K}}{300.7 \text{ K} \cdot 345 \text{ mL}} = 354 \text{ mmHg}$$

27.   The volume of the gas in the cylinder following the compression can be calculated using the general gas law:

$$P_2 = \dfrac{P_1V_1T_2}{T_1V_2} = \dfrac{1.00 \text{ atm} \cdot 0.400 \text{ L} \cdot 350 \text{ K}}{300 \text{ K} \cdot 0.050 \text{ L}} = 9.33 \text{ atm}$$

29.   The pressure of 1.00 g of gaseous water may be calculated with the ideal gas law:

$$1.00 \text{ g } H_2O \cdot \dfrac{1 \text{ mol } H_2O}{18.02 \text{ g } H_2O} = 0.0555 \text{ mol } H_2O$$

Rearranging $PV = nRT$ to solve for P, we obtain:

$$P = \dfrac{nRT}{V} = \dfrac{(0.0555 \text{ mol})(0.082057 \text{ L} \cdot \text{atm/K} \cdot \text{mol})(423 \text{ K})}{10.0 \text{ L}} = 0.193 \text{ atm}$$

31.   The volume of the flask may be calculated by realizing that the gas will expand to fill the flask.

$$4.4 \text{ g } CO_2 \cdot \frac{1 \text{ mol } CO_2}{44.0 \text{ g } CO_2} = 0.10 \text{ mol } CO_2$$

$$P = 730 \text{ mmHg} \cdot \frac{1 \text{ atm}}{760 \text{ mmHg}} = 0.96 \text{ atm}$$

$$V = \frac{(0.10 \text{ mol})(0.082057 \text{ L} \cdot \text{atm/K} \cdot \text{mol})(300 \text{ K})}{0.96 \text{ atm}} = 2.6 \text{ L}$$

33.   The number of moles of CO can be calculated with the ideal gas law:

$$n = \frac{41.8 \text{ mm Hg} \cdot \dfrac{1 \text{ atm}}{760 \text{ mmHg}} \cdot 150. \text{ L}}{0.082057 \dfrac{\text{L} \cdot \text{atm}}{\text{mol} \cdot \text{K}} \cdot 298.2 \text{ K}} = 0.337 \text{ mol CO}$$

35.   Calculate the number of moles of compound:

$$n = \frac{PV}{RT} = \frac{(1.00 \text{ atm})(0.820 \text{ L})}{(0.082057 \text{ L} \cdot \text{atm/K} \cdot \text{mol})(276 \text{ K})} = 0.0362 \text{ mol}$$

So $1.00 \text{ g} = 0.0362 \text{ mol}$ and $\dfrac{1.00 \text{ g}}{0.0362 \text{ mol}} = 27.6 \text{ g/mol}$

The atomic weight of $B = 10.8$ and that of $H = 1.01$. Calculate the molar masses of the suggested compounds.

Compound	Molar Mass
a. B_4H_{10}	53.3 g
b. B_2H_6	27.7 g
c. B_5H_9	63.1 g

The gaseous compound must be B_2H_6.

37.   Using the ideal gas law, we can calculate the moles of gas represented by 0.293 g of the compound.

$$n = \frac{374 \text{ mm Hg} \cdot \dfrac{1 \text{ atm}}{760 \text{ mmHg}} \cdot 0.185 \text{ L}}{0.082057 \dfrac{\text{L} \cdot \text{atm}}{\text{mol} \cdot \text{K}} \cdot 296.2 \text{ K}} = 3.75 \times 10^{-3} \text{ mol}$$

So the molar mass of the compound is: $\dfrac{0.293 \text{ g}}{3.75 \times 10^{-3} \text{ mol}}$ = 78.2 g/mol

The compound is 92.26 % C and 7.74 % H. The empirical formula would then be:

Carbon: 92.26 g • $\dfrac{1 \text{ mol C}}{12.011 \text{ g C}}$ = 7.68 mol C

Hydrogen: 7.74 g • $\dfrac{1 \text{ mol H}}{1.008 \text{ g H}}$ = 7.68 mol H

With a molar mass of about 78 grams and an empirical formula of CH (formula mass $\approx$ 13), the molecular formula would be C_6H_6 (benzene).

39. Using the ideal gas law, we calculate the number of moles of gas represented by 0.0872 g of the boron hydride:

$$n = \dfrac{191 \text{ mm Hg} \cdot \dfrac{1 \text{ atm}}{760 \text{ mmHg}} \cdot 0.159 \text{ L}}{0.082057\dfrac{\text{L} \cdot \text{atm}}{\text{mol} \cdot \text{K}} \cdot 297.2 \text{ K}} = 1.64 \times 10^{-3} \text{ mol}$$

The molar mass of the compound is $\dfrac{0.0872 \text{ g}}{1.64 \times 10^{-3} \text{ mol}}$ = 53.2 g/mol

The compound is 18.9 % H and 81.1 % B. The empirical formula would be:

Hydrogen: 18.9 g H • $\dfrac{1 \text{ mol H}}{1.008 \text{ g H}}$ = 18.75 mol H

Boron: 81.1 g B • $\dfrac{1 \text{ mol B}}{10.81 \text{ g B}}$ = 7.50 mol B

The ratio of boron to hydrogen is 1 : 2.5 or (to an integral number of atoms B_2H_5). The empirical formula weight ($\approx$ 26.6), indicating that the molecular formula is B_4H_{10}.

41. Write the ideal gas law as: Molar Mass = $\dfrac{dRT}{P}$ where d = density in grams per liter.

Solving for d, we obtain: $\dfrac{(\text{Molar Mass}) \cdot P}{R \cdot T}$ = d

The average molecular weight for air is approximately 29.0 g.

$$\dfrac{(29.0 \text{ g/mol})(0.20 \text{ mmHg} \cdot 1 \text{ atm}/760 \text{ mmHg})}{(0.082057 \dfrac{\text{L} \cdot \text{atm}}{\text{K} \cdot \text{mol}})(250 \text{ K})} = 3.7 \times 10^{-4} \text{ g/L} = d$$

43.   Molar mass = $\dfrac{(1.25 \text{ g/L})(0.082057 \text{ L} \cdot \text{atm/K} \cdot \text{mol})(298 \text{ K})}{(195 \text{ mmHg} \cdot 1 \text{ atm/760 mmHg})}$ = 119

The molar mass of chloroform, $CHCl_3$, to three significant figures is 119.

45.   Solve the ideal gas law for volume to obtain

(1) $V_1 = \dfrac{n_1RT_1}{P_1}$

We are interested in the mass (and therefore moles) of oxygen needed to fill this balloon to the same volume, measured at the same pressure but a different temperature. We could express this as:

(2) $V_1 = \dfrac{n_2RT_2}{P_1}$

In equations (1) and (2) we have the common term V_1 so we may write:

(3) $\dfrac{n_1RT_1}{P_1} = \dfrac{n_2RT_2}{P_1}$

Since both R and P_1 are constant in these conditions, equation (3) may be simplified:

$n_1T_1 = n_2T_2$

Calculating the moles of oxygen represented by 12.0 g gives:

$$n_1 = 12.0 \text{ g } O_2 \cdot \dfrac{1 \text{ mol } O_2}{32.00 \text{ g } O_2} = 0.375 \text{ mol } O_2$$

Substituting for T_1 (300. K) and T_2 (400. K) and solving for n_2 we obtain:

$$\dfrac{(0.375 \text{ mol } O_2) \cdot (300. \text{ K})}{400. \text{ K}} = 0.281 \text{ mol } O_2 \text{ or } 9.00 \text{ g } O_2$$

47.   To determine which cylinder contains the greater mass of oxygen **without** performing a calculation, rearrange the ideal gas law--solving for n to obtain: $n = \dfrac{PV}{RT}$

Since P, V, and R are identical for the two cylinders, we note that the number of moles, n, is inversely proportional to T or : $n \propto \dfrac{1}{T}$.

The absolute temperature of cylinder A is greater than the absolute temperature of cylinder B, therefore **n is greater for cylinder B.**

49.   Avogadro's law states that equal volumes of two gases, measured under the same temperature and pressure conditions, contain equal numbers of molecules (and numbers of moles). We can set up a ratio between the mass of these two gases and their molar masses.

$$\frac{0.34 \text{ g A}}{M_A} = \frac{0.48 \text{ g B}}{M_B}$$

Gas B is ozone (M = 48 g). Solving for the molar mass of A we obtain

$$\frac{0.34 \text{ g of A}}{M_A} = \frac{0.48 \text{ g of B}}{48 \text{ g}}$$

and M_A = 34 g. H_2S has a molar mass of 34 and is therefore a likely choice.

51.   We begin by noting that for each mole of water, one mole of gaseous hydrogen will be formed. We can calculate the number of moles of water using the ideal gas law:

$$n = \frac{P \cdot V}{R \cdot T} = \frac{2.0 \text{ atm} \cdot 250. \text{ L}}{0.082057\frac{\text{L} \cdot \text{atm}}{\text{mol} \cdot \text{K}} \cdot 393 \text{ K}} = 15.5 \text{ mol H}_2\text{O}$$

which gives rise to 15.5 mol H_2 [16 mol to 2 significant figures]

Calculating the mass of H_2 gives:

$$15.5 \text{ mol H}_2 \cdot \frac{2.02 \text{ g H}_2}{1 \text{ mol H}_2} = 31 \text{ g H}_2 \text{ (to two significant figures)}$$

53. N_2H_4 (g) + O_2 (g) → N_2 (g) + 2 H_2O (g)

$$10. \text{ kg N}_2\text{H}_4 \cdot \frac{1.0 \times 10^3 \text{ g N}_2\text{H}_4}{1.0 \text{ kg N}_2\text{H}_4} \cdot \frac{1 \text{ mol N}_2\text{H}_4}{32.0 \text{ g N}_2\text{H}_4} \cdot \frac{1 \text{ mol O}_2}{1 \text{ mol N}_2\text{H}_4}$$

$$= 3.1 \times 10^2 \text{ mole O}_2$$

$$P(O_2) = \frac{n(O_2) \cdot R \cdot T}{V} = \frac{(3.1 \times 10^2 \text{ mol})(0.082057 \text{ L} \cdot \text{atm/K} \cdot \text{mol})(299 \text{ K})}{450 \text{ L}}$$

$$P(O_2) = 17 \text{ atm}$$

The pressure in the tank would have to be increased by 1 atm to pump the oxygen out of the tank. So the total tank pressure would be 18 atm.

55.   Calculate the number of moles of UF_6:

$$1.0 \times 10^3 \text{ g U} \cdot \frac{1 \text{ mol U}}{238 \text{ g U}} = 4.2 \text{ mol U} = 4.2 \text{ mol } UF_6$$

Substituting into the ideal gas law we obtain:

$$P = \frac{n \cdot R \cdot T}{V} = \frac{(4.2 \text{ mol})(0.082057 \text{ L} \cdot \text{atm/K} \cdot \text{mol})(335 \text{ K})}{3.0 \times 10^2 \text{ L}} = 0.38 \text{ atm}$$

57.   The CO produced upon heating the oxide represents:

$$n = \frac{369 \text{ mm Hg} \cdot \dfrac{1 \text{ atm}}{760 \text{ mmHg}} \cdot 0.155 \text{ L}}{0.082057 \dfrac{\text{L} \cdot \text{atm}}{\text{mol} \cdot \text{K}} \cdot 300.2 \text{ K}} = 3.5 \times 10^{-3} \text{ mol CO}$$

This amount of CO would have a mass of :

$$3.5 \times 10^{-3} \text{ mol CO} \cdot \frac{28.01 \text{ g CO}}{1 \text{ mol CO}} = 0.0856 \text{ g CO}$$

The mass of Cr in the oxide is : $(0.112 - 0.0856)$ or 0.026 g which corresponds to :

$$0.026 \text{ g Cr} \cdot \frac{1 \text{ mol Cr}}{51.996 \text{ g Cr}} = 5.0 \times 10^{-4} \text{ mol Cr}$$

The ratio of Cr to CO is then:

$$\frac{3.5 \times 10^{-3} \text{ mol CO}}{5.0 \times 10^{-4} \text{ mol Cr}} = \frac{6 \text{ mol CO}}{1 \text{ mol Cr}} \text{ giving an empirical formula of } Cr(CO)_6$$

59.   The amount of methane produced is:

$$n = \frac{208 \text{ mm Hg} \cdot \dfrac{1 \text{ atm}}{760 \text{ mmHg}} \cdot 0.125 \text{ L}}{0.082057 \dfrac{\text{L} \cdot \text{atm}}{\text{mol} \cdot \text{K}} \cdot 297.2 \text{ K}} = 1.40 \times 10^{-3} \text{ mol } CH_4$$

The amount of CH_3 in the compound is equal to this number of moles, with a mass of:

$$1.40 \times 10^{-3} \text{ mol } CH_3 \cdot \frac{15.0 \text{ g } CH_3}{1 \text{ mol } CH_3} = 0.0210 \text{ g } CH_3$$

The mass of Al in the compound is:

$$(0.0338 \text{ g} - 0.0210 \text{ g}) \quad \text{or} \quad 0.0128 \text{ g Al} \quad \text{or} \quad 4.73 \times 10^{-4} \text{ mol Al}$$

The ratio of CH_3 to Al is then: $\dfrac{1.40 \times 10^{-3} \text{ mol } CH_3}{4.73 \times 10^{-4} \text{ mol Al}}$ or $\dfrac{3 \text{ mol } CH_3}{1 \text{ mol Al}}$

61.   The amount of $SClF_5$ represented by 0.265 g is:

$$0.265 \text{ g} \cdot \frac{1 \text{ mol } SClF_5}{162.5 \text{ g } SClF_5} = 1.63 \times 10^{-3} \text{ mol } SClF_5$$

The amount of H_2 present is:

$$n = \frac{95.1 \text{ mm Hg} \cdot \dfrac{1 \text{ atm}}{760 \text{ mmHg}} \cdot 0.275 \text{ L}}{0.082057 \dfrac{\text{L} \cdot \text{atm}}{\text{mol} \cdot \text{K}} \cdot 318.2 \text{ K}} = 1.32 \times 10^{-3} \text{ mol } H_2$$

Noting the ratios of moles of $SClF_5$ to H_2 we note the $SClF_5$ is the limiting reagent.

$$1.63 \times 10^{-3} \text{ mol } SClF_5 \cdot \frac{1 \text{ mol } S_2F_{10}}{2 \text{ mol } SClF_5} \cdot \frac{254.1 \text{ g } S_2F_{10}}{1 \text{ mol } S_2F_{10}} = 0.207 \text{ g } S_2F_{10}$$

The pressure in the flask due solely to the S_2F_{10} is:

$$P = \frac{nRT}{V} = \frac{(8.15 \times 10^{-4} \text{ mol})(0.082057 \dfrac{\text{L} \cdot \text{atm}}{\text{K} \cdot \text{mol}})(318.2 \text{ K})}{0.275 \text{ L}}$$

$$= 0.0774 \text{ atm or } 58.8 \text{ mmHg}$$

63.   Calculate the number of moles of H_2 and Ar:

$$1.0 \text{ g } H_2 \cdot \frac{1 \text{ mol } H_2}{2.02 \text{ g } H_2} = 0.50 \text{ mol } H_2 \quad \text{and} \quad 8.0 \text{ g Ar} \cdot \frac{1 \text{ mol Ar}}{40.0. \text{ g Ar}} = 0.20 \text{ mol Ar}$$

Substituting into the ideal gas law we obtain:

$$P = \frac{n \cdot R \cdot T}{V} = \frac{(0.50 \text{ mol} + 0.20 \text{ mol})(0.082057 \dfrac{\text{L} \cdot \text{atm}}{\text{K} \cdot \text{mol}})(300 \text{ K})}{3.0 \text{ L}}$$

$$P = 5.7 \text{ atm}$$

65.   (a) The ratio of moles of oxygen to moles of cyclopropane is:

$$\frac{\text{moles cyclopropane}}{\text{moles oxygen}} = \frac{170 \text{ mmHg}}{570 \text{ mmHg}} = \frac{1}{3.35}$$

(b) The mass of cyclopropane assuming the tank contains 160 g O_2:

$$160 \text{ g } O_2 \cdot \frac{1 \text{ mol } O_2}{32.0 \text{ g } O_2} = 5.0 \text{ mol } O_2, \text{ and using the ratio from above,}$$

$$5.0 \text{ mol } O_2 \cdot \frac{1 \text{ mol } C_3H_6}{3.35 \text{ mol } O_2} \cdot \frac{42.1 \text{ g } C_3H_6}{1 \text{ mol } C_3H_6} = 63 \text{ g } C_3H_6$$

67. $P(B_2H_6) + P(O_2) = 10. \text{ mmHg}$

The reaction mixture must have 3 moles of O_2 for each mole of B_2H_6 (the stoichiometric ratio dictated by the balanced equation) so we can calculate the mole fraction of each. **Assume** we had 1 mol B_2H_6 and 3 mol O_2. The mole fraction of B_2H_6 would be:

$$X(B_2H_6) = \frac{1 \text{ mol } B_2H_6}{1 \text{ mol } B_2H_6 + 3 \text{ mol } O_2} = 1/4$$

Note that this mole fraction would be independent of the **actual** number of moles of each reactant present.

$P(B_2H_6) = P(\text{Total}) \cdot X(B_2H_6) = 10. \text{ mmHg} \cdot 1/4 = 2.5 \text{ mmHg}$

and $P(O_2) = (10. - 2.5)$ or 7.5 mmHg

69.   The moles represented by 4.25 g of dimethylhydrazine :

$$4.25 \text{ g } (CH_3)_2N_2H_2 \cdot \frac{1 \text{ mol } (CH_3)_2N_2H_2}{60.10 \text{ g } (CH_3)_2N_2H_2} = 0.0707 \text{ mol } (CH_3)_2N_2H_2$$

From the balanced equation, we determine the moles of product gases:

$$0.0707 \text{ mol } (CH_3)_2N_2H_2 \cdot \frac{3 \text{ mol } N_2}{1 \text{ mol } (CH_3)_2N_2H_2} = 0.212 \text{ mol } N_2$$

$$0.0707 \text{ mol } (CH_3)_2N_2H_2 \cdot \frac{4 \text{ mol } H_2}{1 \text{ mol } (CH_3)_2N_2H_2} = 0.283 \text{ mol } H_2$$

$$0.0707 \text{ mol } (CH_3)_2N_2H_2 \cdot \frac{2 \text{ mol } CO_2}{1 \text{ mol } (CH_3)_2N_2H_2} = 0.141 \text{ mol } CO_2$$

According to Dalton's Law, these gases will act independently of each other to exert a pressure. Hence we can add the number of moles of the three gases and substitute into the ideal gas law to obtain:

$$P = \frac{nRT}{V} = \frac{(0.212+0.283+0.141 \text{ mol}) \cdot 0.082057 \frac{L \cdot atm}{K \cdot mol} \cdot 300 \text{ K}}{250. \text{ L}}$$

$$= 0.0626 \text{ atm}$$

71. Begin by calculating the total number of gaseous products which result from the reaction of 1.0 g of nitroglycerine:

$$1.0 \text{ g nitroglycerine} \cdot \frac{1 \text{ mol nitroglycerine}}{227 \text{ g nitroglycerine}} \cdot \frac{14.5 \text{ mol gaseous products}}{2 \text{ mol nitroglycerine}}$$

$$= 0.032 \text{ mol products}$$

Note that the third term arises from adding the moles of all gaseous products.

$$V = \frac{nRT}{P} = \frac{(0.032 \text{ mol})(0.082057 \frac{L \cdot atm}{K \cdot mol})(773 \text{ K})}{1.0 \text{ atm}} = 2.0 \text{ L}$$

The volume occupied by the product gases at 298 K and 1.0 atm may be calculated

$$\frac{V}{T} = \frac{V}{T} \quad \text{(P and n are constant)}$$

$$\frac{2.0 \text{ L}}{773 \text{ K}} = \frac{V}{298 \text{ K}} \quad \text{and solving for V gives: } 0.78 \text{ L}$$

73. According to the stoichiometry of the decomposition equation, each mole of $KClO_3$ produces 1.5 moles of O_2. Begin by calculating the number of moles of $KClO_3$.

$$1.56 \text{ g KClO}_3 \cdot \frac{1 \text{ mol KClO}_3}{122.5 \text{ g KClO}_3} = 0.0127 \text{ mol KClO}_3$$

The amount of oxygen that can be expected is:

$$0.0127 \text{ mol KClO}_3 \cdot \frac{3 \text{ mol O}_2}{2 \text{ mol KClO}_3} = 0.0191 \text{ mol O}_2$$

Since the gas is collected over water, we must subtract the vapor pressure of water from the total pressure(barometric pressure).

Pressure O_2 = 742 mmHg - 18.7 mmHg = 723 mmHg or 0.952 atm

Substituting into the ideal gas law:

$$V = \frac{nRT}{P} = \frac{(0.0191 \text{ mol})(0.082057 \frac{L \cdot atm}{K \cdot mol})(294 \text{ K})}{0.952 \text{ atm}} = 0.484 \text{ L}$$

75.   From Study Question 61 we note the pressure of S_2F_{10} = 58.8 mmHg. The equation indicates that for each mole of S_2F_{10}, two moles of HCl (g) are formed. The pressure attributable to that gas would be (2 x 58.8 mmHg) 117.6 mmHg. The last contributor to pressure in the flask would be unreacted H_2.

Given (from Study Question 61) that 1.63×10^{-3} mol $SClF_5$ react, 8.15×10^{-4} mol H_2 would react (the balanced equation shows a stoichiometry of 2:1) leaving:

$$(1.32 \times 10^{-3} \text{ mol } H_2 - 8.15 \times 10^{-4} \text{ mol of } H_2) \text{ or } 5.05 \times 10^{-4} \text{ mol of } H_2$$

$$P = \frac{nRT}{V} = \frac{(5.05 \times 10^{-4} \text{ mol})(0.082057 \frac{L \cdot atm}{K \cdot mol})(318.2 \text{ K})}{0.275 \text{ L}} = 0.0479 \text{ atm}$$
$$\text{or } 36.4 \text{ mmHg}$$

The total pressure is then: 58.8 mm S_2F_{10}

117.6 mm HCl

<u>36.4 mm H_2</u>

212.8 mm   or 213 mm to three significant figures

77.   a. Kinetic energy depends only on the temperature so the average kinetic energies of these two gases are equal.

b. Since the kinetic energies are equal, we can state:

$$KE(H_2) = KE(CO_2)$$

$$1/2 \text{ m}(H_2) \cdot \overline{V}^2 (H_2) = 1/2 \text{ m}(CO_2) \cdot \overline{V}^2(CO_2)$$

Where m = mass of a molecule and $\overline{V}$ = average velocity of a molecule

Now the molar mass of H_2 = 2.0 g   and the molar mass of CO_2 = 44 g

So $\qquad m(H_2) \cdot \overline{V}^2(H_2) = m(CO_2) \cdot \overline{V}^2(CO_2)$

and $\qquad \dfrac{\overline{V}^2(H_2)}{\overline{V}^2(CO_2)} = \dfrac{m(CO_2)}{m(H_2)}$

and $\qquad \dfrac{\overline{V}_{H_2}}{\overline{V}_{CO_2}} = \sqrt{\dfrac{m_{CO_2}}{m_{H_2}}} = \sqrt{\dfrac{44}{2.0}} = 4.7$

The hydrogen molecules have an average velocity which is 4.7 times the average velocity of the CO_2 molecules.

c. Since the temperatures and the volumes are equal for these two gas samples, the pressure is proportional to the amount of gas present.

$$V_A = \dfrac{n_A R T_A}{P_A} \qquad \text{and} \quad V_B = \dfrac{n_B R T_B}{P_B} \quad \text{now } T_A = T_B \text{ and } V_A = V_B \text{ so}$$

$$\dfrac{n_A R}{P_A} = \dfrac{n_B R}{P_B} \qquad \text{or} \qquad \dfrac{n_A}{P_A} = \dfrac{n_B}{P_B}$$

Since the pressure in Flask B (2 atm) is twice that of Flask A (1 atm), there are two times as many moles (and molecules) of gas in Flask B (CO_2) as there are in Flask A (H_2).

79.   Average speed of xenon atoms at 25 °C is:

$$\overline{u} = \sqrt{\dfrac{3RT}{M}} = \sqrt{\dfrac{(3)(8.314 \text{ J/K} \cdot \text{mol})(298 \text{ K})}{0.13129 \text{ kg/mol}}}$$

$$\overline{u} = \sqrt{5.72 \times 10^4 \text{ J/kg}} \qquad \text{and since } 1 \text{ J} = 1 \dfrac{\text{kg} \cdot \text{m}^2}{\text{s}^2}$$

$$\overline{u} = \sqrt{5.72 \times 10^4 \text{ m}^2/\text{s}^2} \qquad \text{or } 2.38 \times 10^2 \text{ m/s}$$

The average speed of helium atoms would be

$$\overline{u} = \sqrt{\dfrac{(3)(8.314 \text{ J/K mol})(298 \text{ K})}{4.00 \times 10^{-3} \text{ kg/mol}}} = 1.36 \times 10^3 \text{ m/s}$$

Note that molar masses are expressed in units of kilograms to allow cancellation of mass units associated with **Joules**.

81.   The following species will have average molecular speeds which are inversely proportional to their molar masses, hence the order is: $CH_2Cl_2 < Kr < N_2 < CH_4$

83.   Ranking the following substances in order of increasing rates of effusion involves an examination of their relative molar masses. According to Graham's law the gases will effuse at a rate which is inversely proportional to the square root of the molar masses.

   Gas: He   Xe   CO C_2H_6

   Molar Mass: 4.00   131   28.0   30.1

   Ranking these in inverse order of molar masses: $Xe < C_2H_6 < CO < He$

85.   Determine the molar mass of a gas which effuses at a rate 1/3 that of He:

$$\frac{\text{Rate of effusion of He}}{\text{Rate of effusion of unknown}} = \sqrt{\frac{\text{M of unknown}}{\text{M of He}}}$$

$$\frac{3}{1} = \sqrt{\frac{\text{M of unknown}}{4.0 \text{ g/mol}}}$$

Squaring both sides gives: $9 = \dfrac{M}{4.0}$ or M = 36 g/mol

87.   The van der Waals Equation may be written

$$\left[P_{obs} + a\left(\tfrac{n}{V}\right)^2\right]\left[V-bn\right] = nRT$$ where a and b are the van der Waals constants for a specific substance Values for Cl_2 are found in Table 12.2.(page 489)

so $P_{obs} = \dfrac{nRT}{(V-bn)} - a\left(\dfrac{n}{V}\right)^2$

$$P_{obs} = \frac{(8.00 \text{ mol})(0.082057 \text{ L} \cdot \text{atm/K} \cdot \text{mol})(300 \text{ K})}{(4.00 \text{ L} - (0.0562 \text{ L/mol} \cdot 8.00 \text{ mol}))}$$

$$- \frac{6.49 \text{ atm} \cdot \text{L}^2}{\text{mol}^2} \cdot \left(\frac{8.00 \text{ mol}}{4.00 \text{ L}}\right)^2$$

$P_{obs} = \dfrac{199 \text{ L} \cdot \text{atm}}{3.55 \text{ L}} - 26.0 \text{ atm}$

   = 55.5 atm - 26.0 atm = 29.5 atm

From the ideal gas law: $P = \dfrac{nRT}{V} = \dfrac{(8.00 \text{ mol})(0.082057 \text{ L} \cdot \text{atm/K} \cdot \text{mol})(300 \text{ K})}{4.00 \text{ L}}$

   P = 49.3 atm

89.   Rank the following pressures in increasing order:

150 kPa, 742 mmHg, 0.89 atm, 650 torr

Convert each of these to a common pressure unit:

150 kPa = 1130 mmHg and 0.89 atm = 676 mm Hg and   650 torr = 650 mmHg

[See Study question 11 for examples of conversions of pressure units.]

Hence the order is: 650 torr < 0.89 atm < 742 mmHg < 150 kPa

91.   The pressure exerted by a gas originally at 25 °C when warmed to 100. °C may be obtained
by noting that pressure increases directly as a function of the absolute temperature:

$$P_2 = \frac{P_1 \cdot T_2}{T_1} = 3.0 \text{ atm} \cdot \frac{373 \text{ K}}{298 \text{ K}} = 3.8 \text{ atm}$$

93.   Assuming that the temperature remains constant as the CO is released into the room, we can
state:

$$P_{tank} \cdot V_{tank} = P_{room} \cdot V_{room}$$

$$45.6 \text{ mmHg} \cdot 56.0 \text{ L} = P_{room} \cdot 2.70 \times 10^4 \text{ L}$$

$$9.46 \times 10^{-2} \text{ mmHg} = P_{room}$$

95.   Rewriting the ideal gas law we obtain:

$$P \cdot V = n \cdot R \cdot T$$

$$P \cdot V = \frac{mass}{Molar\ mass} \cdot R \cdot T$$

Rearranging this equation gives: Molar mass $\cdot$ P $= \frac{mass}{V} \cdot R \cdot T$

or noting that D $= \frac{mass}{V}$ we write: Molar mass $= \frac{D \cdot R \cdot T}{P}$

Converting 331 mmHg to atm yields:

$$331 \text{ mmHg} \cdot \frac{1 \text{ atm}}{760 \text{ mmHg}} = 0.436 \text{ atm}$$

$$\text{Molar mass} = \frac{0.855 \frac{g}{L} \cdot 0.082057 \frac{L \cdot atm}{K \cdot mol} \cdot 273.2 \text{ K}}{0.436 \text{ atm}} = 44.0 \frac{g}{mol}$$

97.   a. $1.0 \text{ L H}_2 \cdot \frac{1 \text{ mol H}_2}{22.4 \text{ L H}_2} = 0.045 \text{ mol H}_2$

b. $1.0 \text{ L Ne} \cdot \frac{1 \text{ mol Ne}}{22.4 \text{ L Ne}} = 0.045 \text{ mol Ne}$

c. $n = \dfrac{PV}{RT} = \dfrac{(1 \text{ atm})(1.0 \text{ L})}{(0.082057 \ \frac{L \cdot atm}{K \cdot mol})(300 \text{ K})} = 0.041 \text{ mol } H_2$

d. $n = \dfrac{PV}{RT} = \dfrac{(800 \text{ mmHg} \cdot \frac{1 \text{ atm}}{760 \text{mmHg}})(1.0 \text{ L})}{(0.082057 \ \frac{L \cdot atm}{K \cdot mol})(273 \text{ K})} = 0.047 \text{ mol } CO_2$

The number of molecules may be calculated by multiplying the number of moles by Avogadro's number. However, the number of moles is proportional to the number of molecules, so it is sufficient to note that 1.0 L of CO_2 at 0 °C and 800 mmHg contains the largest number of molecules (of this group of gases) and 1.0 L of H_2 at 27 °C and 760 mmHg contains the smallest number of molecules.

99.   The amount of N_2 can be calculated.

$P(\text{Total}) = P(H_2O) + P(N_2)$

731.0 mmHg = 18.7 mmHg + $P(N_2)$

712.3 mmHg = $P(N_2)$ = 0.937 atm

$n(N_2) = \dfrac{(0.937 \text{ atm})(0.325 \text{ L})}{(0.082057 \ \frac{L \cdot atm}{K \cdot mol})(294 \text{ K})} = 1.26 \times 10^{-2} \text{ mol } N_2$

According to the equation in which sodium nitrite reacts with sulfamic acid, one mole of $NaNO_2$ produces one mole of N_2.

$1.26 \times 10^{-2} \text{ mol } N_2 \cdot \dfrac{1 \text{ mol } NaNO_2}{1 \text{ mol } N_2} \cdot \dfrac{69.00 \text{ g } NaNO_2}{1 \text{ mol } NaNO_2} = 0.870 \text{ g } NaNO_2$

Weight percentage of $NaNO_2 = \dfrac{0.870 \text{ g } NaNO_2}{1.012 \text{ g sample}} \times 100 = 86.0\% \ NaNO_2$

101.   The velocities of two species are inversely proportional to the square roots of their molar masses, so we write:

$$\dfrac{V_{rms} \text{ Xe}}{482 \text{ m/s}} = \sqrt{\dfrac{32.0 \text{ g/mol}}{131 \text{ g/mol}}}$$

and solving for V_{rms} Xe yields V= 238 m/s

103.   Calculate the number of moles of air with the ideal gas law:

$$n = \frac{P \cdot V}{R \cdot T} = \frac{1.0 \text{ atm} \cdot 0.500 \text{ L}}{0.082057 \frac{L \cdot atm}{K \cdot mol} \cdot 323 \text{ K}} = 0.019 \text{ mol air}$$

If the mole fraction of oxygen in the air is 0.20, then the number of moles of oxygen is :

$$0.019 \text{ mol air} \cdot \frac{20 \text{ mol O}_2}{100 \text{ mol air}} = 0.0038 \text{ mol O}_2$$

According to the balanced equation, this amount of oxygen would require:

$$0.0038 \text{ mol O}_2 \cdot \frac{2 \text{ mol C}_8\text{H}_{18}}{25 \text{ mol O}_2} \cdot \frac{114.2 \text{ g C}_8\text{H}_{18}}{1 \text{ mol C}_8\text{H}_{18}} = 0.034 \text{ g C}_8\text{H}_{18}$$

105.   a. Writing the ideal gas law as $M = \frac{dRT}{P}$ we can solve for the molar mass (M).

Note that density must be expressed in units of g/L, so we convert

$$\frac{92 \text{ g}}{1 \text{ m}^3} \cdot \frac{1 \text{ m}^3}{1000 \text{ L}} = 0.092 \text{ g/L}$$

Converting pressure to appropriate units: $42 \text{ mmHg} \cdot \frac{1 \text{ atm}}{760 \text{ mmHg}} = 0.055 \text{ atm}$

$$M = \frac{(0.092 \text{ g/L})(0.082057 \frac{L \cdot atm}{K \cdot mol})(210 \text{ K})}{0.055 \text{ atm}} = 28.7 \text{ g/mol (or 29 g/mol)}$$

   b. Mole fraction of O_2 and N_2:

The <u>average</u> molar mass reflects a contribution from the two gases.

Mass (from O_2) + Mass (from N_2) = 28.7 g/mol

$X(O_2) \cdot 32.0$ g/mol + $X(N_2) \cdot 28.0$ g/mol = 28.7 g/mol (29 to two sig. figures)

Since $X(O_2) + X(N_2) = 1$ then $X(N_2) = 1 - X(O_2)$

$X(O_2) \cdot 32.0$ g/mol + $(1 - X(O_2)) \cdot 28.0$ g/mol = 28.7 g/mol

$X(O_2) = 0.18$ and $X(N_2) = 0.82$

107. $Ni(CO)_x + 2 HCl \text{ (aq)} \rightarrow NiCl_2 \text{ (aq)} + H_2\text{(g)} + x CO\text{(g)}$
         0.125 g

                                        565 mL, 25 °C

                                        P_T = 121 mm

151

The total amount of hydrogen and carbon monoxide gas may be calculated with the ideal gas law :

$$n = \frac{121 \text{ mm Hg} \cdot \frac{1 \text{ atm}}{760 \text{ mmHg}} \cdot 0.565 \text{ L}}{0.082057 \frac{\text{L} \cdot \text{atm}}{\text{mol} \cdot \text{K}} \cdot 298.2 \text{ K}} = 3.68 \times 10^{-3} \text{ mol}$$

The number of moles of H_2 + moles of CO = 3.68×10^{-3}.

The stoichiometry of the equation indicates that $n(H_2) = n(Ni)$.

So $n(H_2) + n(CO) = 3.68 \times 10^{-3}$ may be written $n(Ni) + n(CO) = 3.68 \times 10^{-3}$

If we let m = mass of CO, we can write:

$$n(Ni) = \frac{(0.125 - m)}{58.69} \text{ or } 2.13 \times 10^{-3} - 1.70 \times 10^{-2} \text{ m}$$

$$\text{and } n(CO) = \frac{m}{28.01} = 3.57 \times 10^{-2} \text{ m}$$

Substituting:

$$(2.13 \times 10^{-3} - 1.70 \times 10^{-2} \text{ m}) + 3.57 \times 10^{-2} \text{ m} = 3.68 \times 10^{-3}$$
$$1.87 \times 10^{-2} \text{ m} = 1.55 \times 10^{-3}$$
$$m = 0.0829 \text{ g}$$

Then $n(Ni) = 2.13 \times 10^{-3} - 1.70 \times 10^{-2} (0.0829) = 7.21 \times 10^{-4}$ moles Ni
and $n(CO) = (0.0829)(3.57 \times 10^{-2}) = 2.96 \times 10^{-3}$ mol CO

$$\frac{\text{mol CO}}{\text{mol Ni}} = \frac{2.96 \times 10^{-3}}{7.21 \times 10^{-4}} \approx 4 \text{ and the compound's formula is } Ni(CO)_4.$$

109. The difference in pressure is:

$$P_{initial} - P_{final} = 158 \text{ mmHg} - 115 \text{ mmHg} = 43 \text{ mmHg} = 0.057 \text{ atm}$$

The number of moles which correspond to this pressure is :

$$n = \frac{P \cdot V}{R \cdot T} = \frac{0.057 \text{ atm} \cdot 1.0 \text{ L}}{0.082057 \frac{\text{L} \cdot \text{atm}}{\text{K} \cdot \text{mol}} \cdot 298 \text{ K}} = 0.0023 \text{ mol } O_2$$

111.   Since the amount of gas is constant throughout the experiment, we can use the combined gas law :

$$\frac{P_1 \cdot V_1}{T_1} = \frac{P_2 \cdot V_2}{T_2}$$

Since Pressure is constant, we can further simplify the expression to:

$$\frac{V_1}{T_1} = \frac{V_2}{T_2}$$

Substituting a value for $T_2 = T_1 + 100.$ and appropriate values for V_1 and V_2 yields:

$$\frac{10.00 \text{ mL}}{T_1} = \frac{13.66 \text{ mL}}{T_1 + 100.}$$

Solving: $(10.00 \text{ mL})(T_1 + 100.) = 13.66 \text{ mL } (T_1)$

$10.00 \, T_1 + 1000 = 13.66 \, T_1$ and

$1000 = 3.66 \, T_1$ and $T_1 = \frac{1000}{3.66}$ or 273

which (according to the problem) was 0.°C.

113.   According to Graham's Law, the two gases effuse at a rate which is inversely proportional to the square root of their molar masses. The rate of effusion of a gas is also inversely proportional to the time required: Rate $\alpha \, \frac{1}{T}$

$$\frac{\text{Rate of effusion of Iodine Gas}}{\text{Rate of effusion of } SCl_xF_y} = \sqrt{\frac{M \text{ of } SCl_xF_y}{M \text{ of } I_2}}$$

$$\frac{\frac{1}{10820 \text{ seconds}}}{\frac{1}{8470 \text{ seconds}}} = \sqrt{\frac{M \text{ of } SCl_xF_y}{253.81 \text{ g/mol } I_2}}$$

$$(0.782)^2 = \frac{M \text{ of } SCl_xF_y}{253.81 \text{ g/mol}}$$

$$156 = M \text{ of } SCl_xF_y$$

115.   Calculate the amount of CaC_2 and H_2O initially present:

$$95.0 \text{ g } CaC_2 \cdot \frac{1 \text{ mol } CaC_2}{64.10 \text{ g } CaC_2} = 1.48 \text{ mol } CaC_2$$

$$65.0 \text{ g } H_2O \cdot \frac{1 \text{ mol } H_2O}{18.02 \text{ g } H_2O} = 3.61 \text{ mol } H_2O$$

Note that CaC_2 is the limiting reagent (Use the moles-available and moles-required ratios). The amount of C_2H_2 produced would be:

$$1.48 \text{ mol } CaC_2 \; \bullet \; \frac{1 \text{ mol } C_2H_2}{1 \text{ mol } CaC_2} \; = \; 1.48 \text{ mol } C_2H_2$$

The pressure exerted by this amount of acetylene would be:

$$P \; = \; \frac{(1.48 \text{ mol})(0.082057 \frac{L \bullet atm}{K \bullet mol})(295.2 \text{ K})}{0.895 \text{ L}} \; = \; 40.1 \text{ atm}$$

117. Calculate the amount of methane produced with the ideal gas law:

$$n = \frac{P \bullet V}{R \bullet T} = \frac{90.6 \text{ mmHg } \bullet \; \frac{1 \text{ atm}}{760 \text{ mmHg}} \; \bullet 0.256 \text{ L}}{0.082057 \frac{L \bullet atm}{K \bullet mol} \; \bullet 303.2 \text{ K}} = 1.23 \times 10^{-3} \text{ mol } CH_4$$

The amount of CH_3 will also be 1.23×10^{-3} mol, so the mass of CH_3 in the sample is:

$$1.23 \times 10^{-3} \text{ mol } CH_3 \; \bullet \; \frac{15.03 \text{ g}}{1 \text{ mol } CH_3} \; = \; 0.0184 \text{ g } CH_3$$

The mass of thallium in the sample is then: 0.102 g - 0.0184 g or 0.0836 g.

This mass corresponds to 4.09×10^{-4} moles (Atomic mass of Tl = 204.38 g)

The ratio of moles of CH_3 to moles of Tl $= \dfrac{1.23 \times 10^{-3}}{4.09 \times 10^{-4}} = 3.00$ so the formula for the compound is $Tl(CH_3)_3$.

119. a. Mass of $NaHCO_3$ and Na_2CO_3:

The amount of HCl present is: $0.01400 \text{ L} \; \bullet \; 1.50 \text{ M HCl} = 0.0210 \text{ mol HCl}$

The balanced equations indicate that the HCl is consumed in the ratio:

$$\text{mol HCl} \; = \; \text{mol } NaHCO_3 \; + \; 2 \text{ mol } Na_2CO_3$$

Since $\text{mol X} = \dfrac{\text{mass X}}{\text{molar mass X}}$ we can write:

$$0.0210 \text{ mol HCl} \; = \; \text{mass } NaHCO_3 \left(\frac{1}{84.0}\right) \; + \; 2 \text{ mass } Na_2CO_3 \left(\frac{1}{106}\right)$$

and according to the data given:

$$\text{mass } NaHCO_3 \; + \; \text{mass } Na_2CO_3 \; = \; 1.500 \text{ g}$$

So mass Na_2CO_3 = 1.500 - mass $NaHCO_3$

Let m = mass $NaHCO_3$ then mass Na_2CO_3 = 1.500 - m

$$0.0210 \text{ mol} = m(\frac{1}{84.0}) + 2(1.500 - m)\frac{1}{106}$$

$$0.0210 \text{ mol} = 0.0119 \text{ m} + 0.0283 - 0.0189 \text{ m}$$

$$\frac{(0.0210 - 0.0283)}{- 0.0070} = m$$

1.05 g = m = mass $NaHCO_3$ and   0.45 g = mass Na_2CO_3

b. Pressure of CO_2 in 0.450 L flask at 27 ° C :

Moles CO_2 = moles $NaHCO_3$ +   moles Na_2CO_3

$$\text{Moles } CO_2 = \frac{1.05 \text{ g } NaHCO_3}{84.00 \text{ g}} + \frac{0.45 \text{ g } Na_2CO_3}{106 \text{ g } Na_2CO_3} = 1.67 \times 10^{-2} \text{ mol } CO_2$$

$$P = \frac{(0.0167 \text{ mol } CO_2)(0.082057 \text{ L} \cdot \text{atm/K} \cdot \text{mol})(300. \text{ K})}{(0.450 \text{ L})} = 0.916 \text{ atm}$$

121.   a. Partial Pressures of Cl_2, F_2 before reaction:

$$P_{Cl_2} = \frac{0.71 \text{ g } Cl_2 \cdot \frac{1 \text{ mol } Cl_2}{70.9 \text{ g } Cl_2} \cdot 0.082057 \text{ L} \cdot \text{atm/K} \cdot \text{mol} \cdot 296.2 \text{ K}}{0.258 \text{ L}}$$

    = 0.943 atm

$$P_{F_2} = \frac{1.00 \text{ g } F_2 \cdot \frac{1 \text{ mol } F_2}{38.00 \text{ g } F_2} \cdot 0.082057 \text{ L} \cdot \text{atm/K} \cdot \text{mol} \cdot 296.2 \text{ K}}{0.258 \text{ L}}$$

    = 2.48 atm

Calculate the limiting reagent:

moles required: $\frac{3 \text{ mol } F_2}{1 \text{ mol } Cl_2}$ moles available: $\frac{0.0263 \text{ mol } F_2}{0.0100 \text{ mol } Cl_2} = 2.63$

So **F_2** is the limiting reagent.

The amount of ClF_3 formed will be :

$$0.0263 \text{ mol } F_2 \cdot \frac{2 \text{ mol } ClF_3}{3 \text{ mol } F_2} = 0.0175 \text{ mol } ClF_3$$

and the pressure will be:

$$P_{ClF_3} = \frac{0.0175 \text{ mol } ClF_3 \cdot 0.082057 \text{ L} \cdot \text{atm/K} \cdot \text{mol} \cdot 296.2 \text{ K}}{0.258 \text{ L}}$$

$$= 1.65 \text{ atm}$$

The stoichiometry indicates that number of moles of Cl_2 consumed is one-third the number of moles of F_2 or 8.77×10^{-3} moles. This unreacted Cl_2 (0.0100 mol - 0.00877 mol) would exert a pressure of 0.116 atm

b. The Lewis dot structure for ClF_3 is:

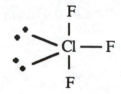

giving a structural pair geometry which is **trigonal bipyramidal**, and a molecular geometry which is **T-shaped**. The Cl atom would be utilizing sp^3d hybrid orbitals.

Outer electrons for fluorine are omitted for clarity's sake.

c.   Enthalpy of reaction

Bond	Energy
Cl - Cl	243 kJ/mol
F - F	159 kJ/mol
Cl - F	255 kJ/mol

Bonds broken: Bonds formed:

1 mol Cl-Cl • 243 kJ/mol = 243 kJ   6 mol Cl-F • 255 kJ/mol = 1530 kJ

3 mol F-F • 159 kJ/mol = <u>477 kJ</u>

                                    720 kJ

Energy of reaction = Bonds broken - Bonds formed

                    = 720 - 1530 = - 810 kJ

18.   Intermolecular force(s) to overcome to:

   (a) melt ice: hydrogen bonds (dipole - dipole)

   (b) melt solid I_2 : induced dipole - induced dipole

   (c) remove the water of hydration from $MnCl_2 \cdot 4\,H_2O$: ion - dipole

   (d) convert liquid NH_3 to NH_3 vapor: hydrogen bonds (dipole - dipole)

20.   **Ion - dipole** forces bind the waters of hydration (dipole) to the $CuSO_4$ species (ions).

22.

Molecules	Intermolecular Force
a. CO_2	induced dipole-induced dipole (London)
b. NH_3	dipole-dipole (hydrogen bonding)
c. $CHCl_3$	dipole-dipole
d. CCl_4	induced dipole-induced dipole (London)

24.   Increasing strength of intermolecular forces:

$Ne \;<\; CH_4 \;<\; CO \;<\; CCl_4$

Neon and methane are nonpolar species and possess only induced dipole-induced dipole interactions. Neon has a smaller molar mass than CH_4, and therefore weaker London (dispersion) forces. Carbon monoxide is a polar molecule. Molecules of CO would be attracted to each other by dipole-dipole interactions. The CCl_4 molecule is a non-polar molecule, but very heavy (when compared to the other three). Hence the larger London forces that accompany larger molecules would result in the strongest attractions of this set of molecules.

   The lower molecular weight molecules with weaker interparticle forces should be gases at 25 °C at 1 atmosphere: Ne, CH_4, CO.

26.

Higher Boiling Substance	Intermolecular Force
a. O_2	induced dipole-induced dipole (London) (greater for heavier molecules)
b. SO_2	dipole-dipole (stronger than London)
c. HF	hydrogen bonding (stronger than dipole-dipole)
d. GeH_4	induced dipole-induced dipole (greater for larger molecules)

28.   The compounds in order of increasing boiling point:

$$CH_4 < CO < NH_3 < SCl_2$$

The nonpolar molecule methane would exhibit only Induced dipole-Induced dipole (London) forces. The polar molecule CO would exhibit dipole-dipole forces. Ammonia would exhibit hydrogen bonding. SCl_2 is polar and with its large molar mass would have the strongest London forces.

30.   Compounds which are capable of forming hydrogen bonds with water are those containing polar O-H bonds and lone pairs of electrons on N,O, or F.

(a) CH_3-O-CH_3 no; no "polar H's" and the C-O bond is not very polar

(b) CH_4 no

(c) HF   yes: lone pairs of electrons on F and a "polar hydrogen".

(d) HCOOH   yes: lone pairs of electrons on O atoms, and a "polar hydrogen"
                     attached to one of the oxygen atoms

(e) I_2 no

(f) CH_3OH yes: "polar H" and lone pairs of electrons on O

32.   Since ion-dipole attractions increase as the distance between centers decreases, the energy of hydration of Cs^+ is less than that of Na^+. The attraction also increases with charge density, hence the order is: $Cs^+ < Na^+ < Mg^{2+}$

34.   a. increase -- The greater the intermolecular forces, the greater the energy (temperature) necessary to "separate" the molecules.

       b. decrease -- The stronger interparticle forces will reduce the number of liquid molecules escaping into the gaseous state -- reducing the vapor pressure.

       c. not change -- Pressure is force per unit area. If you increase the surface area of the liquid, you also increase the force so the pressure does not change.

36.   The heat energy necessary to vaporize 1.00 kg of CCl_3F:

$$1.00 \text{ kg } CCl_3F \cdot \frac{1.00 \times 10^3 \text{ g} CCl_3F}{1.00 \text{ kg } CCl_3F} \cdot \frac{1 \text{ mol } CCl_3F}{137.4 \text{ g } CCl_3F} \cdot \frac{24.8 \text{ kJ}}{1 \text{ mol } CCl_3F} = 180. \text{ kJ}$$

38.   The heat energy necessary to convert 250. mL of CH_3OH from liquid to vapor:

$$250.\text{ mL} \cdot \frac{0.787\text{ g }CH_3OH}{1\text{ mL }CH_3OH} \cdot \frac{1\text{ mol }CH_3OH}{32.04\text{ g }CH_3OH} \cdot \frac{38.0\text{ kJ}}{1\text{ mol }CH_3OH} = 233\text{ kJ}$$

40.   a. The vapor pressure of water at 60 °C from Figure 13.14 is approximately 150 mmHg.
          From Appendix F, the vapor pressure is 149.4 mmHg.

       b. At 94 °C the equilibrium vapor pressure of water will be 600 mmHg.

       c. The vapor pressure for water at 70 °C is approximately 225 mmHg and for ethyl
          alcohol at 70 °C is approximately 525 mmHg.

42.   The vapor pressure of $(C_2H_5)_2O$ at 30. °C is **600 mmHg**. (From Figure 13.14).
       Calculate the amount of $(C_2H_5)_2O$ to furnish this vapor pressure at 30.°C.

$$n = \frac{PV}{RT} = \frac{600\text{ mm} \cdot \frac{1\text{ atm}}{760\text{ mm}} \cdot 0.1\text{ L}}{0.082057\ \frac{L \cdot atm}{K \cdot mol} \cdot 303\text{ K}} = 3.2 \times 10^{-3}\text{ mol}$$

The total mass of $(C_2H_5)_2O$ present (1.0 g) corresponds to 0.014 mol.
The amount of $(C_2H_5)_2O$ in the liquid state is (0.014 - 0.0032) 0.0107 mol or 0.79 g.
So at 30.°C approximately 0.21 g is in the gaseous state. As the flask is cooled from
30.°C to 0 °C, **some of the gaseous ether will condense** to form liquid ether.

44.   (a) The normal boiling point of benzene is that temperature at which the vapor pressure of
       the liquid equals atmospheric pressure. Hence according to the data given, the boiling
       point for benzene is 80.1 °C.

(b) Plot of the vapor pressure of benzene with the data supplied:

Vapor Pressure Curve for Benzene

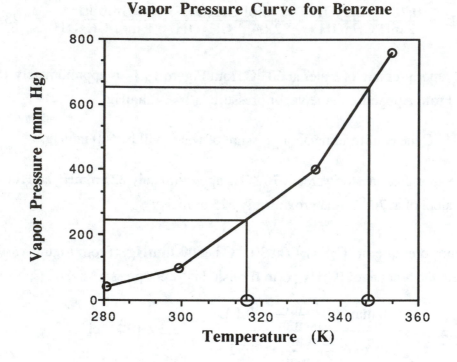

Note that at a vapor pressure of about 250 mmHg, the temperature corresponds to approximately 315 K (42 °C), while the temperature corresponding to 650 mmHg is approximately 347 K (74 °C).

(c) Using the Clausius-Clapeyron equation (found on page 522 of the text) we can substitute two T,P pairs from the data given. I chose P_1 = 100 mm, T_1 = 26.1 °C
and P_2 = 760 mm, T_2 = 80.1 °C.

$$\ln\left(\frac{P_2}{P_1}\right) = \frac{\Delta H_{vap}}{R}\left[\frac{1}{T_1} - \frac{1}{T_2}\right]$$

$$\ln\left(\frac{760 \text{ mm Hg}}{100 \text{ mm Hg}}\right) = \frac{\Delta H_{vap}}{8.3143 \times 10^{-3} \frac{kJ}{K \cdot mol}}\left[\frac{1}{299.3} - \frac{1}{353.3}\right]$$

Solving for ΔH_{vap} = 33.0 kJ/mol

46.   No, there are no conditions at which CF_4 will be a liquid at room temperature since room temperature is greater than the critical temperature.

160

48.   a. The length of the cell edge:

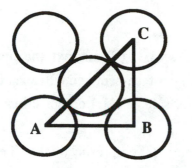

Since the lead atoms in a face-centered cubic arrangement touch along the diagonal ($\overline{AC}$) and since the angles in this lattice are 90°, we can calculate the cell edge ($\overline{AB}$ or $\overline{BC}$) using the Pythagorean theorem:

$$\overline{AC^2} = \overline{AB^2} + \overline{BC^2}$$

The length $\overline{AC}$ corresponds to 1 diameter and 2 radii (or a total of 4 radii)

$$(4 \text{ radii})^2 = 2 \text{ edge}^2$$
$$(4 \cdot 175.0 \text{ pm})^2 = 2 \text{ edge}^2$$
$$495.0 \text{ pm} = \text{edge}$$

b. The volume of the cell:

$$\text{edge}^3 \text{ or } (495.0 \text{ pm})^3 = 1.213 \times 10^8 \text{ pm}^3$$

or in units of cubic centimeters:

$$1.213 \times 10^8 \text{ pm}^3 \cdot \left(\frac{1 \text{ cm}}{1 \times 10^{10} \text{ pm}}\right)^3 = 1.213 \times 10^{-22} \text{ cm}^3$$

c. The density of the lead is:

\# atoms/unit cell = 6 atoms (faces) x 1/2 + 8 atoms (corners) x 1/8 = 4

$$\frac{4 \text{ atoms Pb}}{1 \text{ unit cell}} \cdot \frac{1 \text{ mol Pb}}{6.022 \times 10^{23} \text{ atoms Pb}} \cdot \frac{207.2 \text{ g Pb}}{1 \text{ mol Pb}} \cdot \frac{1}{1.213 \times 10^{-22} \text{ cm}^3}$$

$$= 11.35 \text{ g/cm}^3$$

50.   Consult Exercise 13.5 for a graphic of the unit cell geometry. The edge of this cell would be: $\frac{1}{3} \cdot (4 \cdot 124.9 \text{ pm})^2 = \text{edge}^2$

The volume of the cell would be: $(288.4 \text{ pm})^3 \cdot \left(\frac{1 \text{ cm}}{1 \times 10^{10} \text{ pm}}\right)^3 = 2.40 \times 10^{-23} \text{ cm}^3$

The number of atoms in the unit cell is: 2 for bcc and 4 for fcc.

The calculated density for a bcc lattice would be:

$$\frac{2 \text{ atoms Cr}}{\text{unit cell}} \cdot \frac{1 \text{ mol Cr}}{6.023 \times 10^{23} \text{ atoms Cr}} \cdot \frac{52.0 \text{ g Cr}}{1 \text{ mol Cr}} \cdot \frac{1}{2.40 \times 10^{-23} \text{ cm}^3}$$

$$= 7.20 \text{ g/cm}^3$$

That density for a fcc lattice would be 7.84 g/cm^3. These data indicate that Cr would be **bcc**.

52. $CaCl_2$ cannot have the NaCl structure. As shown in Figure 13.31 of your text, the cubic structure possesses 4 net lattice ions (occupied by anions) per face-centered lattice and 4 octahedral holes (occupied by cations). This is suitable for salts of a 1:1 composition.

54.   The ionic radii for Rb^+ and Cl^- are reported as 166 pm and 167 pm, respectively. The unit cell would have alternating rubidium and chloride ions. The edges of the unit cell would contain a net of one of each type ion. Their diameters would give a length of 666 pm. The volume of this cell would be

$$v = l \times l \times l = (666 \text{ pm})^3 = 2.95 \times 10^8 \text{ pm}^3 = 2.95 \times 10^{-22} \text{ cm}^3$$

One unit cell would contain 4 ion-pairs. The mass of these ion-pairs is:

$$\frac{120.9 \text{ g RbCl}}{1 \text{mol RbCl}} \cdot \frac{1 \text{ mol RbCl}}{6.022 \times 10^{23} \text{ ion pairs}} \cdot 4 \text{ ion pairs} = 8.030 \times 10^{-22} \text{ g RbCl}$$

The density would be: $D = \dfrac{M}{V} = \dfrac{8.030 \times 10^{-22} \text{ g RbCl}}{2.95 \times 10^{-22} \text{ cm}^3} = 2.72 \text{ g/cm}^3$

56.   Solid's density = 0.77 g/cm^3 Edge of unit cell = 4.086×10^{-8} cm

Volume of unit cell = $(4.086 \times 10^{-8} \text{ cm})^3 = 6.822 \times 10^{-22}$ cm^3

Mass = Density $\cdot$ Volume = 0.77 g/cm^3 $\cdot$ 6.822×10^{23} cm^3 = 5.3×10^{-23} g

1 LiH ion pair has a mass of:

$$\frac{7.94 \text{ LiH}}{1 \text{ mol LiH}} \cdot \frac{1 \text{ mol LiH}}{6.022 \times 10^{23} \text{ LiH ion pairs}} = 1.32 \times 10^{-23} \text{ g/LiH ion pair}$$

$$5.3 \times 10^{-23} \text{ g/unit cell} \cdot \frac{1 \text{ LiH ion pair}}{1.32 \times 10^{-23} \text{ g}} = 4 \text{ ion pairs/ unit cell}$$

The **face-centered cubic lattice is appropriate** for LiH.

58.   The attractive forces between the cation and anion depend inversely upon the distance between the charged species. Hence as the cation size increases from Li^+ to Cs^+, the distance between these ions and the I^- ion increases, resulting in a decrease in lattice energy.

162

60.   Since the iodide ion is larger than the fluoride ion, the ion distances would be greater in
        CsI, and we expect **CsF to have a higher melting point than CsI**.

62.   The heat evolved when 15.5 g of benzene freezes at 5.5 °C :

$$15.5 \text{ g benzene} \cdot \frac{1 \text{ mol benzene}}{78.1 \text{ g benzene}} \cdot \frac{9.94 \text{ kJ}}{1 \text{ mol benzene}} = 1.97 \text{ kJ}$$

        The quantity of heat needed to remelt this 15.5 g sample of benzene would be 1.97 kJ.

64.   (1) Heat evolved when 10.0 g of freon is cooled from 40 °C to - 30°C :

$$q = 10.0 \text{ g } CCl_2F_2 \cdot \frac{0.61 \text{ J}}{\text{g} \cdot \text{K}} \cdot (-70 \text{ K}) = -427 \text{ J} \text{ cooling gas to boiling point}$$

        (2) Heat evolved when freon condenses from gaseous to liquid state at the boiling point:

$$q = 10.0 \text{ g } CCl_2F_2 \cdot \frac{-165 \text{ kJ}}{\text{g}} = -1650 \text{ kJ or } -1.65 \times 10^3 \text{ J}$$

        (3) Heat evolved as the liquid freon cools from - 30°C to - 40°C:

$$\text{Heat evolved} = 10.0 \text{ g } CCl_2F_2 \cdot \frac{0.97 \text{ J}}{\text{g} \cdot \text{K}} \cdot (-10 \text{ K}) = -97 \text{ J}$$

        The quantity of heat evolved is then $(427 + 1.65 \times 10^3 + 97)$ Joules or 2174 J.
        The data for the specific heat capacities restrict the answer to 2200 Joules.

66.   a. At 1.0 atm pressure and room temperature xenon is in the **gaseous phase**.
        b. At 0.75 atm pressure and -114 °C xenon is in the **liquid phase**.
        c. The vapor pressure of 380 mmHg (0.5 atm) is found at a temperature of
                approximately **- 117 °C**.
        d. At - 122 °C the vapor pressure of the solid is approximately **0.25 atm** .
        e. The **solid** phase of xenon is more dense than the liquid phase. The liquid-solid
        interface has a positive slope. Remember the liquid-solid interface for water has a negative
        slope--and that liquid water is more dense than solid water.

68.   The observation that cooking oil floats on water, instead of forming a homogeneous solution with water indicates dissimilar interparticle forces to those present in water. Hence we conclude that the **amount of hydrogen bonding in cooking oil is minuscule**.

70.   The alcohol 1-propanol possesses the ability to form hydrogen bonds with the concomitant higher boiling point. The ether lacks the "polar" O-H bond found in the alcohol and has the lower boiling point since the interparticle forces between ether molecules are much weaker induced dipole-induced dipole forces.

72.   Differences in intermolecular forces for diethyl ether, ethyl alcohol, and water:
>   water -- hydrogen bonds
>   ethyl alcohol -- hydrogen bonds
>   diethyl ether -- dipole- dipole forces

74.   Volume of room $= 3.0 \times 10^2$ cm $\bullet$ 2.5×10^2 cm $\bullet$ 2.5×10^2 cm $= 1.9 \times 10^7$ cm^3
Convert volume to L:

$$1.9 \times 10^7 \, cm^3 \bullet \frac{1 \, L}{1.0 \times 10^3 \, cm^3} = 1.9 \times 10^4 \, L$$

To produce a pressure of 60 mmHg, calculate the amount of ethanol required.

$$P = 60 \, mm \bullet \frac{1 \, atm}{760 \, mm} = 7.9 \times 10^{-2} \, atm$$

$$V = 1.9 \times 10^4 \, L \qquad and \; T = 25 + 273 = 298 \, K$$

$$\frac{PV}{RT} = \frac{7.9 \times 10^{-2} \, atm \; \bullet 1.9 \times 10^4 \, L}{0.082057 \frac{L \bullet atm}{K \bullet mol} \bullet 298 \, K} = 6.1 \times 10^1 \, mol \; ethanol$$

$$6.1 \times 10^1 \; mol \; ethanol \bullet \frac{46.1 \, g \; ethanol}{1 \, mol \; ethanol} = 2.8 \times 10^3 \, g \; ethanol$$

This mass of ethanol would occupy a volume of:

$$2.8 \times 10^3 \, g \; C_2H_5OH \bullet \frac{1 \, cm^3}{0.79 \, g} = 3.6 \times 10^3 \, cm^3$$

As only 1 L of C_2H_5OH (1.0×10^3 cm^3) was introduced into the room, **all the ethanol would evaporate**.

76.   The **viscosity of ethylene glycol would be predicted to be greater** than that of ethanol since the glycol possesses two O-H groups per molecule while ethanol possesses one.

78.   For the unit cell shown there are 8 corner atoms (1/8 in the cell), 6 face atoms (1/2 in the cell), and 4 atoms wholly within the cell, for a total of **8 carbon atoms**.

a. The volume of the unit cell may be calculated if we first calculate the masses of the atoms involved:

$$\frac{12.01 \text{ g C atom}}{1 \text{ mol C atom}} \cdot \frac{1 \text{ mol C atom}}{6.022 \times 10^{23} \text{ C atom}} \cdot \frac{8 \text{ atoms}}{1 \text{ unit cell}} = 1.60 \times 10^{-22} \frac{\text{g C atom}}{\text{unit cell}}$$

$$D = \frac{M}{V} \quad \text{and} \quad V = \frac{M}{D} = \frac{1.60 \times 10^{-22} \text{ g/unit cell}}{3.51 \text{ g/cm}^3} = 4.55 \times 10^{-23} \text{ cm}^3/\text{unit cell}$$

b. The length of an edge: Volume $= l^3$

$$4.55 \times 10^{-23} \text{ cm}^3 = l^3$$

$$3.57 \times 10^{-8} \text{ cm} = l$$

80.   Ice has a relatively high vapor pressure. The net effect of this property is that hail stones sublime in the freezing compartment.

82.   An examination of Figure 13.48 shows that the solid/liquid line for CO_2 has a positive slope (unlike that of water). Hence an increase in pressure on the solid (as the blades pressed upon the "ice") would not cause the solid to melt. Additionally since the solid is more dense than the liquid, the solid would not float on liquid CO_2. Hence, skating would NOT BE EASY!

84.   If one counts the titanium atoms in the unit cell pictured, one notes a Ti atom at the corner of the cube. Each of these eight atoms is shared by seven other unit cells, hence the number of Ti atoms contributed by this source is (8 x $\frac{1}{8}$) or 1 atom.

The lattice also possesses one Ti atom in the center, providing an additional Ti atom to the count for a total of **two** Ti atoms (and TiO_2) units per unit cell.

86.   For a simple cubic unit cell, each corner is occupied by an atom or ion. Each of these is contained within EIGHT unit cells contributing 1/8 to each. Within one unit cell, therefore, there is (8 x $\frac{1}{8}$) 1 atom or ion. The volume occupied by that one net atom would be equal to 4/3 πr^3 with r representing the radius of the spherical atom or ion. The volume of the unit cell may be calculated by noting that the length of one side of the cell (an edge) corresponds to two radii (2r)-- since the spheres touch. The volume of this cube is therefore $(2r)^3$. The empty space within the cell is therefore:

   $(2r)^3$ - 4/3 πr^3 and the fraction of space unoccupied is:

$$\frac{(8 - 4/3 \pi)r^3}{8r^3} = \frac{8 - 4/3 \pi}{8} = 0.476 \text{ or approximately 48 \%}$$

88.   Consult problem 48 for an explanation of the edge calculation:

Radius of gold atom = 144.0 pm therefore the length of edge = 407.3 pm

The volume of the cell would be :

$(407.3 \text{ pm})^3 \cdot (\frac{1 \text{ cm}}{1 \times 10^{10} \text{ pm}})^3 = 6.757 \times 10^{-23} \text{ cm}^3$

The density of Au is 19.32 g/cm^3, so the mass of a unit cell is:

19.32 $g/cm^3 \cdot 6.757 \times 10^{-23} \text{ cm}^3 = 1.305 \times 10^{-21}$ g

$\frac{1.305 \times 10^{-21} \text{ g Au}}{\text{unit cell}} \cdot \frac{1 \text{ unit cell}}{4 \text{ atoms Au}} \cdot \frac{1 \text{ mol Au}}{196.97 \text{ g Au}} = 1.657 \times 10^{-24} \frac{\text{mol Au}}{\text{atom Au}}$

and taking the reciprocal gives 6.039 x $10^{23} \frac{\text{atom Au}}{\text{mol Au}}$.

90.   a. The electron dot structure for SO_2:

$\ddot{\text{O}}=\ddot{\text{S}}-\ddot{\text{O}}\colon$

(i) The OSO angle is approximately 120°. (Slightly less due to the lone pair on S).

(ii) The structural pair geometry is trigonal planar and the molecular geometry is bent.

(iii) The S atom is using sp^2 hybrid orbitals.

166

b. The forces binding SO_2 molecules to each other are dipole-dipole forces since SO_2 molecules are polar.

c. Listed in order of increasing intermolecular forces:

CH_4 < NH_3 < SO_2 < H_2O

For H_2O hydrogen bonding is possible. SO_2 is a relatively heavy molecule that will exhibit dipole-dipole forces. The lighter NH_3 molecule will have the hydrogen bonding forces that would be absent in CH_4.

d. Since the melting point of SO_2 is higher than the triple point, SO_2 would have a phase diagram somewhat similar to that of CO_2 (see Figure 13.48 in the text) and **solid SO_2 would be more dense than the liquid**.

e. Enthalpy change for SO_2 (g) $\rightarrow$ SO_3 (g):

$$\Delta H_{rxn} = [\Delta H_f\, SO_3\, (g)] - [\Delta H_f\, SO_2\, (g) + \Delta H_f\, O_2\, (g)]$$
$$= (-395.72 \text{ kJ/mol}) - (-296.83 \text{ kJ/mol})$$
$$= -98.89 \text{ kJ/mol}$$

Enthalpy change for H_2SO_4 formation:

$$\Delta H_{rxn} = [\Delta H_f\, H_2SO_4\, (aq)] - [\Delta H_f\, SO_3\, (g) + \Delta H_f\, H_2O\, (l)]$$
$$= -909.27 \text{ kJ/mol} - [-395.72 \text{ kJ/mol} + (-285.83 \text{ kJ/mol})]$$
$$= -227.72 \text{ kJ/mol}$$

92.   a. Metric tons of $CuFeS_2$ to produce 1 metric ton of Cu:

$$1000. \text{ kg Cu} \cdot \frac{1 \times 10^3 \text{ g Cu}}{1 \text{ kg Cu}} \cdot \frac{1 \text{ mol Cu}}{63.546 \text{ g Cu}} \cdot \frac{1 \text{ mol } CuFeS_2}{1 \text{ mol Cu}}$$

$$\cdot \frac{183.51 \text{ g } CuFeS_2}{1 \text{ mol } CuFeS_2} \cdot \frac{1 \text{ kg } CuFeS_2}{1 \times 10^3 \text{ g } CuFeS_2} \cdot \frac{1 \text{ metric ton } CuFeS_2}{1 \times 10^3 \text{ kg } CuFeS_2}$$

$$= 2.898 \text{ metric ton } CuFeS_2$$

b. $$1000. \text{ kg } CuFeS_2 \cdot \frac{64.13 \text{ kg S}}{183.51 \text{ kg } CuFeS_2} \cdot \frac{64.066 \text{ kg } SO_2}{32.066 \text{ kg S}} \cdot \frac{1 \text{ metric ton } SO_2}{1 \times 10^3 \text{ kg } SO_2}$$

$$= 0.698 \text{ metric ton } SO_2$$

c. The radius of a copper atom:

Given that copper crystallizes as a face-centered cube, there are 4 Cu atoms per unit cell. (See problems 48, 50, and 88 for similar calculations.) Given that the atomic mass for Cu is 63.5436 g, we can calculate the volume of a unit cell:

$$\frac{63.546 \text{ g Cu}}{1 \text{ mol Cu atoms}} \cdot \frac{1 \text{ mol Cu atoms}}{6.023 \times 10^{23} \text{ Cu atoms}} \cdot \frac{4 \text{ Cu atoms}}{1 \text{ unit cell}} = 4.22 \times 10^{-22} \frac{\text{g Cu}}{\text{unit cell}}$$

Using the density of the solid Cu:

$$4.22 \times 10^{-22} \frac{\text{g Cu}}{\text{unit cell}} \cdot \frac{1 \text{ cm}^3 \text{ Cu}}{8.95 \text{ g Cu}} = 4.22 \times 10^{-23} \frac{\text{cm}^3 \text{ Cu}}{\text{unit cell}}$$

Since the volume is 4.72×10^{-23} cm^3, the length of one edge is

$$\sqrt[3]{4.72 \times 10^{-23} \text{cm}^3} = 3.61 \times 10^{-8} \text{ cm}$$

and the diagonal distance $= \sqrt{2}$ (edge) [see p 530 in text]

diagonal distance $= (1.414)(3.61 \times 10^{-8} \text{ cm}) = 5.11 \times 10^{-8} \text{ cm}$

or expressed in pm: $5.11 \times 10^{-8} \text{ cm} \cdot \frac{1 \times 10^{10} \text{ pm}}{1 \text{ cm}} = 511 \text{ pm}$

The diagonal distance $= 4 \cdot$ radius thus

511 pm $= 4 \cdot$ radius

128 pm $=$ radius

CHAPTER 14: Solutions and Their Behavior

13. For a solution containing 45.0 g of $C_2H_4(OH)_2$ in 500.0 g of water, the molality is:

$$45.0 \text{ g } C_2H_4(OH)_2 \cdot \frac{1 \text{ mol } C_2H_4(OH)_2}{62.07 \text{ g } C_2H_4(OH)_2} \cdot \frac{1}{0.500 \text{ kg}} = 1.45 \text{ molal}$$

The mole fraction of glycol is: $500.0 \text{ g water} \cdot \dfrac{1 \text{ mol water}}{18.015 \text{ g water}} = 27.75 \text{ mol } H_2O$

and the moles of glycol calculated above are: $0.725 \text{ mol } C_2H_4(OH)_2$

giving a mf glycol of $\dfrac{0.725 \text{ mol } C_2H_4(OH)_2}{0.725 \text{ mol } C_2H_4(OH)_2 + 27.75 \text{ mol } H_2O} = 0.0255$

and a mf water of (1 - 0.0255) or 0.9745

The weight percent of glycol is:

$$\frac{45.0 \text{ g } C_2H_4(OH)_2}{45.0 \text{ g } C_2H_4(OH)_2 + 500.0 \text{ g } H_2O} \cdot 100 = 8.26 \text{ \% glycol}$$

15. Complete the following transformations for
NaCl:

Weight percent:

$$\frac{0.25 \text{ mol NaCl}}{1 \text{ kg solvent}} \cdot \frac{58.44 \text{ g NaCl}}{1 \text{ mol NaCl}} = \frac{14.6 \text{ g NaCl}}{1 \text{ kg solvent}}$$

$$\frac{14.6 \text{ g NaCl}}{1000 \text{ g solvent} + 14.6 \text{ g NaCl}} \cdot 100 = 1.4 \text{ \% NaCl}$$

Mole fraction:

$1000 \text{ g } H_2O = 55.51 \text{ mol } H_2O$

$$X_{NaCl} = \frac{0.25 \text{ mol NaCl}}{55.51 \text{ mol } H_2O + 0.25 \text{ mol NaCl}} = 4.5 \times 10^{-3}$$

C_2H_5OH:

Molality:

$$\frac{5.0 \text{ g } C_2H_5OH}{100 \text{ g solution}} \cdot \frac{1 \text{ mol } C_2H_5OH}{46.07 \text{ g } C_2H_5OH} \cdot \frac{100 \text{ g solution}}{95 \text{ g solvent}} \cdot \frac{1000 \text{ g solvent}}{1 \text{ kg solvent}}$$

$$= 1.1 \text{ molal}$$

Mole fraction:

$$\frac{5.0 \text{ g } C_2H_5OH}{1} \cdot \frac{1 \text{ mol } C_2H_5OH}{46.07 \text{ g } C_2H_5OH} = 0.11 \text{ mol } C_2H_5OH$$

and for water : $\frac{95 \text{ g } H_2O}{1} \cdot \frac{1 \text{ mol } H_2O}{18.02 \text{ g } H_2O} = 5.28 \text{ mol } H_2O$

$$X_{C_2H_5OH} = \frac{0.11 \text{ mol } C_2H_5OH}{5.28 \text{ mol } H_2O + 0.11 \text{ mol } C_2H_5OH} = 0.020$$

$C_{12}H_{22}O_{11}$:

Weight percent:

$$\frac{0.10 \text{ mol } C_{12}H_{22}O_{11}}{1 \text{ kg solvent}} \cdot \frac{342.3 \text{ g } C_{12}H_{22}O_{11}}{1 \text{ mol } C_{12}H_{22}O_{11}} = \frac{34.2 \text{ g } C_{12}H_{22}O_{11}}{1 \text{ kg solvent}}$$

$$\frac{34.2 \text{ g } C_{12}H_{22}O_{11}}{1000 \text{ g } H_2O + 34.2 \text{ g } C_{12}H_{22}O_{11}} \times 100 = 3.3 \% \ C_{12}H_{22}O_{11}$$

Mole fraction:

$$X_{C_{12}H_{22}O_{11}} = \frac{0.10 \text{ mol } C_{12}H_{22}O_{11}}{55.51 \text{ mol } H_2O + 0.10 \text{ mol } C_{12}H_{22}O_{11}} = 1.8 \times 10^{-3}$$

17.   a. To prepare a solution that is 0.200 m $NaNO_3$:

$$\frac{0.200 \text{ mol } NaNO_3}{1 \text{ kg } H_2O} \cdot \frac{0.500 \text{ kg } H_2O}{1} \cdot \frac{84.99 \text{ g } NaNO_3}{1 \text{ mol } NaNO_3} = 8.50 \text{ g } NaNO_3$$

The mole fraction of $NaNO_3$ in the resulting solution:

$$\frac{500. \text{ g } H_2O}{1} \cdot \frac{1 \text{ mol } H_2O}{18.02 \text{ g } H_2O} = 27.8 \text{ mol } H_2O$$

$$X (NaNO_3) = \frac{0.100 \text{ mol } NaNO_3}{0.100 \text{ mol } NaNO_3 + 27.8 \text{ mol } H_2O} = 0.00359$$

19.   The mole fraction (mf) of CH_3OH is 0.093 which means the mf (water) = 0.907.

The number of moles of water corresponding to this mole fraction is:

$$500. \text{ g water} \cdot \frac{1 \text{ mol water}}{18.02 \text{ g water}} = 27.8 \text{ mol } H_2O$$

Since 27.8 mol H_2O corresponds to a mf of 0.907 we can write:

$$\frac{\text{mol } H_2O}{\text{mol } H_2O + \text{mol } CH_3OH} = \frac{27.8}{27.8 + \text{mol } CH_3OH} = 0.907 \text{ or}$$

$$27.8 = 0.907(27.8 + \text{mol } CH_3OH) \text{ giving: } 2.85 = \text{mol } CH_3OH$$

which would correspond to a mass of:

$$2.85 \text{ mol } CH_3OH \cdot \frac{32.04 \text{ g } CH_3OH}{1 \text{ mol } CH_3OH} = 91.2 \text{ g } CH_3OH$$

The molality of the solution would be: $\dfrac{2.85 \text{ mol } CH_3OH}{0.500 \text{ kg } H_2O} = 5.70 \text{ molal}$

21.   For a solution containing 15.0 g of $C_2H_4(OH)_2$ in 775 g of water, the molality is:

$$15.0 \text{ g } C_2H_4(OH)_2 \cdot \frac{1 \text{ mol } C_2H_4(OH)_2}{62.07 \text{ g } C_2H_4(OH)_2} \cdot \frac{1}{0.775 \text{ kg}} = 0.312 \text{ molal}$$

The mf of $C_2H_4(OH)_2$ is:

$$15.0 \text{ g } C_2H_4(OH)_2 \cdot \frac{1 \text{ mol } C_2H_4(OH)_2}{62.07 \text{ g } C_2H_4(OH)_2} = 0.242 \text{ mol glycol}$$

$$775 \text{ g water} \cdot \frac{1 \text{ mol water}}{18.02 \text{ g water}} = 43.0 \text{ mol } H_2O$$

giving a mf (glycol): $\dfrac{0.242 \text{ mol } C_2H_4(OH)_2}{0.242 \text{ mol } C_2H_4(OH)_2 + 43.0 \text{ mol } H_2O} = 0.00559$ and

a mf (water) = (1 - 0.00559) = 0.994

23.   For the compound Na_2CO_3:

$$0.0125 \text{ molal solution} = \frac{x \text{ mol } Na_2CO_3}{0.250 \text{ kg } H_2O} \text{ and}$$

$$x = 3.13 \times 10^{-3} \text{ mol } Na_2CO_3 \text{ or } 0.331 \text{ g } Na_2CO_3$$

Since 250. g H_2O corresponds to 13.9 mol H_2O,

$$\text{the mf}(Na_2CO_3) = \frac{3.13 \times 10^{-3} \text{ mol } Na_2CO_3}{3.13 \times 10^{-3} \text{ mol } Na_2CO_3 + 13.9 \text{ mol } H_2O} = 2.25 \times 10^{-4}$$

For the compound CH_3OH:

13.5 g CH_3OH = 0.421 mol and 150. g H_2O = 8.32 mol

The molality of the solution is: $\dfrac{0.421 \text{ mol}}{0.150 \text{ kg}} = 2.81$ molal

The mf (methanol) of the solution is: $\dfrac{0.421}{0.421 + 8.32} = 0.0482$

For the compound KNO_3:

The mf (water) = 1 - 0.0934 = 0.9066

and 555 g water = $555 \text{ g } H_2O \cdot \dfrac{1 \text{ mol } H_2O}{18.02 \text{ g } H_2O} = 30.8 \text{ mol } H_2O$

We can write: $\dfrac{\text{mol } H_2O}{\text{mol } H_2O + \text{mol } KNO_3} = 0.9066$ and substituting

$\dfrac{30.8}{30.8 + x} = 0.9066$ so x = 3.17 and 3.17 mol of $KNO_3 = 321 \text{ g } KNO_3$

For the molality:

$$\dfrac{3.17 \text{ mol of } KNO_3}{0.555 \text{ kg } H_2O} = 5.72 \text{ molal}$$

25.   a. Calculate the molality of a solution of 12.0 M HCl with a density of 1.18 g/cm^3:

$$\dfrac{1.18 \text{ g HCl}}{1 \text{ mL}} \cdot \dfrac{1000 \text{ mL}}{1 \text{ L}} = 1180 \text{ g/L}$$

The mass of HCl in 12.0 M HCl: $\dfrac{12.0 \text{ mol HCl}}{1 \text{ L}} \cdot \dfrac{36.5 \text{ g HCl}}{1 \text{ mol HCl}} = 438 \text{ g HCl/L}$

The mass of water in the solution is: 1180 g solution - 438 g HCl = 742 g H_2O

The molality of HCl $= \dfrac{12.0 \text{ mol HCl}}{0.742 \text{ kg}} = 16.2$ molal

   b. Calculate the weight percent of HCl:

$$\dfrac{438 \text{ g HCl}}{1180 \text{ g solution}} \text{ x } 100 = 37.1 \text{ \% HCl}$$

27.   a. Mole fraction of NaOH:

$$\dfrac{10.7 \text{ mol NaOH}}{1 \text{ kg solvent}} \cdot \dfrac{1 \text{kg solvent}}{1000 \text{ g solvent}} \cdot \dfrac{18.02 \text{ g solvent}}{1 \text{ mol solvent}} = \dfrac{10.7 \text{mol NaOH}}{55.5 \text{ mol } H_2O}$$

$$\dfrac{10.7 \text{ mol NaOH}}{55.5 \text{ mol } H_2O + 10.7 \text{ mol NaOH}} = 0.162 \text{ mf NaOH}$$

b. Weight percentage of NaOH:

$$\frac{10.7 \text{ mol NaOH}}{1000 \text{ g solvent}} \cdot \frac{40.0 \text{ g NaOH}}{1 \text{ mol NaOH}} = \frac{428 \text{ g NaOH}}{1000 \text{ g solvent}}$$

The mass of solution would be (428 g + 1000. g) 1428 g.

$$\frac{428 \text{ g NaOH}}{1428 \text{ g solution}} \times 100 = 30.0\% \text{ NaOH}$$

c. Molarity of the solution:

$$\frac{10.7 \text{ mol NaOH}}{1428 \text{ g solution}} \cdot \frac{1.33 \text{ g NaOH}}{1 \text{ cm}^3 \text{ solution}} \cdot \frac{1000 \text{ cm}^3}{1 \text{ L solution}} = 9.97 \text{ M NaOH}$$

29. The molality of the $CaCl_2$ solution:

$$\frac{2.00 \text{ g CaCl}_2}{0.750 \text{ kg solvent}} \cdot \frac{1 \text{ mol CaCl}_2}{111.0 \text{ g CaCl}_2} = 0.0240 \text{ molal CaCl}_2$$

One mol $CaCl_2$ provides 3 mol of ions (1 Ca^{2+} and 2 Cl^-). The total molality would be
(3 x 0.0240) 0.0721 molal.

31. The concentration of ppm expressed in grams is:

$$1.9 \text{ ppm} = \frac{1.9 \text{ g solute}}{1.0 \times 10^6 \text{ g solvent}} = \frac{1.9 \text{ g solute}}{1.0 \times 10^3 \text{ kg solvent}} \text{ or } \frac{0.0019 \text{ g solute}}{1 \text{ kg water}}$$

$$\frac{0.0019 \text{ g Li}^+}{1 \text{ kg water}} \cdot \frac{1 \text{ mol Li}^+}{6.939 \text{ g Li}^+} = 2.7 \times 10^{-4} \text{ molal Li}^+$$

33. Explain the following observations:

a. Octane is miscible in CCl_4 since both are nonpolar liquids.

b. Methyl alcohol mixes with water since both are polar liquids, and can hydrogen bond.

c. NaBr is poorly soluble in diethyl ether since NaBr is ionic (very polar) while diethyl
   ether is only slightly polar.

35. The enthalpy of formation (ΔH_f) for NH_4Cl:

$\Delta H°_{solution} = \Delta H_f° \text{ NH}_4\text{Cl (1m)} - \Delta H_f° \text{ NH}_4\text{Cl (s)}$

$+ 14.7 \text{ kJ/mol} = \Delta H_f° \text{ NH}_4\text{Cl (1 m)} - (- 314.4 \text{ kJ/mol})$

$- 299.7 \text{ kJ/mol} = \Delta H_f° \text{ NH}_4\text{Cl (1 m)}$

37. The data indicate increased solubility at higher temperatures so the enthalpy of solution for KNO_3 is **positive** (endothermic).

$$\Delta H_{solution}\ KNO_3 = \Delta H_f\ KNO_3\ (aq, 1\ m) - \Delta H_f\ KNO_3\ (s)$$
$$= -459.7\ kJ/mol - (-494.6\ kJ/mol)$$
$$= 34.9\ kJ/mol$$

39. To calculate the concentration of CO_2 at various pressures, use Henry's Law:

Molality $= K \cdot P$ where K is the Henry's Law constant for a given gas.

From Exercise 14.3 we find that K for CO_2 is 4.48×10^{-5} molal/mmHg.

At a pressure of 1 atm (760 mmHg):

Molality $= 4.48 \times 10^{-5}$ molal/mmHg $\cdot$ 760 mmHg $= 0.0340$ molal

At a pressure of 1/3 atm:

Molality $= 4.48 \times 10^{-5}$ molal/mmHg $\cdot$ 253 mmHg $= 0.0113$ molal

41. To calculate the vapor pressure, we must first calculate the number of moles of each substance:

$$1.80 \times 10^3\ g\ H_2O\ \rightarrow\ 100.\ mol\ H_2O$$
$$620.\ g\ of\ (CH_2OH)_2 \rightarrow 10.0\ mol\ (CH_2OH)_2$$

The vapor pressure can be calculated :

$$P_{water} = X_{water}\ P°_{water}$$

$$P_{water} = \frac{100.\ mol}{(100. + 10.0)\ mol} \cdot 760\ mm = 691\ mmHg$$

43. Using Raoult's Law, we know that the vapor pressure of pure water (P°) multiplied by the mole fraction(X) of the solute gives the vapor pressure of the solvent above the solution (P).

$$P_{water} = X_{water}\ P°_{water}$$

The vapor pressure of pure water at 90 °C is 525.8 mmHg (from Appendix D).

Since the P_{water} is given as 457 mmHg, the mole fraction of the water is:

$$\frac{457\ mmHg}{525.8\ mmHg} = 0.869$$

The 2.00 kg of water correspond to a mf of 0.869. This mass of water corresponds to:

$$2.00 \times 10^3 \text{ g H}_2\text{O} \cdot \frac{1 \text{ mol H}_2\text{O}}{18.02 \text{ g H}_2\text{O}} = 111 \text{ mol water.}$$

Representing moles of ethylene glycol as x we can write:

$$\frac{\text{mol H}_2\text{O}}{\text{mol H}_2\text{O} + \text{mol C}_2\text{H}_4(\text{OH})_2} = \frac{111}{111 + x} = 0.869$$

$$\frac{111}{0.869} = 111 + x \ ; \ \ 16.7 = x \text{ (mol of ethylene glycol)}$$

$$16.7 \text{ mol C}_2\text{H}_4(\text{OH})_2 \cdot \frac{62.07 \text{ g C}_2\text{H}_4(\text{OH})_2}{1 \text{ mol C}_2\text{H}_4(\text{OH})_2} = 1.04 \times 10^3 \text{ g C}_2\text{H}_4(\text{OH})_2$$

45.   The molar mass of the solute may be calculated by using Raoult's Law:

$$P_{\text{benzene}} = 121.8 \text{ mmHg at } 30 \text{ °C} \qquad P_{\text{solution}} = 113.0 \text{ mmHg at } 30 \text{ °C}$$

$$P_{\text{solution}} = P_{\text{benzene}} \cdot X_{\text{benzene}}$$

$$113.0 \text{ mm} = 121.8 \text{ mm} \cdot X_{\text{benzene}}$$

$$0.9278 = X_{\text{benzene}}$$

$$0.9278 = \frac{\text{mol C}_6\text{H}_6}{\text{mol C}_6\text{H}_6 + \text{mol unknown}}$$

Calculating the amount of C_6H_6:

$$100. \text{ g C}_6\text{H}_6 \cdot \frac{1 \text{ mol C}_6\text{H}_6}{78.11 \text{ g C}_6\text{H}_6} = 1.28 \text{ mol C}_6\text{H}_6$$

Substituting into the mf expression above

$$0.9278 = \frac{1.28 \text{ mol C}_6\text{H}_6}{1.28 + x} \text{ and solving gives: } x = 0.0996 \text{ mol}$$

Since 10.0 g corresponds to 0.0996 mol, we can calculate the molar mass:

10.0 g = 0.0996 mol   and 100. g = 1 mol

47.   Calculate the boiling point of a solution containing 0.200 mol of a solute in 100. g C_6H_6

$$\Delta T_{bp} = K_{bp} \cdot m$$

The molality of the solution is $\dfrac{0.200 \text{ mol solute}}{0.100 \text{ kg } C_6H_6} = 2.00$ molal

K_{bp} for benzene is +2.53 °C/molal , so

$$\Delta T_{bp} = +2.53 \text{ °C/molal} \cdot 2.00 \text{ molal} = 5.06 \text{ °C}.$$

Hence the solution boils at 80.10 °C + 5.06 = 85.16 °C.

49.   The boiling point of a solution can be calculated if the molality of the solution is known.

$$0.515 \text{ g } C_{12}H_{10} \cdot \dfrac{1 \text{ mol } C_{12}H_{10}}{154.2 \text{ g } C_{12}H_{10}} = 3.34 \times 10^{-3} \text{ mol } C_{12}H_{10}$$

$$\dfrac{3.34 \times 10^{-3} \text{ mol } C_{12}H_{10}}{15.0 \text{ g } CHCl_3} \cdot \dfrac{1000 \text{ g } CHCl_3}{1 \text{ kg } CHCl_3} = 0.223 \text{ molal solution}$$

$$\Delta T_{bp} = K_{bp} \cdot m$$
$$= +3.63 \text{ °C/molal} \cdot 0.223 \text{ molal or } +0.808 \text{ °C}$$

Thus the solution should boil at (61.7 + 0.808) °C or 62.5 °C.

51.   The boiling point has been elevated by 4.3 °C

The molality of the solution is :

$$\Delta T_{bp} = K_{bp} \cdot m$$
$$4.3 \text{ °C} = +0.512 \text{ °C/molal} \cdot m \quad \text{and} \quad 8.4 = m$$

Since molality is defined as:

$$m = \dfrac{\text{moles solute}}{\text{kg solvent}} \quad \text{then} \quad 8.4 = \dfrac{\text{moles ethylene glycol}}{0.750 \text{ kg water}}$$

$$6.3 = \text{mol ethylene glycol}$$

$$6.3 \text{ mol } C_2H_4(OH)_2 \cdot \dfrac{62.07 \text{ g } C_2H_4(OH)_2}{1 \text{ mol } C_2H_4(OH)_2} = 391 \text{ g } C_2H_4(OH)_2 \text{ or to two s.f. 390 g}$$

The mf(glycol) is:

$$750. \text{ g } H_2O \cdot \dfrac{1 \text{ mol } H_2O}{18.02 \text{ g } H_2O} = 41.6 \text{ mol } H_2O$$

$$\dfrac{6.3 \text{ mol glycol}}{6.3 \text{ mol glycol} + 41.6 \text{ mol } H_2O} = 0.13$$

53.   The **solution with the highest boiling point** will have the **greatest number of particles** in solution.

If we assume total dissociation of the solutes given the molality of particles for the solution will be:

$$0.10 \text{ m NaCl} \rightarrow 0.10 \text{ m Na}^+ + 0.10 \text{ m Cl}^- = 0.20 \text{ m}$$

$$0.10 \text{ m sugar} \rightarrow \text{(covalently bonded specie)} = 0.10 \text{ m}$$

$$0.080 \text{ m CaCl}_2 \rightarrow 0.080 \text{ m Ca}^{2+} + 0.16 \text{ m Cl}^- = 0.24 \text{ m}$$

In order of increasing boiling point: $0.10 \text{ m sugar} < 0.10 \text{ m NaCl} < 0.080 \text{ m CaCl}_2$

55.   From Figure 14.11 the solubility of NaCl at 100 °C is approximately 39.1 g/100. g H_2O.

$$\text{molality} = \frac{39.1 \text{ g NaCl}}{100 \text{ g } H_2O} \cdot \frac{1 \text{ mol NaCl}}{58.44 \text{ g NaCl}} \cdot \frac{1000 \text{ g } H_2O}{1 \text{ kg } H_2O} = 6.69 \text{ molal NaCl}$$

NaCl is expected to produce 2 ions/ formula unit, giving an effective molality of 13.4.

The change in the boiling point of this solution is:

$$\Delta T_{bp} = K_{bp} \cdot m = +0.512 \text{ °C/molal} \cdot 13.4 \text{ molal} = +6.85 \text{ °C.}$$

The solution will begin to boil at $(100.00 + 6.85)$ or 106.85 °C.

57.   Calculate the molar mass of hexachlorophene if 0.640 g of the solid dissolved in 25.0 g chloroform gives a solution boiling at 61.93 °C.

$$\Delta T_{bp} = K_{bp} \cdot m \quad \text{and} \quad m = \frac{\Delta T_{bp}}{K_{bp}}$$

$$\frac{(61.93 \text{ °C} - 61.70 \text{ °C})}{3.63 \text{ °C/molal}} = 0.0634 \text{ molal}$$

The number of moles of hexachlorophene is: $0.0634 \text{ molal} \cdot 0.0250 \text{ kg} = 1.58 \times 10^{-3} \text{ mol}$

The molar mass is then $\dfrac{0.640 \text{ g}}{1.58 \times 10^{-3} \text{ mol}} = 404 \text{ g/mol}$

59.   Determine the molar mass and formula of a compound if 0.255 g of the solid dissolved in
11.12 g benzene gives a solution boiling at 80.26 °C.

$$\Delta T_{bp} = K_{bp} \bullet m \quad \text{and} \quad m = \frac{\Delta T_{bp}}{K_{bp}}$$

$$\frac{(80.26 \text{ °C} - 80.10 \text{ °C})}{2.53 \text{ °C/molal}} = 0.063 \text{ molal}$$

The number of moles of the compound is: 0.063 molal • 0.01112 kg = 7.0 x 10^{-4} mol
The molar mass is then $\frac{0.255 \text{ g}}{7.0 \text{ x } 10^{-4} \text{ mol}}$ = 360 g/mol. The empirical formula, $C_{10}H_8Fe$

has a formula mass of 184.0. With a molar mass of 360 the molecular formula of the
compound is 2 x $C_{10}H_8Fe$ or $C_{20}H_{16}Fe_2$.

61.   To lower the freezing point 31.0 °C :
$$\Delta T_{fp} = K_{fp} \bullet m$$
$$-31.0 \text{ °C} = -1.86 \text{ °C/molal} \bullet m$$
$$16.7 = m$$

A solution containing 16.7 mol of ethylene glycol per kilogram of water would be
adequate. The mass of glycol per quart of water would be:

$$\frac{16.7 \text{ mol glycol}}{1000 \text{ g H}_2\text{O}} \bullet \frac{946 \text{ g H}_2\text{O}}{1 \text{ quart H}_2\text{O}} \bullet \frac{62.07 \text{ g glycol}}{1 \text{ mol glycol}} = \frac{979 \text{ g glycol}}{1 \text{ quart H}_2\text{O}}$$

63.   Grams of NaCl to be added to 3.0 kg of water to form a mixture freezing at - 10. °C:
To lower the freezing point 10. °C:
$$- 10. \text{ °C} = - 1.86 \text{ °C/molal} \bullet m$$
$$5.38 \text{ molal} = m$$

A solution containing 5.38 moles of NaCl per kg of water will freeze at -10. °C. The mass
of NaCl for 3.0 kg of water is:

$$\frac{5.38 \text{ mol NaCl}}{1 \text{ kg H}_2\text{O}} \bullet \frac{3 \text{ kg H}_2\text{O}}{1} \bullet \frac{58.44 \text{ g NaCl}}{1 \text{ mol NaCl}} = 940 \text{ g NaCl if NaCl were not}$$

ionic. Since NaCl furnishes two particles/formula unit, then half the NaCl (470 g) is
required.

65.   Calculate the molar mass of erythritol if 2.50 g of the compound in 50.0 g of water freezes
        at - 0.773 °C:

$$m = \frac{\Delta t_{fp}}{K_{fp}} = \frac{(0.000 \text{ °C} - 0.773 \text{°C})}{-1.86 \text{ °C/molal}} = 0.416 \text{ molal}$$

$$\text{moles erythritol} = \frac{0.416 \text{ mol erythritol}}{1 \text{ kg water}} \cdot 0.0500 \text{ kg water} = 2.08 \times 10^{-2} \text{ mol}$$

$$\text{Therefore the molar mass for erythritol is: } \frac{2.50 \text{ g}}{2.08 \times 10^{-2} \text{ mol}} = 120. \text{ g/mol}$$

67.   Determine the molar mass for benzaldehyde:

$$m = \frac{\Delta t_{fp}}{K_{fp}} = \frac{- 0.049 \text{ °C}}{- 1.86 \text{ °C/molal}} = 0.026 \text{ molal}$$

$$\text{moles benzaldehyde} = \frac{0.026 \text{ mol benzaldehyde}}{1 \text{ kg water}} \cdot 0.040 \text{ kg water} = 1.1 \times 10^{-3} \text{ mol}$$

$$\text{Therefore the molar mass for benzaldehyde is: } \frac{0.112 \text{ g}}{1.1 \times 10^{-3} \text{ mol}} = 110 \text{ g/mol}$$

69.   Solutions given in order of increasing melting point:

        The solution with the greatest **number** of particles will have the lowest melting
        point.

        The total molality of solutions is then:
        (a) 0.1 m sugar   x 1 particle/formula unit = 0.1 m
        (b) 0.1 m NaCl   x 2 particles/formula unit = 0.2 m   [Na^+ , Cl^-]
        (c) 0.08 m $CaCl_2$ x 3 particles/formula unit = 0.24 m   [Ca^{2+}, 2 Cl^-]
        (d) 0.04 m Na_2SO_4 x 3 particles/formula unit = 0.12 m   [2 Na^+, SO_4^{2-}]

        The melting points would increase in the order: $CaCl_2$ < NaCl < Na_2SO_4 < sugar

71.   When meat is placed in a strong salt solution, the concentration of salt in the solution is
        greater than the concentration of salt in the meat--osmosis causes water to flow from the
        meat into the salt in an attempt to "equalize the concentration" of the salt solutions.

73.   Solving the equation $\Pi = MRT$ for M yields:

$$M = \frac{\Pi}{R \cdot T} = \frac{7.53 \text{ atm}}{0.0821 \frac{L \cdot atm}{K \cdot mol} \cdot 310. \text{ K}} = 0.296 \text{ M solute particles}$$

75.   a. Weight percentage of NaCl: $\dfrac{\text{mass of NaCl}}{\text{mass of salt + ice}} = \dfrac{1130 \text{ g}}{8390 \text{ g}} \times 100 = 13.5\%$

         b. Mole fraction of NaCl $= \dfrac{\text{mol NaCl}}{\text{mol } H_2O + \text{mol NaCl}}$

$$= \frac{19.3 \text{ mol NaCl}}{402.9 \text{ mol } H_2O + 19.3 \text{ mol NaCl}} = 0.0457 \text{ mf NaCl}$$

         c. Molality of the solution $= \dfrac{\text{\# mol NaCl}}{\text{kg } H_2O} = \dfrac{19.3 \text{ mol NaCl}}{7.26 \text{ kg } H_2O}$ or 2.66 m NaCl

77.   a. The molality of a solution containing 300. g of silver nitrate in 100. g of water:

$$300. \text{ g } AgNO_3 \cdot \frac{1 \text{ mol } AgNO_3}{169.9 \text{ g } AgNO_3} = 1.77 \text{ mol } AgNO_3$$

$$\text{molality} = \frac{1.77 \text{ mol } AgNO_3}{0.100 \text{ kg water}} = 17.7 \text{ molal}$$

         b. The mf of $AgNO_3$:

$$100. \text{ g } H_2O \cdot \frac{1 \text{ mol } H_2O}{18.02 \text{ g } H_2O} = 5.55 \text{ mol } H_2O$$

$$\text{mf } AgNO_3 = \frac{1.77 \text{ mol } AgNO_3}{1.77 \text{ mol } AgNO_3 + 5.55 \text{ mol } H_2O} = 0.241$$

         c. The weight percent of $AgNO_3$:

$$\frac{300. \text{ g } AgNO_3}{300. \text{ g } AgNO_3 + 100. \text{ g } H_2O} \times 100 = 75.0 \% \text{ } AgNO_3$$

79.   Calculate the partial pressure of each gas:

$$P(N_2) = P_T \cdot X(N_2) = 760 \text{ mmHg} \cdot 0.783 = 595 \text{ mmHg or } 6.0 \times 10^2$$

$$P(O_2) = P_T \cdot X(O_2) = 760 \text{ mmHg} \cdot 0.210 = 160 \text{ mmHg or } 1.6 \times 10^2$$

The molality is then:

N_2: 8.4×10^{-7} molal/mmHg $\cdot$ 6.0×10^2 mmHg $= 5.0 \times 10^{-4}$ molal

O_2: 1.7×10^{-6} molal/mmHg $\cdot$ 1.6×10^2 mmHg $= 2.7 \times 10^{-4}$ molal

81.   Calculate the molar mass by freezing point depression:

$$m = \frac{\Delta T_{fp}}{K_{fp}} = \frac{(0.00 \text{ °C} - 1.28 \text{ °C})}{-1.86 \text{ °C/molal}} = 0.688 \text{ molal}$$

Since KX provides two particles/formula unit, the number of moles of KX is half this number or 0.344.

$$\text{moles KX} = \frac{0.344 \text{ mol KX}}{1 \text{ kg water}} \cdot 0.100 \text{ kg water} = 3.44 \times 10^{-2} \text{ mol}$$

Therefore the molar mass for KX is: $\dfrac{4.00 \text{ g}}{3.44 \times 10^{-2} \text{ mol}} = 116 \text{ g/mol}$

Subtracting the mass of potassium from this molar mass yields:

$116 - 39.1 = 77$ g/mol X   thus the element X must be **Br**.

84.   The solution with the greater vapor pressure of water will be the one containing the least number of solute particles. The solutions given are:

(a) 33.0 g sugar in 180. g H_2O

$$33.0 \text{ g sugar} \cdot \frac{1 \text{ mol sugar}}{342 \text{ g sugar}} = 0.0965 \text{ mol}$$

$$\text{molality} = \frac{0.0965 \text{ mol sugar}}{0.180 \text{ kg water}} = 0.536 \text{ molal}$$

(b) 5.85 g NaCl in 180. g H_2O

$$5.85 \text{ g NaCl} \cdot \frac{1 \text{ mol NaCl}}{58.44 \text{ g NaCl}} = 0.100 \text{ mol}$$

$$\text{molality} = \frac{0.100 \text{ mol NaCl}}{0.180 \text{ kg water}} = 0.556 \text{ molal}$$

The ionic compound NaCl dissociates in water, producing two particles per formula unit.
The net molality of ions in the NaCl solution is therefore 2 x 0.556 molal.
The sugar solution will therefore have a higher vapor pressure of water.
The NaCl solution (having the greater molality) will have the higher boiling point.

85.   Examine the number of particles formed when 1 formula unit of each solute dissolves.

a. 0.35 m $C_2H_4(OH)_2$ Covalently bonded substance exists as 1 particle in solution

b. 0.50 m sugar   Covalently bonded substance exists as 1 particle in solution

c. 0.20 m KBr   forms 2 particles/formula [K^+, Br^-]

d. 0.20 m Na_2SO_4 forms 3 particles/formula [2 Na^+, SO_4^{2-}]

The "total" molality of each may be found by multiplying the number of particles/formula
unit by the molality of each solution.

a. 0.35 m x 1 = 0.35 m   b. 0.50 m x 1 = 0.50 m

c. 0.20 m x 2 = 0.40 m   d. 0.20 m x 3 = 0.60 m

From Raoult's Law, we note that the larger the mole fraction (greater the concentration) of
solute, the smaller the mole fraction of solvent. Knowing that the vapor pressure of water
may be calculated $P(H_2O) = X(H_2O) \cdot P_t°H_2O$, we can list the solutions in increasing
order of vapor pressure by listing the solutions in decreasing solute concentration.
Increasing vapor pressure :

$$0.20 \text{ m } Na_2SO_4 < 0.50 \text{ m sugar} < 0.20 \text{ m KBr} < 0.35 \text{ m } C_2H_4(OH)_2$$

87.   To calculate the boiling point of the solution requires that we know the molality. In Study
question 61, this molality is found to be 16.7 molal. With this information we can calcualte
the boiling point:

$$\Delta t_{bp} = K_{bp} \cdot m = +0.512 °C/\text{molal} \cdot 16.7 \text{ molal} = 8.55 °C$$
$$\text{New boiling point} = 100.00 °C + 8.55 °C \text{ or } 108.55 °C.$$

88.   To calculate the molarity of the solution we note that:

$$3.0 \text{ g NaCl} \cdot \frac{1 \text{ mol NaCl}}{58.4 \text{ g NaCl}} = 0.051 \text{ mol NaCl}$$

We are not provided with the density of sea water, so we note that 100 g of sea water
(which contains 3.0 g NaCl) will also contain 97 g of H_2O.

182

The molality of the solution is: $\dfrac{0.051 \text{ mol NaCl}}{0.097 \text{ kg}}$ = 0.53 molal

For dilute solutions we will assume that molality is ≈ molarity. The molarity is then 0.53 M. Substituting into the osmotic pressure expression:

$$\Pi = M \bullet R \bullet T = \dfrac{0.53 \text{ mol NaCl}}{L} \bullet 0.082057 \dfrac{L \bullet atm}{K \bullet mol} \bullet 298 \text{ K}$$

$$\Pi = 12.9 \text{ atm (if we assume that NaCl remains as ONE UNIT)}$$

Since NaCl dissociates into two particles/formula unit, the effective molarity will be twice as great as we calculated, and the osmotic pressure will be 2 x 12.9 or approximately 26 atm (to two significant figures).

90.   a. The vapor pressure of a solution of 15.0 g urea in 0.500 kg water:

$$15.0 \text{ g urea} \bullet \dfrac{1 \text{ mol urea}}{60.06 \text{ g urea}} = 0.250 \text{ mol urea}$$

$$500. \text{ g H}_2\text{O} \bullet \dfrac{1 \text{ mol H}_2\text{O}}{18.02 \text{ g H}_2\text{O}} = 27.8 \text{ mol water}$$

$$\text{MF (water)} = \dfrac{27.8 \text{ mol water}}{27.8 \text{ mol water} + 0.250 \text{ mol urea}} = 0.991$$

$$P\ (\text{H}_2\text{O}) = 0.991 \bullet 17.5 \text{ mmHg} = 17.3 \text{ mmHg}$$

b. The vapor pressure of a solution of 15.0 g sugar in 0.500 kg water:

$$15.0 \text{ g sugar} \bullet \dfrac{1 \text{ mol sugar}}{342 \text{ g sugar}} = 0.0439 \text{ mol sugar}$$

$$\text{MF (water)} = \dfrac{27.8 \text{ mol water}}{27.8 \text{ mol water} + 0.0439 \text{ mol sugar}} = 0.998$$

$$P\ (\text{H}_2\text{O}) = 0.998 \bullet 17.5 \text{ mmHg} = 17.5 \text{ mmHg}$$

c. Mass of sugar in 0.500 kg water to produce a vapor pressure of 17.3 mmHg:

The amount of sugar necessary will be 0.250 mol (as in the urea solution).

The mass of sugar corresponding to this amount of sugar is:

$$\dfrac{342 \text{ g sugar}}{1 \text{ mol sugar}} \bullet 0.250 \text{ mol sugar} = 85.5 \text{ g sugar}$$

92.   a. The solution with the **greater number** of particles will have the higher boiling point. Hence a 0.10 m NaCl solution will boil at a higher temperature than 0.15 m sugar since the ionic substance, NaCl, will furnish twice as many particles per formula unit as will the covalently bonded sugar molecule.

b. Since the enthalpy of solution of NaOH is strongly **exo**thermic, the solid will be **less** soluble as the temperature goes up.

c. The vapor pressure of a solution decreases as the number of particles increase. As NH_4NO_3 dissolves, two ions are released per formula unit-- the solution containing 0.30 mol NH_4NO_3 per kilogram of solvent will contain 0.60 mol of ions. For Na_2SO_4 each formula unit releases three ions. So the 0.15 molal Na_2SO_4 solution contains 0.45 mol of ions. **Thus the NH_4NO_3 solution contains more particles and will have the lower vapor pressure**.

94.   To calculate the vapor pressure of the mixture, you must know the mole fractions of each component of the mixture.

$$\text{Mf (benzene)} = \frac{2.0 \text{ mol}}{2.0 + 1.0 \text{ mol}} = \frac{2}{3}$$

$$\text{Mf (toluene)} = 1.00 - \frac{2}{3} = \frac{1}{3}$$

$$P_T = P_{benzene} + P_{toluene} = X_{benzene}P°_{benzene} + X_{toluene}P°_{toluene}$$

$$P_T = \frac{2}{3} \cdot (75 \text{ mmHg}) + \frac{1}{3} \cdot (22 \text{ mmHg})$$

$$= 50. \text{ mmHg} + 7.33 \text{ mmHg} = 57 \text{ mmHg}$$

96. $10.0 \text{ g } CH_2Cl_2 \cdot \dfrac{1 \text{ mol } CH_2Cl_2}{84.93 \text{g } CH_2Cl_2} = 0.118 \text{ mol } CH_2Cl_2$

$1.00 \text{ g } C_6H_{12} \cdot \dfrac{1 \text{ mol } C_6H_{12}}{84.16 \text{ g } C_6H_{12}} = 0.0119 \text{ mol } C_6H_{12}$

At 25 °C, $P°(C_6H_{12}) = 166 \text{ mmHg}$ and $P°(CH_2Cl_2) = 435 \text{ mmHg}$

$$P_T = P(C_6H_{12}) + P(CH_2Cl_2)$$

$$P_T = X(C_6H_{12}) \cdot P°(C_6H_{12}) + X(CH_2Cl_2) \cdot P°(CH_2Cl_2)$$

$$= \frac{0.0119}{0.130} \cdot (166 \text{ mm}) + \frac{0.118}{0.130} \cdot (435 \text{ mm})$$

$$= 15.2 \text{ mm} + 395 \text{ mm} = 410. \text{ mmHg}$$

98.   Determine the vapor pressure of an alcohol-water mixture at 25 °C: $\Delta T_{fp} = -10.0$ °C

The molality of the C_2H_5OH solution must be

$$-10.0 \text{ °C} = -1.86 \text{ °C/molal} \cdot m$$

$$5.38 \text{ molal } C_2H_5OH = m$$

$$\frac{5.38 \text{ mol } C_2H_5OH}{1000 \text{ g } H_2O} \quad \text{or} \quad \frac{5.38 \text{ mol } C_2H_5OH}{55.6 \text{ mol } H_2O}$$

$$X(C_2H_5OH) = \frac{5.38 \text{ mol}}{(55.6 + 5.38) \text{ mol}} = 0.0883$$

$$X(H_2O) = \frac{55.6 \text{ mol}}{(55.6 + 5.38) \text{ mol}} = 0.912$$

$$P_T = P(H_2O) + P(C_2H_5OH)$$

$$P_T = X(H_2O) \cdot P°(H_2O) + X(C_2H_5OH) \cdot P°(C_2H_5OH)$$
$$= (0.912)(24 \text{ mmHg}) + (0.0883)(60. \text{ mmHg})$$
$$= 22 \text{ mmHg} + 5.3 \text{ mmHg} = 27 \text{ mmHg}$$

99.   0.146 g of BxFy in 10.0 g of benzene freezes at 4.77 °C.
$$\Delta T_{fp} = 4.77 - 5.50 = -0.73 \text{ °C}$$
The molality of the solution is: $-0.73 \text{ °C} = K_{fp} \cdot m$
$$-0.73 \text{ °C} = -4.90 \text{ °C/molal} \cdot m$$
$$0.149 = m$$

$$\frac{0.149 \text{ mol } B_xF_y}{1 \text{ kg } C_6H_6} \cdot \frac{0.0100 \text{ kg } C_6H_6}{1} = 0.00149 \text{ mol } B_xF_y$$

This amount of B_xF_y (1.49×10^{-3} mol) has a mass of 0.146 g.
So 1.49×10^{-3} mol = 0.146 g or   1 mol = 98 g

Now we need to know the relative number of boron and fluorine atoms which have a molar mass of 98 g. Examine the percentage composition: 22.1% B means that the compound is 77.9% F by weight.

Calculate the empirical formula.

A 100 g sample of B_xF_y would contain 22.1 g B and 77.9 g F.

$$22.1 \text{ g B} \cdot \frac{1 \text{ mol B}}{10.81 \text{ g B}} = 2.05 \text{ mol B}$$

and $77.9 \text{ g F} \cdot \dfrac{1 \text{ mol F}}{19.00 \text{ g F}} = 4.10 \text{ mol F}$

So the ratio of F to B is 2:1. With the empirical formula, BF_2, and the molar mass 98 g, we can postulate a molecular formula.

$$\frac{98 \text{ g } B_xF_y}{1 \text{ mol } B_xF_y} \cdot \frac{1 \text{ empirical formula}}{48.8 \text{ g } B_xF_y} = \frac{2.0 \text{ empirical formulas}}{1 \text{ mol } B_xF_y}$$

The molecular formula must be B_2F_4.

101.   a. Compound contains 73.87% C, 8.21% H, and 17.92% Cr

Calculate the empirical formula.

$$73.87 \text{ g C} \cdot \frac{1 \text{ mol C}}{12.011 \text{ g C}} = 6.150 \text{ mol C}$$

$$8.21 \text{ g H} \cdot \frac{1 \text{ mol H}}{1.008 \text{ g H}} = 8.14 \text{ mol H}$$

$$17.92 \text{ g Cr} \cdot \frac{1 \text{ mol Cr}}{52.00 \text{ g Cr}} = 0.3446 \text{ mol Cr}$$

The ratio of atoms is: $\dfrac{6.150 \text{ mol C}}{0.3446 \text{ mol Cr}} = 17.85$ $\dfrac{8.14 \text{ mol H}}{0.3446 \text{ mol Cr}} = 23.63$

or to the nearest integral values: $CrC_{18}H_{24}$.

b. $\Pi = 3.17$ mmHg or 4.17×10^{-3} atm

$T = 298$ K   and $V = 0.100$ L   mass $= 5.00 \times 10^{-3}$ g

$\Pi = M \cdot R \cdot T$ and since $M = \dfrac{n}{V}$

$\Pi = \dfrac{n}{V} \cdot R \cdot T$

$n = \dfrac{\Pi \cdot V}{R \cdot T} = \dfrac{4.17 \times 10^{-3}\text{atm} \cdot 0.100 \text{ L}}{0.082057 \dfrac{\text{L} \cdot \text{atm}}{\text{K} \cdot \text{mol}} \cdot 298 \text{ K}} = 1.71 \times 10^{-5}$ mol

So 5.00 mg of the sample corresponds to 1.71×10^{-5} mol.

The molar mass is:

$$5.00 \times 10^{-3} \text{ g} = 1.71 \times 10^{-5} \text{ mol}$$

$$\frac{5.00 \times 10^{-3} \text{ g}}{1.71 \times 10^{-5}\text{mol}} = 293 \text{ g/mol}$$

Using the empirical formula from part (a), we calculate an empirical mass of 292.

The molecular formula must be $CrC_{18}H_{24}$.

20.　　a. $2 O_3 (g) \rightarrow 3 O_2 (g)$

$$\text{Reaction Rate} = -\frac{1}{2} \cdot \frac{\Delta[O_3]}{\Delta t} = +\frac{1}{3} \cdot \frac{\Delta[O_2]}{\Delta t}$$

b. $2 HOF (g) \rightarrow 2 HF (g) + O_2 (g)$

$$\text{Reaction Rate} = -\frac{1}{2} \cdot \frac{\Delta[HOF]}{\Delta t} = +\frac{1}{2} \cdot \frac{\Delta[HF]}{\Delta t} = + \frac{\Delta[O_2]}{\Delta t}$$

c. $N_2 (g) + 3 H_2 (g) \rightarrow 2 NH_3 (g)$

$$\text{Reaction Rate} = -\frac{\Delta[N_2]}{\Delta t} = -\frac{1}{3} \cdot \frac{\Delta[H_2]}{\Delta t} = +\frac{1}{2} \cdot \frac{\Delta[NH_3]}{\Delta t}$$

22. Plot the data for the hypothetical reaction $A \rightarrow 2 B$

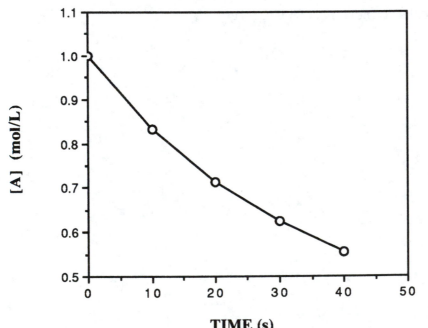

a.　Rate = $\f(\Delta[A],\Delta t)$　　 = $\f((0833 - 1.000),10.0 - 0.00)$　 = $- \f(0.167,10.0)$

= $- 0.0167 \frac{mol}{L \cdot s}$

$$= \frac{(0.714 - 0.833)}{20.0 - 10.00} = -\frac{0.119}{10.0} = -0.0119 \frac{mol}{L \cdot s}$$

$$= \frac{(0.625 - 0.714)}{30.0 - 20.00} = -\frac{0.089}{10.0} = -0.0089 \frac{mol}{L \cdot s}$$

$$= \frac{(0.555 - 0.625)}{40.0 - 30.00} = -\frac{0.070}{10.0} = -0.0070 \frac{mol}{L \cdot s}$$

The rate of change decreases from one time interval to the next due to a continuing decrease in the amount of reacting material (A).

b. Since each A molecule forms 2 molecules of B

T	[A]	$[B] = 2([A]_0 - [A])$
10.0 s	0.833	$2(0.167) = 0.334$
20.0 s	0.714	$2(0.286) = 0.572$

$$\text{Rate} = \frac{\Delta[B]}{\Delta t} = \frac{(0.572 - 0.334)}{20.0 - 10.00} = \frac{0.238}{10.0} \text{ or } 0.0238 \frac{mol}{L \cdot s}$$

The appearance of B is twice as great as the disappearance of A.

24.   a. Rate expression for CO (g) + NO_2 (g) $\rightarrow$ CO_2 (g) + NO (g) given :

   Below 500 K the reaction is second order in NO_2 and zero order in CO.

   $$\text{Rate} = k[NO_2]^2[CO]^0 \text{ or } k[NO_2]^2$$

   b. Viewing the rate expression, one can see that reducing the "value of the NO_2 term" by a factor of two will reduce the rate by a factor of four. Said another way, halving the concentration of NO_2 will result in a rate which is one-fourth of the original rate.

26.   Reaction order of each reagent and the overall reaction:

	Order in A	Order in B	Overall
a. Rate = $k[A][B]^2$	1	2	3
b. Rate = $k[A][B]$	1	1	2
c. Rate = $k[A]$	1	0	1
d. Rate = $k[A]^3[B]$	3	1	4

28.   If doubling [B] increases the rate by a factor of four, the dependence on [B] is second order. The lack of effect on the rate by changing the concentration of A implies a zero order dependence.

   The rate expression is: (c) Rate = $k[B]^2$

30.   The rate expression indicates that the reaction is first order in the Pt complex.
        The initial rate of reaction when the $[Pt(NH_3)_2Cl_2]$ = 0.020 M

$$Rate = k[Pt(NH_3)_2Cl_2]$$

$$Rate = 0.090 \text{ hr}^{-1} \cdot \frac{0.020 \text{ mol}}{L} = 1.8 \times 10^{-3} \frac{mol}{L \cdot hr}$$

32.   a. After you do part b, the rate expression may then be written as: Rate = k $[CO][NO_2]$

        b. Using the kinetic data, note that the rate of the second experiment is two times that of the
        first experiment. Note that this corresponds to a doubling in $[NO_2]$--hence the reaction is
        first order in $[NO_2]$. If you compare the fourth and fifth experiments, note that the [CO]
        increases by a factor of 1.5 , and the rate increases by a factor of $\frac{10.2}{6.8} = 1.5$.
        The reaction is therefore first order with respect to CO, and second order overall.

        c. Using the Rate expression from part a, substitute some of the data from the table.

$$Rate = k \,[CO][NO_2]$$

$$3.4 \times 10^{-8} \frac{mol}{L \cdot hr} = k \cdot 5.1 \times 10^{-4} \frac{mol}{L} \cdot 0.35 \times 10^{-4} \frac{mol}{L} \text{ and}$$

$$\frac{3.4 \times 10^{-8} \frac{mol}{L \cdot hr}}{1.785 \times 10^{-8} \frac{mol^2}{L^2}} = 1.9 \frac{L}{mol \cdot hr}$$

34.   Comparing the first and second experiments--you find that doubling $[Br_2]$ shows no
        change in rate. Comparing the first and third experiments,--you find that doubling $[H^+]$
        the rate doubles. Comparing the first and last experiments--you find that increasing the
        [acetone] by 1.3, changes the rate by a factor of 1.3 .

        a. Rate = $k[CH_3COCH_3] [H^+]$ indicating that the order in acetone = 1,
           order in H^+ = 1, in Br_2 = 0, for an overall order = 2

        b. Substituting data from the first experiment into the rate expression, we solve for k:

$$5.7 \times 10^{-5} \text{ M/sec} = k \cdot 0.30 \cdot 0.05 \text{ M}^2$$
$$k = 4 \times 10^{-3} \text{ M}^{-1}\text{sec}^{-1}$$

c. With $[H^+] = 0.050$ M and $[CH_3COCH_3] = 0.10$ M we substitute these values (and that of k) into the rate expression:

$$\text{Rate} = 4 \times 10^{-3} \text{ M}^{-1}\text{sec}^{-1} \cdot 0.10 \text{ M} \cdot 0.050 \text{ M} = 2 \times 10^{-5} \text{ M/sec}$$

36. Compare the first and second experiments: when [B] is halved, the rate is quartered.
    Compare the first and third experiments: when the original [A] is tripled, the rate triples.

   a. $\text{Rate} = k[A]_0[B]_0^2$ order: in A = 1, in B = 2, overall = 3

   b. 2.0×10^2 M/min $= k(0.10)(0.20)^2 \text{ M}^3$
      $$k = 5.0 \times 10^4 \text{ M}^{-2} \text{ min}^{-1}$$

   c. $\text{Rate} = (5.0 \times 10^4 \text{ M}^{-2} \text{ min}^{-1})(0.20 \text{ M})(0.20 \text{ M})^2 = 4.0 \times 10^2 \text{ M min}^{-1}$

   d. $6.0 \times 10^2 \text{ M min}^{-1} = (5.0 \times 10^4 \text{ M}^{-2}\text{min}^{-1})(0.30 \text{ M})[B]_0^2$
      $$[B]_0 = 2.0 \times 10^{-1} \text{ M}$$

38. The transformation of cyclopropane to propene is described as first order.
    Hence we use the following form of the rate expression:

$$\ln\left(\frac{[\text{cyclopropane}]}{[\text{cyclopropane}]_0}\right) = -kt$$

Substituting

$$\ln\left(\frac{[0.025]}{[0.050]}\right) = -5.4 \times 10^{-2} \text{ hr}^{-1} \cdot t$$

$$13 \text{ hr} = t$$

This **time is also the half-life** for the conversion of cyclopropane to propene.

40. The reaction is first order: $t_{1/2} = \dfrac{0.693}{k} = \dfrac{0.693}{3.33 \times 10^{-6} \text{ hr}^{-1}} = 2.08 \times 10^5 \text{ hr}$

Time for concentration to drop from 1.0 M to 0.20 M:

$$\ln\left(\frac{0.20}{1.00}\right) = -3.33 \times 10^{-6} \text{ hr}^{-1} \cdot t \quad \text{and solving: } t = 4.8 \times 10^5 \text{ hr}$$

42.   If $t_{1/2} = 37.9$ sec, then $k = 1.83 \times 10^{-2}$ sec^{-1}

$$\ln\left[\frac{[PH_3]_t}{[PH_3]_0}\right] = \ln\left(\frac{1}{4}\right) = -(1.83 \times 10^{-2})\,t$$

$$t = 7.58 \times 10^1 \text{ sec}$$

Alternately, for $[PH_3]$ to decrease to $\frac{1}{4}[PH_3]_0$ requires 2 half-lives

$[(\frac{1}{2})^2 = \frac{1}{4}]$ so $t = 2\,t_{1/2} = 75.8$ sec

44.   The reaction is first order with $t_{1/2} = 40.$ min

   a. The rate constant for the reaction

$$\ln\left(\frac{1}{2}\right) = -k \cdot 40.\text{ min and solving for } k \text{ yields}$$

$$1.7 \times 10^{-2} \text{ min}^{-1} = k$$

   b. Concentration of M-SCH$_3$ after 2.0 hours:

$\ln[M\text{-}SCH_3] - \ln[M\text{-}SCH_3]_0 = -kt$ rearranging and substituting yields

$\ln[M\text{-}SCH_3] = \ln[6.0\times10^{-3}] - (1.7 \times 10^{-2}/\text{min})(120 \text{ min})$

$\ln[M\text{-}SCH_3] = -5.12 - 2.08 = -7.20$

Taking the antilogarithm of both sides: $[M\text{-}SCH_3] = 7.5 \times 10^{-4}$ M

   c. Time for M-SCH$_3$ concentration to decrease from 6.0×10^{-3} M to 1.0×10^{-4} M.

$$\ln\left(\frac{1.0 \times 10^{-4}}{6.0 \times 10^{-3}}\right) = -1.7 \times 10^{-2} \text{ min}^{-1} \cdot t$$

$$\frac{\ln(1.67 \times 10^{-2})}{-1.7 \times 10^{-2} \text{ min}^{-1}} = t = 2.4 \times 10^2 \text{ min}$$

46.   For the reaction HOF (g) $\rightarrow$ HF (g) + 1/2 O$_2$ (g)

the half-life [for HOF (g)] at room temperature is 30. minutes. If the partial pressure of HOF is initially 100. mm, after half of the HOF has decomposed, **its partial pressure will be 50. mm**. The HF formed will contribute 50. mm and the O$_2$ an additional 25 mm for a **total pressure of 125 mm**.

The first order kinetics for this decomposition allow us to calculate the rate constant, k.

$$\ln\left(\frac{[HOF]}{[HOF]_0}\right) = -kt$$

The [HOF] will be 1/2 of the original value, and substituting yields:

$$\ln (0.5) = -k \, (30. \text{ min}) \text{ and solving for k yields } k = 0.0231 \text{ min}^{-1}$$

Substitution of the rate constant into the concentration/time equation permits us to calculate the fraction of HOF remaining (X) after 45 minutes.

$$\ln X = -(0.0231 \text{min}^{-1})(45 \text{ min}) \qquad \text{and} \qquad X = 0.35$$

After 45 minutes the partial pressure of HOF is then 35% of its original value or 35 mm. The HF pressure will be 65 mm (100 - 35) and the O_2 pressure approximately 33 mm ($1/2 \, O_2$) for a total pressure ($HOF + HF + O_2$) of 133 mm.

48.   The rate law ($k[C_4H_6]^2$) indicates a **second order** dependence. The concentration/time equation may be written: $\quad \dfrac{1}{[C_4H_6]} - \dfrac{1}{[C_4H_6]_0} = kt$

The half-life of the reaction if $k = 0.014 \dfrac{L}{mol \cdot s}$ and $[C_4H_6]_0 = 0.016$ M is:

$$\frac{1 \text{ L}}{0.008 \text{ mol}} - \frac{1 \text{ L}}{0.016 \text{ mol}} = 0.014 \frac{L}{mol \cdot s} \cdot t$$

$$\frac{62.5 \dfrac{L}{mol}}{0.014 \dfrac{L}{mol \cdot s}} = t \quad \text{and} \quad \textbf{4.5 x 10}^3 \textbf{ s = t}$$

Time for $[C_4H_6]$ to drop to 0.0016 M:

$$\frac{1 \text{ L}}{0.0016 \text{ mol}} - \frac{1 \text{ L}}{0.016 \text{ mol}} = 0.014 \frac{L}{mol \cdot s} \cdot t$$

$$\frac{5.63 \text{ x } 10^2 \text{ L/mol}}{0.014 \text{ L/mol} \cdot s} = t = 4.0 \text{ x } 10^4 \text{ seconds}$$

50.   a. Plot of ln[sucrose] and $\dfrac{1}{[\text{sucrose}]}$ versus time.

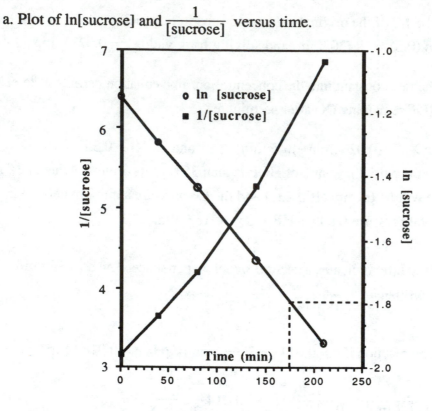

b. Since the reaction is first order with respect to sucrose, the rate
expression may be written: Rate = k [sucrose]. The rate constant can be calculated using
two data points: Using the first two points yields:

$$\ln \left(\frac{[A]}{[A]_0}\right) = - kt \quad \text{and substituting :} \quad \ln \left(\frac{0.274}{0.316} \right) = - k(39 \text{ min})$$

$$\frac{\ln (0.867)}{39 \text{ min}} = - k \quad \text{and } 3.66 \times 10^{-3} \text{ min}^{-1} = k$$

c. Using the graph of ln[sucrose] vs time, an estimate at 175 minutes yields:

ln[sucrose] = -1.8 corresponding to [sucrose] = 0.165 M

52.   For the reaction of an organic bromide(A) vs. an organic amine(B):

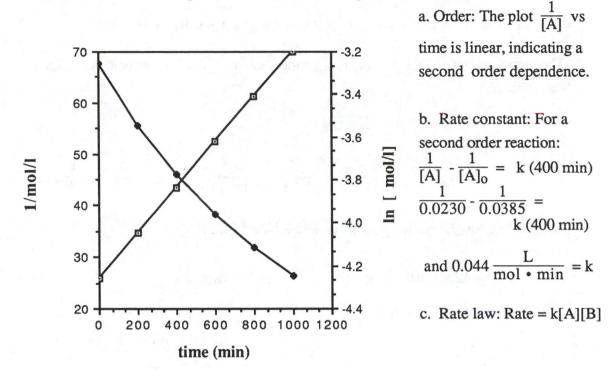

a. Order: The plot $\frac{1}{[A]}$ vs time is linear, indicating a second order dependence.

b. Rate constant: For a second order reaction:

$$\frac{1}{[A]} - \frac{1}{[A]_0} = k \,(400 \text{ min})$$

$$\frac{1}{0.0230} - \frac{1}{0.0385} = k \,(400 \text{ min})$$

$$\text{and } 0.044 \,\frac{L}{mol \cdot min} = k$$

c. Rate law: Rate = k[A][B]

d. $t_{1/2} = \dfrac{1}{k[A_0]} = \dfrac{1}{(0.044 \,\frac{L}{mol \cdot min})(0.0385 \text{ M})} = 590 \text{ min}$

54.   The reaction coordinate for the reaction $A + B \rightarrow C + D$

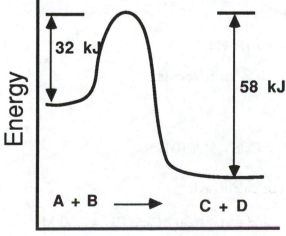

The reaction $A + B \rightarrow C + D$ is exothermic.

Note that the energy level of the products $C + D$ is lower than that of the reactants, $A + B$. As the reaction proceeds the energy corresponding to the difference between these is (32- 58) kJ. So as the reaction proceeds, about 26 kJ of energy are liberated.

56.   The E* for the reaction N_2O_5 (g) → 2 NO_2 (g) + 1/2 O_2 (g)

Given k at 25 °C = 3.46 x 10^{-5} s^{-1} and k at 55 °C = 1.5 x 10^{-3} s^{-1}

The rearrangement of the Arrhenius equation shown in your text as equation 15.6 is helpful here.

$$\ln \frac{k_2}{k_1} = -\frac{E^*}{R} \left(\frac{1}{t_2} - \frac{1}{t_1}\right)$$

$$\ln \frac{1.5 \times 10^{-3} \, s^{-1}}{3.46 \times 10^{-5} \, s^{-1}} = -\frac{E^*}{8.31 \times 10^{-3} \, kJ/mol \cdot K} \left(\frac{1}{328} - \frac{1}{298}\right)$$

and solving for E* yields a value of **102 kJ/mol for E***.

58.   Elementary Step   Rate law

a. NO (g) + NO_3 (g) → 2 NO_2 (g)   Rate = k[NO][NO_3]

                                                                                        Reaction is bimolecular

b. Cl (g) + H_2 (g) → HCl (g) + H (g)   Rate = k[Cl][H_2]

                                                                                        Reaction is bimolecular

c. $(CH_3)_3CBr$ (aq) → $(CH_3)_3C^+$ (aq) + Br^- (aq)   Rate = k[$(CH_3)_3CBr$]

                                                                                        Reaction is unimolecular

60.   a. The rate determining step in a mechanism is the slowest step. For the steps given,
        Step 2 is the rate determining step.

        b. Rate = k[O_3][O] is the rate expression for Step 2.

        c. **Step 1 is unimolecular** and **Step 2 is bimolecular**.

62.   For the reaction A + 2B → AB_2

        a. The rate expression for this reaction is: Rate = k[A][B]^2

        b. The rate constant for the reaction can be calculated:

             For rate = 2.0 x 10^{-5} mol/L • s   and [A] = 0.30 M and [B] = 0.30 M

Substituting into the rate expression above:

2.0×10^{-5} mol/L $\cdot$ s $= k \cdot (0.30)(0.30)^2$ and solving for k yields

$7.4 \times 10^{-4} \dfrac{L^2}{mol^2 \cdot s} = k$

64.   a. True--The concentration of a catalyst may appear in the rate expression. The function may be to provide an alternative pathway or to assist in optimizing molecular orientations.

b. False--While a catalyst is involved in one or more steps of a process, the catalyst is regenerated in subsequent steps, being available for further reaction.

c. False--Homogeneous catalysts are always in the same phase as the reactants, but heterogeneous catalysts are in different phases. As an example, the Pt gauze used in the production of nitric acid is in the solid state, while the reactants are not.

66.   Reactions which *appear* to involve a catalyst are identified by remembering that catalysts are substances that **are not changed (consumed)** during the course of a reaction. An examination of the four reactions given shows that equations (a) and (c) appear to involve a catalyst:

For (a) the catalyst is H^+ which is in the same phase as the reactants and products--hence it is a homogeneous catalyst.

For (c) the catalyst is Pt. Pt is in the solid state while the hydrogen, oxygen, and water are in the gaseous state. Pt is a heterogeneous catalyst.

68.   a. To determine the rate law for the reaction between methyl alcohol (A) and the organic halide (B):

Note that while the [A] remains constant in the first two experiments and the [B] is doubled, the rate also increases by a factor of two. This indicates first order dependence on B. Note that in the difference between the second and third experiments is a doubling of the [A] while [B] is constant. Note that the initial rate varies as the **square** of the doubling, indicating second order dependence. Hence the rate expression may be written:

Rate $= k [A]^2 [B]$

b. To calculate the rate of change of C if [A] = 0.15 M and [B] = 0.10 M, calculate the value for the rate constant, k , under any of the three given sets of conditions.

$$1.32 \times 10^{-4} = k \cdot (0.10)^2 \cdot (0.05)$$ and solving for k gives 0.264.

Substituting this value for k along with the stated concentrations gives:

$$\text{Rate} = 0.264 \cdot (0.15)^2 \cdot (0.10) = 5.94 \times 10^{-4} \frac{mol}{L \cdot sec}$$

69.   a. The rate expression for the decomposition of nitryl fluoride can be determined:

Examine the first and second columns of concentration data. Note that in Experiments 1 and 2 the $[F_2]$ is constant when the concentration of $[NO_2]$ doubles. Since the rate of experiment 2 is twice as great as the rate of experiment 1, we conclude that the dependence on $[NO_2]$ is first order. Compare experiments 3 and 4. Note that the rate of experiment 4 is twice that of experiment 3, corresponding to a doubling of the concentration of F_2. {Note that in these two experiments, the $[NO_2]$ is constant.} This is also a first order dependence. We also note that the rate seems to be unaffected by the concentration of product $[NO_2F]$. From these observations we can write the rate expression as :   Rate = k $[NO_2]$ $[F_2]$.

b. The order of reaction with respect to each component (as shown in part a) is :

First order with respect to F_2 ; First order with respect to NO_2 ; and second order overall.

c. To determine the value for the rate constant k, substitute values for the components of the rate expression.

$$\text{Rate} = k\ [NO_2]\ [F_2]$$

$$2 \times 10^{-4} \frac{mol}{L \cdot s} = k \cdot (1 \times 10^{-3} \frac{mol}{L}) \cdot (5 \times 10^{-3} \frac{mol}{L})$$

$$\frac{2 \times 10^{-4} \frac{mol}{L \cdot s}}{(1 \times 10^{-3}) \cdot (5 \times 10^{-3})} = 4 \times 10^{1} \frac{L}{mol \cdot s}$$

71.   To determine the $t_{1/2}$ for the first order decomposition of SO_2Cl_2 we need to determine the period of time required for the concentration of SO_2Cl_2 to be halved.

Pick a point on the graph.

e.g. $\ln[SO_2Cl_2]$ = 1.3 at time = 2.0 hrs

if $\ln[SO_2Cl_2]$ = 1.3 then $[SO_2Cl_2]$ = 3.67

Determine the ln for $[SO_2Cl_2]$ = 1.83 (that is 1/2 the concentration at t = 2.0 hrs):

$\ln[SO_2Cl_2]$ = 0.607

Consulting the graph, you will see that the time corresponding to ln = 0.607 is 6 hours. Since our t_0 was 2.0 the elapsed time was (6 - 2) or **4 hours**.

72.   For the first order process we can calculate the rate constant, k, from the equation :

$$\ln \left(\frac{[A]}{[A]_0} \right) = -kt$$

Since we know that the concentration has **decreased by 20%** in 6 hours,

$$\ln \left(\frac{[0.8]}{[1.0]} \right) = -k \cdot (6 \text{ hours}) \text{ and } k = 3.7 \times 10^{-2} \text{ hr}^{-1}$$

To calculate the half-life, $t_{1/2}$, we use a modified form of the previous expression since [A] is exactly 1/2 the original concentration of A.

$$t_{1/2} = \frac{0.693}{k} = \frac{0.693}{3.7 \times 10^{-2} \text{ hr}^{-1}} = 1.9 \times 10^1 \text{ hr}$$

74.   The Arrhenius equation may be written:

$$k = A \cdot e^{-\frac{E^*}{RT}}$$

If we assume that A will be constant for the two pathways, we can judge the effect of changing the activation energy by substitution of some values into the Arrhenius equation. Pick an arbitrary value for E*, say 50 kJ/mol, at a temperature of 298 K.

$$k_1 = A \cdot e^{-\frac{50 \text{ kJ/mol}}{8.31 \times 10^{-3} \frac{kJ}{K \cdot mol} \cdot 298 \text{ K}}}$$

and $k_1 = e^{-20.2}$ or a value of approximately 1.70×10^{-9}

Now solve for k using $E^* = -45$ kJ/mol

$$k_2 = A \cdot e^{-\dfrac{45\text{ kJ/mol}}{8.31 \times 10^{-3}\,\dfrac{kJ}{K \cdot mol} \cdot 298\text{ K}}}$$

and $k_2 = e^{-18.2}$ or a value of approximately 1.28×10^{-8}

The ratio of these two values is: $\dfrac{k_2}{k_1} = \dfrac{1.28 \times 10^{-8}}{1.70 \times 10^{-9}} = 7.5$

Note that this ratio of the rate constants is good for 298 K. The ratio is dependent upon temperature. For example, using the two activation energies chosen here (50 and 45 kJ/mol) at 1000 K, the ratio of the k's is approximately 1.8. The important point to be made is that lower activation energies will result in increased reaction rates and the ratio of the rates varies with temperature.

78.   a. The slow reaction is unimolecular and the fast reaction is bimolecular.

b. The lack of effect of the concentration of L on the reaction rate indicates that L doesn't participate in the rate-determining (slow) step. The doubling in reaction rate when the $[Ni(CO)_4]$ is doubled is indicative of first order dependence on the nickel compound. Both facts are consistent with the proposed mechanism.

c. With first order kinetics one can calculate the concentration after some time t has passed.

$$\ln\left(\frac{[Ni(CO)_4]}{[Ni(CO)_4]_0}\right) = -kt$$

$$\ln\left(\frac{[Ni(CO)_4]}{0.025}\right) = -(9.3 \times 10^{-3}\text{ s}^{-1})(300\text{ s})$$

Note 5.0 minutes has been expressed as 300 seconds.

$$\ln\left(\frac{[Ni(CO)_4]}{0.025}\right) = -2.79$$

$$\frac{[Ni(CO)_4]}{0.025} = 0.0614$$

and $[Ni(CO)_4] = 1.5 \times 10^{-3}$

d. The formation of [Ni(CO)$_4$] may be written: Ni + 4 CO $\rightarrow$ Ni(CO)$_4$

$$\frac{0.125 \text{ g Ni}}{1} \cdot \frac{1 \text{ mol}}{58.69 \text{ g Ni}} = 2.13 \times 10^{-3} \text{ mol Ni}$$

From the ideal gas law:

$$n_{CO} = \frac{1.50 \text{ atm} \cdot 0.750 \text{ L}}{0.082057 \frac{\text{L} \cdot \text{atm}}{\text{K} \cdot \text{mol}} \cdot 295.2 \text{ K}} = 0.0464 \text{ mol CO}$$

The amount of CO needed is 4 x 2.13 x 10^{-3} or 8.52 x 10^{-3} mol. An examination of the moles-available and moles-required ratios indicates that Ni is the limiting reagent.

The amount of [Ni(CO)$_4$] formed is: 2.13 x 10^{-3} mol $\cdot \dfrac{170.7 \text{ g}}{1 \text{ mol}}$ = 0.364 g Ni(CO)$_4$

The amount of excess CO is (4.64 x 10^{-2} - 8.52 x 10^{-3}) = 3.79 x 10^{-2} mol

The pressure exerted by the excess CO at 29 °C is:

$$P = \frac{(3.79 \times 10^{-2} \text{ mol})(0.082057 \frac{\text{L} \cdot \text{atm}}{\text{K} \cdot \text{mol}})(302.2 \text{ K})}{0.75 \text{ L}} = 1.3 \text{ atm}$$

e. Enthalpy change for:

Ni(CO)$_4$ (g) $\rightarrow$ Ni (s) + 4 CO (g)

ΔH_{rxn} = (0 + 4 mol $\cdot$ - 110.525 kJ/mol) - (1 mol $\cdot$ - 602.91 kJ/mol)

= 160.81 kJ

CHAPTER 16: Chemical Equilibria: General Concepts

7. Equilibrium constant expressions:

   a. $K_c = \dfrac{[H_2O]^2[O_2]}{[H_2O_2]^2}$

   b. $K_c = \dfrac{[PCl_5]}{[PCl_3][Cl_2]}$

   c. $K_c = [CO]^2$

   d. $K_c = \dfrac{[H_2S]}{[H_2]}$

9. The equilibrium constant expressions are as follows:

   a. $K_p = P_{Cl_2}$

   b. $K = \dfrac{[Cu^{2+}] \cdot [Cl^-]^4}{[CuCl_4^{2-}]}$

   c. $K_p = \dfrac{P_{NO}^2 \cdot P_{O_2}}{P_{NO_2}^2}$

   d. $K_p = \dfrac{P_{CO_2} \cdot P_{H_2}}{P_{CO} \cdot P_{H_2O}}$

   e. $K = \dfrac{[Mn^{2+}] \cdot P_{Cl_2}}{[H_3O^+]^4 \cdot [Cl^-]^2}$

11. Comparing the two equilibria:

$$SO_2\,(g) + \tfrac{1}{2}O_2\,(g) \Leftrightarrow SO_3\,(g) \qquad K_1$$

$$2\,SO_3\,(g) \Leftrightarrow 2\,SO_2\,(g) + O_2\,(g) \qquad K_2$$

the expression that relates K_1 and K_2 is (e) : $K_2 = \dfrac{1}{K_1^2}$

13. a. The equilibrium expression for the "standard" equation is:
$$K_c = \dfrac{[H_2S]}{[H_2]} = 7.6 \times 10^5$$

For the equation given in this problem, it is:
$$K_c^* = \dfrac{[H_2S]^8}{[H_2]^8}$$

To determine the difference between these two expressions, assume the concentration of elemental hydrogen in the "standard" equation is 25 mol/L. Solving for the H_2S concentration we get 1.90×10^7 M.

Substituting the concentrations obtained in our assumptions into the second equation yields

$$K_c^* = \frac{[1.90 \times 10^7]^8}{[25]^8} = 1.1 \times 10^{47}$$

The relationship between K_c^* and K_c may be seen by dividing K_c^* by K_c:

$$\frac{K_c^*}{K_c} = \frac{1.1 \times 10^{47}}{7.6 \times 10^5} = 1.5 \times 10^{41}$$

That is $K_c^* = (K_c)^8$. In general the change of all coefficients by a common factor (8) results in a new equilibrium constant which is the old equilibrium constant (K_c) raised to the power of the change.

b. The value K_c would not differ from K_p since $K_p = K_c(RT)^{\Delta n}$, and for this reaction $\Delta n = 0$.

15. The equilibrium constant for the reaction:

$$CoO\ (s) + H_2\ (g) \Leftrightarrow Co\ (s) + H_2O\ (g) \quad is\ K_1 = \frac{[H_2O]}{[H_2]} = 67 \text{ and}$$

the equilibrium constant for the reaction:

$$CoO\ (s) + CO\ (g) \Leftrightarrow Co\ (s) + CO_2\ (g) \quad is\ K_2 = \frac{[CO_2]}{[CO]} = 490$$

For the reaction : $CO_2\ (g) + H_2\ (g) \Leftrightarrow CO\ (g) + H_2O\ (g)$ the equilibrium constant expression is:

$$K_3 = \frac{[CO][H_2O]}{[CO_2][H_2]}$$

Multiplication of K_1 by the reciprocal of K_2 yields K_3:

$$\frac{[H_2O]}{[H_2]} \cdot \frac{[CO]}{[CO_2]} = 67 \cdot \frac{1}{490} = 0.14$$

Similarly K_4 can be obtained by multiplying $K_2 \cdot \frac{1}{K_1}$ [or $K_4 = \frac{1}{K_3}$]

$$K_4 = \frac{[CO_2]}{[CO]} \cdot \frac{[H_2]}{[H_2O]} = 490 \cdot \frac{1}{67} = 7.3$$

17. If the equilibrium constant for the reaction: $N_2O_4 (g) \Leftrightarrow 2\, NO_2 (g)$ is 0.15 then the constant for : $NO_2 \Leftrightarrow \frac{1}{2} N_2O_4(g)$ would be:

$$K_2 = \frac{1}{\sqrt{K_1}} = \frac{1}{\sqrt{0.15}} = 2.6$$

19. The equilibrium constant expression for the reaction is: $K_c = \frac{[H_2O][CO]}{[H_2][CO_2]}$

Substitution of the first data set yields: $Kc = \frac{[9.40][9.40]}{[80.52]\,[0.69]} = 1.6$

Similarly substitution of the remaining data sets yields values of 1.57, 1.58, and 1.60 for an average **Kc = 1.6**. At this temperature, **the products are slightly favored**.

For the reaction: $H_2O (g) + CO (g) \Leftrightarrow H_2 (g) + CO_2 (g)$ the equilibrium constant is the **reciprocal** of the original equilibrium constant.

$$K_{rev} = \frac{1}{K} = \frac{1}{1.6} = 0.63$$

21. For the reaction: $H_2 (g) + CO_2 (g) \Leftrightarrow H_2O (g) + CO (g)$
a. Calculate K at 986 °C if at equilibrium : $P(CO) = P(H_2O) = 11.2$ atm
$$P(H_2) = P(CO_2) = 8.8 \text{ atm}$$

$$K_p = \frac{P_{CO} \cdot P_{H_2O}}{P_{CO_2} \cdot P_{H_2}} = \frac{(11.2 \text{ atm})^2}{(8.8 \text{ atm})^2} = 1.6$$

b. For this reaction $K_p = K_c$ ($\Delta n = 0$)
This means simply that the **molar concentrations** and **partial pressures** are **interchangeable** for this system. The concentrations of H_2 and CO_2 (8.8 M) may be substituted into the equilibrium constant expression to obtain values of 11 mol/L for both CO and H_2O.

23.   a. K_c for the equation: H_2O (g) $\Leftrightarrow$ H_2 (g) $+ \frac{1}{2} O_2$ (g)

If $[H_2O]_i = 2.0$ M and 10. % dissociates to H_2 and O_2.

	H_2O	**H_2**	**O_2**
Initial	2.0 M	0	0
Change	- 0.20	+ 0.20	+ 0.10
Equilibrium	1.8	+ 0.20	+ 0.10

and $K_c = \dfrac{[H_2][O_2]^{1/2}}{[H_2O]} = \dfrac{(0.20)(0.10)^{1/2}}{(1.8)} = 0.035$

25. K_c for the equation: PCl_5 (g) $\Leftrightarrow$ PCl_3 (g) $+ Cl_2$ (g)

If $[PCl_5]_i = 0.84$, $[PCl_3]_i = 0.18$, and $[PCl_5]_{eq} = 0.72$

	PCl_5	**PCl_3**	**Cl_2**
Initial	0.84 M	0.18	0
Change	-x	+x	+x
Equilibrium	0.84 - x	0.18 + x	+x

Since $0.84 - x = 0.72$, then $x = 0.12$ and $K = \dfrac{(0.030)(0.12)}{0.72} = 5.0 \times 10^{-2}$

27. K_p at 25 °C for the system: $N_2H_6CO_2$ (s) $\Leftrightarrow$ 2 NH_3 (g) $+ CO_2$ (g)

Total Pressure = 0.116 atm = P(NH_3) + P(CO_2)

The reaction produces 2 mol NH_3 and 1 mol O_2 for each mol of ammonium carbamate that sublimes. The partial pressure of ammonia will be twice as great as that of carbon dioxide.

Pressure = 0.116 atm = 2 • P(CO_2) + P(CO_2)

0.116 atm = 3 • P(CO_2)

and P(NH_3) = 2 • P(CO_2) = 0.0773 atm.

Substitution into the equilibrium constant expression yields:

$K_p = P_{CO_2} • P_{NH_3}^2 = (0.0387) • (0.0773)^2 = 2.31 \times 10^{-4}$

29.   Determine the value of the reaction quotient:

$$Q = \frac{[N_2O_4]}{[NO_2]^2} = \frac{\frac{1.5 \times 10^{-3}\text{mol}}{10.\ L}}{\left(\frac{2.0 \times 10^{-3}\text{mol}}{10.\ L}\right)^2} = 3750$$

Since Q is greater than K (170), the system is **not at equilibrium**. Since Q is greater than K, the system must "go to the left" to attain equilibrium--**producing more NO_2**.

31.   For the system $COBr_2\ (g) \Leftrightarrow CO\ (g) + Br_2\ (g)$, $K_c = 0.190$

$$Q = \frac{[CO] \cdot [Br_2]}{[COBr_2]} = \frac{(0.402)^2}{(0.950)} = 1.70 \times 10^{-1}$$

Since Q < K the system is **not at equilibrium**, and $[COBr_2]$ will decrease.

33.   For the system n-butane $\Leftrightarrow$ isobutane $K_c = 2.5$
       Equilibrium concentrations may be found:

	n-butane	isobutane
Initial	0.034	0
Change	- x	+ x
Equilibrium	0.034 - x	x

$$K_c = \frac{x}{0.034 - x} = 2.5 \text{ and } x = 2.4 \times 10^{-2}$$

and [isobutane] $= 2.4 \times 10^{-2}$ M, [n-butane] $= 0.034 - x = 9.7 \times 10^{-3}$ M

35.   The equilibrium constant expression is:

$$K_P = \frac{P_{NO_2}^2}{P_{N_2O_4}} = 0.15 = \frac{P_{NO_2}^2}{0.85}$$

and taking the square root of both sides :

$$P_{NO_2} = 0.36$$

The total pressure at equilibrium is $0.36 + 0.85 = 1.21$ atm

37.   The equilibrium expression for the equation is: $K_c = \dfrac{[Cl_2O]}{[Cl_2]^2}$

If x represents the molar concentration (the volume of the flask is 1.0 L), the amount of Cl_2 needed to form x mol/L would be 2x. The equilibrium concentration of Cl_2 would be 0.15 - 2x. The proper expression would be:

$$(b) \ K_c \ = \ \frac{x}{(0.15 - 2x)^2}$$

39.   For the system $CaCO_3$ (s) $\Leftrightarrow$ CaO (s) + CO_2 (g), $K_c = 0.10$.

With $K_c = [CO_2] = 0.10$, the molar concentration of $CO_2 = 0.10$. The mass of CO_2 at

equilibrium in a 1.0 L container: $\dfrac{0.10 \ mol \ CO_2}{L} \cdot \dfrac{1.0 \ L}{1} \cdot \dfrac{44 \ g \ CO_2}{1 \ mol \ CO_2} = 4.4 \ g \ CO_2$

41.   For the system : 2 HCN (g) $\Leftrightarrow$ H_2 (g) + C_2N_2 (g) $K_p = 4.00 \times 10^{-4}$
$P(C_2N_2)$ if $[HCN]_i = 25.0$ atm:
Assume that the equilibrium pressure of C_2N_2 (and H_2) is x atm.

	HCN	H₂	C₂N₂
Initial	25.0	0	0
Change	-2x	+ x	+ x
Equilibrium	25.0 - 2x	x	x

$$K_p \ = \ \frac{P_{C_2N_2} \cdot P_{H_2}}{P_{HCN}^2} \ = \ \frac{(x)(x)}{(25.0 - 2x)^2} \ = 4.0 \times 10^{-4}$$

Taking the square root of both sides we get:
$$\frac{x}{25.0 - 2x} \ = \ 0.0200$$

$$x \ = \ 0.0200 \ (25.0 - 2x) \ = \ 0.500 - 0.0400 \ x$$
$$1.04 \ x \ = \ 0.500$$
$$x \ = \ 0.481 = P(H_2) = \ P(C_2N_2)$$

The equilibrium pressure of HCN is 24.0 atm.

43.   a. Given: $K_c = \dfrac{[CO][Br_2]}{[COBr_2]} = 0.190$ at 73 °C, calculate equilibrium concentrations.

$$\frac{1.06 \text{ g COBr}_2}{8.0 \text{ L}} \cdot \frac{1 \text{ mol COBr}_2}{187.8 \text{ g COBr}_2} = 7.05 \times 10^{-4} \text{ M COBr}_2$$

Substituting into K_c : $\dfrac{[CO][Br_2]}{[COBr_2]} = \dfrac{x^2}{7.05 \times 10^{-4} - x} = 0.190$

Since $K_c \cdot 100 \gg [COBr_2]$ we must solve via the quadratic equation :
$$[CO] = [Br_2] = 7.029 \times 10^{-4} \text{ M}$$

and $[COBr_2] = (7.055 \times 10^{-4} \text{ M} - 7.029 \times 10^{-4} \text{ M}) = 2.6 \times 10^{-6} \text{ M}$

b. The total pressure is:

$P_T = P(CO) + P(Br_2) + P(COBr_2)$

$\quad = [2(7.0 \times 10^{-4}) + 2.6 \times 10^{-6}] \cdot 0.082057 \dfrac{\text{L} \cdot \text{atm}}{\text{K} \cdot \text{mol}} \cdot 346.2 \text{ K}$

$\quad = 0.040 \text{ atm}$

45.   a. The decomposition expression may be written:
$$K_P = \frac{P_{PCl_3} \cdot P_{Cl_2}}{P_{PCl_5}} = 11.5$$

At equilibrium $P(PCl_5) = 1.50$ atm

Then $P(PCl_3) = P(Cl_2) = x$ and $\dfrac{x^2}{1.50} = 11.5$; solving gives: $x = 4.15$ atm

b. Total pressure at equilibrium

$P_T = P(PCl_3) + P(Cl_2) + P(PCl_5)$

$\quad = 4.15 \text{ atm} + 4.15 \text{ atm} + 1.50 \text{ atm}$

$\quad = 9.80 \text{ atm}$

c. Fraction of PCl_5 dissociated:

Since the stoichiometry tells us that for each mole of PCl_3 (or Cl_2) present at equilibrium one mole of PCl_5 dissociated, the initial pressure of PCl_5 was: 4.15 atm + 1.50 atm = 5.65 atm

$$\frac{P_{PCl_3}}{P_{PCl_5}} = \frac{4.15}{5.65} = 0.735$$

47. K_c for the system $ZnS\ (s) \Leftrightarrow Zn^{2+}\ (aq) + S^{2-}\ (aq)$ is 1.1×10^{-21}

$$K_c = [Zn^{2+}][S^{2-}] = 1.1 \times 10^{-21}$$

[Reminder: Solids are omitted from equilibrium constant expressions.]

As ZnS dissolves, zinc ions and sulfide ions are formed in equimolar amounts. Let **x** represent the amount of ZnS which dissolves. The equilibrium concentration of zinc and sulfide ions will be **x** molar.

$$K_c = [Zn^{2+}][S^{2-}] = x^2 = 1.1 \times 10^{-21}$$
$$x = 3.3 \times 10^{-11}$$
$$[Zn^{2+}] = 3.3 \times 10^{-11}\ M$$

49.   Effect upon the quantities associated with the HBr equilibrium when the following changes occur:

Change	$[Br_2]$	[HBr]	K_c
H_2 is added to the container	**decrease**	**increase**	**no change**

Equilibrium shifts to reduce the added H_2, thereby consuming Br_2 and producing additional HBr.

Temperature of the gases in the container is increased	**increase**	**decrease**	**decrease**

An increase in temperature favors the endothermic process. Hence "left-hand" species are favored. The equilibrium constant shifts to favor that process.

Volume of the flask is increased	**no change**	**no change**	**no change**

Since there is no difference in the total number of gaseous molecules on either side of the equation, a volume change (pressure change) should not affect the equilibrium.

51.   Predict the changes which will occur on the equilibrium:
$$H_2\ (g) + S\ (s) \Leftrightarrow H_2S\ (g) + 20.6\ kJ$$
when:

a. Sulfur is added--No change

Adding solid has no effect on the concentration of S--that's a function of the solid's density.

b. More $H_2(g)$ added--Equilibrium shifts to right.

Such a shift would consume added hydrogen.

c. Volume is decreased--No change

If the volume's decreased, the pressures of both gases will be affected equally and no shift in the equilibrium position is anticipated.

d. Temperature is raised--Equilibrium shifts to left.

Increased temperature favors endothermic processes -- in this case the decomposition of H_2S.

53. K_c for n-butane $\Leftrightarrow$ iso-butane is 2.5.

a. Equilibrium concentration if 0.50 mol/L of iso-butane is added:

	n-butane	iso-butane
Original concentration	1.0	2.5
Change immediately after addition	1.0	2.5 + 0.50
Change (going to equilibrium)	+ x	- x
Equilibrium concentration	1.0 + x	3.0 - x

$$K_c = \frac{[\text{iso-butane}]}{[\text{n-butane}]} = \frac{3.0 - x}{1.0 + x} = 2.5$$

$$3.0 - x = 2.5 (1.0 + x)$$

$$\text{and } 0.14 = x$$

The equilibrium concentrations are:

[n-butane] = 1.0 + x = 1.1 M and [iso-butane] = 3.0 - x = 2.9 M

b.Equilibrium concentrations if 0.50 mol /L of n-butane is added:

	n-butane	iso-butane
Original concentration	1.0	2.5
Change immediately after addition	1.0 + 0.50	2.5
Change (going to equilibrium)	- x	+ x
Equilibrium concentration	1.5 - x	2.5 + x

$$K_c = \frac{[\text{iso-butane}]}{[\text{n-butane}]} = \frac{2.5 + x}{1.5 - x} = 2.5$$

$$2.5 + x = 2.5 (1.5 - x)$$

$$\text{and } x = 0.36$$

The equilibrium concentrations are:

$$[\text{n-butane}] = 1.5 - x = 1.1 \text{ M} \quad \text{and} \quad [\text{iso-butane}] = 2.5 + x = 2.9 \text{ M}$$

55.   For the system $PCl_5 (g) \Leftrightarrow PCl_3 (g) + Cl_2 (g)$, if $[PCl_3] = x$ the equilibrium expression may be written as:

      b. $K_c = \dfrac{(2 + x) \, x}{(2 - x)}$

57. K_p for the system $N_2H_6CO_2 (s) \Leftrightarrow 2 NH_3 (g) + CO_2 (g)$ is 2.31×10^{-4}.

At equilibrium the pressure of CO_2 may be represented as $\mathbf{x}$, and pressure of NH_3 by $\mathbf{2x}$.

The equilibrium constant expression may be written:

$$K_p = P_{CO_2} \cdot P_{NH_3}^2 = (x) \cdot (2x)^2 = 2.31 \times 10^{-4}$$

$$4x^3 = 2.31 \times 10^{-4}$$

$$x = 3.87 \times 10^{-2} \text{ atm}$$

The pressure of NH_3 in the absence of added CO_2 will be $(2x)$ or 7.73×10^{-2} atm.

If 0.05 atm of CO_2 is added to the equilibrium system:

	NH_3	CO_2
Pressure at equilibrium (atm)	0.0773	0.0387
Change immediately after addition		0.05
Change (going to equilibrium)	- 2x	- x
Equilibrium pressure	0.0773 - 2x	0.0887 - x

$$K_p = (0.0773 - 2x)^2(0.0887 - x) = 2.31 \times 10^{-4}$$

$$x = 1.14 \times 10^{-2} \text{ atm}$$

[Note: An exact solution of this cubic equation is somewhat tedious. One rather straightforward solution of this problem may be achieved with an iterative approach on a computer. The author used a spreadsheet on a microcomputer to obtain the answer in about 10 iterations.]

The pressure of CO_2: 0.0887 - 0.0114 = 0.0773 atm [greater than 0.05 atm].

The pressure of NH_3 [0.0774 - (2 • 0.0114)] will be less than the pressure of NH_3 (7.74×10^{-2} atm) in the absence of added CO_2.

59.   a. The rate determining step in a mechanism is the <u>slow</u> step--elementary step 2 is rate determining.

    The rate law for this step would be: Rate $= k$ [CHCl$_3$][Cl]

    b. The first step controls the production of Cl.

    The equilibrium expression for this first step may be written:
$$K = \frac{[Cl]^2}{[Cl_2]} = \text{ or rearranging}$$

$$[Cl] = \sqrt{K \cdot [Cl_2]} = K^{1/2} \cdot [Cl_2]^{1/2}$$

Substituting into the rate law for the slow step gives:

$$\text{Rate} = k[CHCl_3] \cdot K^{1/2} \cdot [Cl_2]^{1/2} \quad \text{or } k_{exp}[CHCl_3][Cl_2]^{1/2}$$
$$\text{where } k_{exp} = k \cdot K^{1/2}$$

61.   For the reaction   A + 2 B + C → D with the mechanism:

        Step 1   A + B ⇔ X   very rapid equilibrium
        Step 2   X + C → Y   slow
        Step 3   Y + B → D   fast

The logic involved here is identical to that in study question 59. The Rate law for Step 2 would be: Rate $= k[X][C]$. Since Step 1 controls the concentration of X, we may substitute for [X] the rate expressioin: Rate $= k'[A][B]$, for an overall rate law of:

        (e) Rate $= k[A][B][C]$

63.   For the system at 700 °C: 2 H$_2$ (g) + S$_2$ (g) ⇔ 2 H$_2$S (g) at equilibrium:

$$[H_2S] = \frac{0.333 \text{ mol}}{23.5 \text{ L}} \qquad [H_2] = \frac{0.512 \text{ mol}}{23.5 \text{ L}} \qquad [S_2] = \frac{2.67 \times 10^{-5} \text{ mol}}{23.5 \text{ L}}$$

$$= 1.42 \times 10^{-2} \text{ M} \qquad = 2.18 \times 10^{-2} \text{ M} \qquad = 1.14 \times 10^{-6} \text{ M}$$

$$Q = \frac{[H_2S]^2}{[H_2]^2 [S_2]} = \frac{(1.42 \times 10^{-2})^2}{(2.18 \times 10^{-2})^2(1.14 \times 10^{-6})} = 3.72 \times 10^5$$

Since K$_c$ is 1.1×10^7, the system is NOT at equilibrium. On proceeding to equilibrium, more H$_2$S is produced, and H$_2$ and S$_2$ are consumed.

65.   For the system $2\ CH_3CO_2H \Leftrightarrow$ dimer, K_p at 25 °C $= 1.3 \times 10^3$.

     a.

	monomer	dimer
Initial pressure (mmHg)	10.	0
(atm)	1.3×10^{-2}	
Equilibrium pressure	x	$\frac{1}{2}(1.3 \times 10^{-2} - x)$

$$K_p = \frac{P(dimer)}{P(monomer)^2} = \frac{\frac{1}{2}(1.3 \times 10^{-2} - x)}{x^2} = 1.3 \times 10^3$$

Note that the magnitude of the equilibrium constant, allows us to assume that almost all of the acetic acid dimerizes.

$$K_p = \frac{6.5 \times 10^{-3} - 0.5x}{x^2} = 1.3 \times 10^3$$

Solving via the quadratic equation yields $x = 2.05 \times 10^{-3}$ (or 2.1×10^{-3} to two significant figures. The amount of acetic acid in the monomeric form is: 2.1×10^{-3} atm or

$$\frac{2.1 \times 10^{-3}\ atm}{1.3 \times 10^{-2}\ atm} \times 100 \quad \text{or 16 \% of the original acid.}$$

The percentage of acid forming the dimer is 84 %.

b. As the temperature goes up, the equilibrium shifts to the left (producing more monomer).

67.   For zinc carbonate dissolving in water, the equilibrium constant expression is:
$$K_c = [Zn^{2+}][CO_3^{2-}] = 1.5 \times 10^{-11} \qquad \text{[Solids are omitted]}$$
At equilibrium $[Zn^{2+}] = [CO_3^{2-}]$ so $[Zn^{2+}]^2 = [CO_3^{2-}]^2 = 1.5 \times 10^{-11}$
$$\text{and}\quad [Zn^{2+}] = [CO_3^{2-}] = 3.9 \times 10^{-6}$$

69.   For the system shown, the relationship between K_P and K_C is:
$$K_P = K_c \cdot (RT)^{\Delta n}$$
Convert the given K_P to the more convenient (in this case) K_c.
Since $\Delta n = (2 - 1)$, $K_c = K_P \cdot (RT)^{-1}$

$$K_c = \frac{1.7 \times 10^{-8}}{(RT)^{\Delta n}} = \frac{1.7 \times 10^{-8}}{\left(0.082057 \frac{L \cdot atm}{K \cdot mol} \cdot 1800\ K\right)} = 1.15 \times 10^{-10}$$

With an original concentration of 0.10 M, if we assume that x mol/L of O_2 dissociates, the amount of O present at equilibrium will be 2x.

$$K_c = \frac{[O]^2}{[O_2]} = \frac{(2x)^2}{(0.10 - x)} = 1.15 \times 10^{-10}$$

The magnitude of K allows us to simplify the denominator to ~ 0.10, and solving gives
$$x = 1.7 \times 10^{-6} \text{ and } [O] = 3.4 \times 10^{-6} \text{ M}.$$

The number of O atoms present:

$$\frac{3.4 \times 10^{-6} \text{ mol}}{1 \text{ L}} \cdot \frac{10. \text{ L}}{1} \cdot \frac{6.022 \times 10^{23} \text{ O atoms}}{1 \text{ mol O atoms}} = 2.0 \times 10^{19} \text{ O atoms}$$

71.   For the system: O_3 (g) + NO (g) $\Leftrightarrow$ O_2 (g) + NO_2 (g), K = 6.0×10^{34}

a. Calculate a reaction quotient:

$$Q = \frac{[O_2][NO_2]}{[O_3][NO]} = \frac{(8.2 \times 10^{-3})(2.5 \times 10^{-4})}{(1.0 \times 10^{-6})(1.0 \times 10^{-5})} = 2.1 \times 10^5$$

Q<<K. The system is **not at equilibrium**. To reach equilibrium more ozone and nitric oxide must react to form oxygen and nitrogen dioxide--products will increase.

b. To determine the effect which heating (or cooling) will have on the system, we must know something about the enthalpy change for the reaction.

[Data from Appendix J]

ΔHreaction$_{(298K)}$ = [ΔH_f NO_2 (g) + ΔH_f O_2 (g)] - [ΔH_f O_3 (g) + ΔH_f NO (g)]

      = (33.18 + 0) - (142.7 + 90.25)

      = (33.18 - 233.0) kJ

      = - 199.8 kJ

The exothermic nature of this reaction (at 298 K) suggests that an increase in temperature would shift the equilibrium to the left, reducing the concentration of O_2 and NO_2.

73.   a. For equilibria involving gases:

$$K_P = K_C \cdot (RT)^{\Delta n} \text{ and } \Delta n = (4 - 2) \text{ or } 2$$

$$K_P = 6.56 \times 10^{-3} \cdot \left[\left(0.082057\frac{L \cdot atm}{K \cdot mol}\right) \cdot (906.2 \text{ K})\right]^2$$

$$= 36.3$$

b. The pressure of NH_3 (in atm) $= 456 \text{ mmHg} \cdot \dfrac{1 \text{ atm}}{760 \text{ mmHg}} = 0.600$ atm

$$K_P = \frac{P_{N_2} \cdot P_{H_2}^{3}}{P_{NH_3}^{2}} = 36.3$$

substituting: $\dfrac{x \cdot (3x)^3}{(0.600 - 2x)^2} = 36.3$

simplifying yields: $\dfrac{27 \, x^4}{(0.600 - 2x)^2} = 36.3$

and taking the square root of both sides gives: $\dfrac{5.196 \, x^2}{(0.600 - 2x)} = 6.025$

Using the quadratic equation gives x $= P(N_2) = 0.27$ atm

$P(H_2) \quad = 3x = 0.81$ atm

$P(NH_3) = (0.600 - 2x) = 0.060$ atm

$P_T = P_{H_2} + P_{NH_3} + P_{N_2} = 1.14$ atm

75.   The pressure of NO_2 initially is:

$$P = \frac{8.90 \text{ g } NO_2}{36 \text{ L}} \cdot \frac{1 \text{ mol } NO_2}{46.01 \text{ g } NO_2} \cdot 0.082057 \frac{L \cdot atm}{K \cdot mol} \cdot 298.2 \text{ K} = 0.13 \text{ atm}$$

At equilibrium $P(NO_2) + P(N_2O_4) = 0.100$ atm

For each 2 moles of NO_2 that combine, 1 mole of N_2O_4 forms.

So at equilibrium: $P(N_2O_4) = x$ and $P(NO_2) = 0.13 - 2x$.

Substituting into the P_T expression

$(0.13 - 2x) + x = 0.100$ atm

$0.13 - x = 0.100$ atm   and $0.03 = x$

$$K_P = \frac{P_{NO_2}^{2}}{P_{N_2O_4}} = \frac{(0.13 - 0.06)^2}{0.03} = 0.2$$

77.   The $[SO_3]$ is: $\dfrac{2.0 \text{ mol}}{50.0 \text{ L}} = 0.040$ M and

$$K_c = \frac{[SO_3]^2}{[SO_2]^2 [O_2]} = 11.39$$

If **x** mol/L of O_2 are produced in going to equilibrium, then **2x** mol/L of SO_2 are produced, and the equilibrium concentration of SO_3 is (0.040 - 2x).

$$K_c = \frac{(0.040 - 2x)^2}{(2x)^2(x)} = 11.39$$

This is a cubic equation. One possible solution to this equation is obtained using a spreadsheet--substituting values of x. Using this technique a value of x = 0.014256 or (to two significant figures) 0.014 mol/L is obtained.

The total pressure for this system is obtained with the ideal gas law:

$$P(T) = P(SO_2) + P(O_2) + P(SO_2) = \frac{n(SO_2) + n(O_2) + n(SO_2)}{V} \cdot R \cdot T$$

Substituting yields:

$$(0.028 \text{ M} + 0.014 \text{ M} + 0.012 \text{ M}) \cdot 0.082057 \frac{L \cdot atm}{K \cdot mol} \cdot 1173 \text{ K} = 0.52 \text{ atm}$$

Partial pressure of SO_3 is: $P(T) \cdot X(SO_3) = 0.52 \text{ atm} \cdot \dfrac{0.012 \text{ M } SO_3}{0.054 \text{ M gas}} = 1.1 \text{ atm}$

79.   a. The initial pressure of SO_2Cl_2 is:

$$\frac{6.70 \text{ g } SO_2Cl_2}{1.00 \text{ L}} \cdot \frac{1 \text{ mol } SO_2Cl_2}{135.0 \text{ g } SO_2Cl_2} \cdot 0.082057 \frac{L \cdot atm}{K \cdot mol} \cdot 648.2 \text{ K} = 2.64 \text{ atm}$$

As the SO_2Cl_2 dissociates in SO_2 and Cl_2 (x mol/L) equilibrium occurs and we write:

$$K_P = \frac{x^2}{(2.64 - x)} = 2.4$$

Solving this equation via the quadration equations gives x = 1.59 (or 1.6) atm

At equilibrium, $P(SO_2) = P(Cl_2) = 1.6 \text{ atm}$ and $P(SO_2Cl_2) = (2.64 - 1.59) = 1.05$ (or 1.1 atm).

The total pressure is: $P(SO_2) + P(Cl_2) + P(SO_2Cl_2) = 1.6 + 1.6 + 1.1 = 4.3 \text{ atm}$

The fraction of SO_2Cl_2 that has dissociated is $\dfrac{1.59 \text{ atm}}{2.64 \text{ atm}} = 0.60$.

b. Pressure of gases at equilibrium if we begin with $P(SO_2Cl_2) = 2.64 \text{ atm}$ and $P(Cl_2) = 1.00 \text{ atm}$

$$K_P = \frac{x \cdot (x + 1.00)}{(2.64 - x)} = 2.4$$

Once again, solving via the quadratic equation gives x = 1.337 (or 1.34 atm)

At equilibrium, $P(SO_2)$ = 1.34 atm; $P(Cl_2)$ = 2.34 atm

and $P(SO_2Cl_2)$ = (2.64 - 1.34) = 1.30 atm

The fraction of SO_2Cl_2 dissociated is: $\frac{1.34}{2.64}$ = 0.51

c. LeChatelier's principle predicts that the addition of chlorine would shift the position of equilibrium to the left, preserving more SO_2Cl_2--a prediction which is confirmed by these calculations.

80.   a. Mass of NOBr obtainable:

$$\frac{3.50 \text{ g NO}}{1} \cdot \frac{1 \text{ mol NO}}{30.01 \text{ g NO}} = 0.117 \text{ mol NO}$$

$$\frac{9.67 \text{ g Br}_2}{1} \cdot \frac{1 \text{ mol Br}_2}{159.8 \text{ g Br}_2} = 0.0605 \text{ mol Br}_2$$

Moles-required ratio: $\frac{2 \text{ mol NO}}{1 \text{ mol Br}_2}$ Moles-available ratio: $\frac{0.117 \text{ mol NO}}{0.0605 \text{ mol Br}_2} =$

$$\frac{1.93 \text{ mol NO}}{1 \text{ mol Br}_2}$$

NO is the Limiting Reagent:

$$\frac{0.117 \text{ mol NO}}{1} \cdot \frac{2 \text{ mol NOBr}}{2 \text{ mol NO}} \cdot \frac{109.9 \text{ g NOBr}}{1 \text{ mol NOBr}} = 12.8 \text{ g NOBr}$$

b. The electron dot structure for

Nitrosyl bromide is: $\ddot{O}=\ddot{N}-\ddot{Br}:$

c. The central atom, N, has three groups (2 atoms and 1 lone pair) around it--hence the structural geometry is **trigonal planar** and the molecular geometry is **bent**.

The molecule is **polar**, as is very often the case with this molecular geometry.

d. Value of K_P for the equilibrium at 25 °C: NOBr (g) $\Leftrightarrow$ NO (g) $+\frac{1}{2}$ Br$_2$ (g)

Given: $P(T) = P(NO) + P(NOBr) + P(Br_2) = 190$ mmHg

If we let x = $P(NOBr)_{initial}$ then $P(NOBr)_{equilibrium}$ = 0.66 x

[The compound is 34 % dissociated.]

The stoichiometry dictates that for each mole of NOBr that dissociates, one mole of NO forms. Hence $P(NO)_{equilibrium} = 0.34\ x$. Note also that the $P(Br_2)_{equilibrium}$ is one-half that of P(NO)--or 0.17 x. Stated mathematically this:

$$P(T) = P(NO) + P(NOBr) + P(Br_2) = 190\ \text{mmHg}$$
$$= 0.34\ x + 0.66\ x + 0.17\ x = 190\ \text{mmHg}$$
$$1.17\ x = 190\ \text{mmHg}$$
$$\text{and}\ x = 162\ \text{mmHg}.$$

The equilibrium pressures are:

P (NO) = 55.2 mmHg	(0.0725 atm)
P (Br$_2$) = 27.6 mmHg	(0.0363 atm)
P (NOBr) = 107 mmHg	(0.141 atm)

Substitution into the K_P expression yields:

$$K_P = \frac{P_{NO} \cdot (P_{Br_2})^{0.5}}{P_{NOBr}} = \frac{(0.0725) \cdot (0.0363)^{0.5}}{(0.141)} = 0.098$$

14.

Conjugate Base of:	Formula	Name
a. HCN	CN^-	Cyanide ion
b. HSO_4^-	SO_4^{2-}	Sulfate ion
c. HF	F^-	Fluoride ion
d. HNO_2	NO_2^-	Nitrite ion
e. HCO_3^-	CO_3^{2-}	Carbonate ion

16.

Conjugate Partner of:	Formula	Name
a. CN^-	HCN	Hydrocyanic acid
b. SO_4^{2-}	HSO_4^-	Hydrogen sulfate ion
c. HI	I^-	Iodide ion
d. S^{2-}	HS^-	Hydrogen sulfide ion
e. HNO_3	NO_3^-	Nitrate ion

18. The equation for potassium carbonate dissolving in water:

$$K_2CO_3 \ (aq) \quad \rightarrow 2 \ K^+ \ (aq) + CO_3^{2-} \ (aq)$$

Soluble salts--like K_2CO_3--dissociate in water. The carbonate ion formed in this process is a base, and reacts with the acid, water.

$$CO_3^{2-} \ (aq) + H_2O \ (l) \rightarrow HCO_3^- \ (aq) + OH^- \ (aq)$$

The production of the hydroxide ion, a strong base, in this second step is responsible for the <u>basic</u> nature of solutions of this carbonate salt.

20. $HPO_4^{2-} \ (aq) + H_2O \ (l) \quad \Leftrightarrow \quad PO_4^{3-} \ (aq) + H_3O^+(aq) \qquad$ (Acid)

$HPO_4^{2-} \ (aq) + H_3O^+ \ (aq) \Leftrightarrow \quad H_2PO_4^- \ (aq) + H_2O \ (l) \qquad$ (Base)

22.

a. $HCOOH \ (aq) + H_2O \ (l) \quad \Leftrightarrow \quad HCOO^- \ (aq) + H_3O^+ \ (aq)$
    acid   base   conjugate   conjugate
                                          of HCOOH   of H_2O

b. $H_2S \ (aq) + NH_3 \ (aq) \quad \Leftrightarrow \quad NH_4^+ \ (aq) + HS^- \ (aq)$
    acid   base   conjugate   conjugate
                                       of NH_3 of H_2S

c. $HSO_4^- \ (aq) + OH^- \ (aq) \quad \Leftrightarrow \quad SO_4^{2-} \ (aq) + H_2O \ (l)$
    acid   base   conjugate   conjugate
                                    of HSO_4^- of OH^-

24.   Since both NH_4Cl and NaH_2PO_4 are soluble in water, we should view this in two steps:
        [1] dissociation into component ions:

$$NH_4Cl \quad \rightarrow \quad NH_4^+ \text{ (aq)} + Cl^- \text{ (aq)}$$
$$NaH_2PO_4 \rightarrow \quad Na^+ \text{ (aq)} + H_2PO_4^- \text{ (aq)}$$

        [2] transfer of proton:

$$NH_4^+ \text{ (aq)} + H_2PO_4^- \text{ (aq)} \rightarrow NH_3 \text{ (aq)} + H_3PO_4 \text{ (aq)}$$

        One might suggest that the proton would be transferred from the ammonium ion to the
        dihydrogen phosphate ion, as shown above. However, examination of Table 17.2 in your
        text shows H_3PO_4 to be more acidic than NH_4^+, making this process unlikely.

26.   The reaction may be written:

$$H_2S \;+\; NaCH_3COO \rightarrow HS^- \;+\; CH_3COOH \;+\; Na^+$$

        Since H_2S is a weaker acid than CH_3COOH, the reaction does not occur extensively.

28.   a. CO_3^{2-} (aq) $+$ H_2S (aq) $\Leftrightarrow$ HCO_3^- (aq) $+$ HS^- (aq)

            H_2S is a stronger acid than HCO_3^- and CO_3^{2-} a stronger base than HS^-, hence
            the equilibrium lies to the **right**.

        b. HCN (aq) $+$ SO_4^{2-} (aq) $\Leftrightarrow$ CN^- (aq) $+$ HSO_4^- (aq)

            The HSO_4^- ion is a stronger acid than HCN and CN^- a stronger base than SO_4^2,
            so the equilibrium lies to the **left**.

        c. CN^- (aq) $+$ NH_3 (aq) $\Leftrightarrow$ HCN (aq) $+$ NH_2^- (aq)

            Since HCN is a stronger acid than NH_3 and NH_2^- is a stronger base than OH^-, the
            equilibrium lies to the **left**.

        d. SO_4^{2-} (aq) $+$ CH_3COOH (aq) $\Leftrightarrow$ HSO_4^- (aq) $+$ CH_3COO^- (aq)

            As in part (c) the stronger acid (HSO_4^-) and base (CH_3COO^-) are on the right side
            of the equilibrium, hence this equilibrium would lie to the **left**.

30.   a. strongest acid: HF   largest K_a
            weakest acid: HS^- smallest K_a

        b. conjugate base of HF: F^-

    c. acid with weakest conjugate base: HF   conjugate of strongest acid → weakest base.

    d. acid with strongest conjugate base: HS⁻   conjugate of weakest acid → strongest base

32.   a. strongest base: NH_3 (largest K_b)   weakest base: C_5H_5N (smallest K_b)

    b. conjugate acid of C_5H_5N: $C_5H_5NH^+$

    c. base with strongest conjugate acid: C_5H_5N conjugate of weakest base → strongest acid

       base with weakest conjugate acid: NH_3 conjugate of strongest base → weakest acid

34.   The substance which has the smallest value for K_a will have the strongest conjugate base. An examination of Table 17.4 shows that--of these three substances--HClO has the smallest K_a, and ClO⁻ will therefore be the strongest conjugate base.

36.   Since pH = 3.40, $[H_3O^+] = 10^{-pH}$ or $10^{-3.40}$ or 4.0×10^{-4}
    Since the $[H_3O^+]$ is greater than 1×10^{-7} (pH<7), the solution <u>is acidic</u>.

38.   pH of a solution of 0.0013 M HNO_3:
    Since HNO_3 is considered a strong acid, a solution of 0.0013 M HNO_3 has
    $[H_3O^+] = 0.0013$ or 1.3×10^{-3}.

$$pH = -\log[H_3O^+] = -\log[1.3 \times 10^{-3}] = 2.89$$

    The hydroxide ion concentration is readily determined since $[H_3O^+] \cdot [OH^-] = 1.0 \times 10^{-14}$

$$[OH^-] = \frac{1.0 \times 10^{-14}}{1.3 \times 10^{-3}} = 7.7 \times 10^{-12}$$

40.   The pH of a solution of $Ca(OH)_2$ can be calculated by remembering that this base dissolves in water to provide two OH⁻ ions for each formula unit.

       So $[Ca(OH)_2] = 0.0015$ M $\Rightarrow [OH^-] = 0.0030$ M

       pOH = $-\log[OH^-] = -\log[0.0030] = 2.52$ and

       pH $= 14.00 - $ pOH $= 11.48$

42.   Interconversions:

	pH	$[H_3O^+]$ M	$[OH^-]$ M	Solution character
a.	1.00	1.0×10^{-1}	1.0×10^{-13}	acidic
b.	10.50	3.2×10^{-11}	3.2×10^{-4}	basic
c.	4.89	1.3×10^{-5}	7.7×10^{-10}	acidic
d.	9.25	5.6×10^{-10}	1.8×10^{-5}	basic
e.	10.36	4.3×10^{-11}	2.3×10^{-4}	basic

44.   a. A solution of HCl with a concentration equal to 5.5×10^{-4} will have a pH equal to:

$$pH = -\log[H_3O^+] = -\log[5.5 \times 10^{-4}] = 3.30$$

b. At a pH = 3.30 the indicator methyl orange would be a red color.

c. At a pH = 3.30 phenolphthalein would be colorless.

46.   a. With a pH = 3.80 the solution has a $[H_3O^+] = 10^{-3.80}$ or 1.6×10^{-4}

b. Writing the equation for the unknown acid, HA, in water we obtain:

$$HA\ (aq) + H_2O\ (l) \Leftrightarrow H_3O^+\ (aq) + A^-\ (aq)$$

$[H_3O^+] = 1.6 \times 10^{-4}$ implying that $[A^-]$ is also 1.6×10^{-4}. Therefore the equilibrium concentration of acid, HA, is $(2.5 \times 10^{-3} - 1.6 \times 10^{-4})$ or $\approx 2.3 \times 10^{-3}$.

$$K_a = \frac{[H_3O^+][A]}{[HA]} = \frac{(1.6 \times 10^{-4})^2}{2.3 \times 10^{-3}} = 1.1 \times 10^{-5}$$

We would classify this acid as a moderately weak acid.

48.   a. If we write the formula for the acid as HA, then [HA] = 0.015 M initially.
            Since pH = 2.67 $[H_3O^+] = 10^{-2.67}$ or 2.1×10^{-3}

b. Substituting into the equilibrium expression for a weak monoprotic acid we obtain:

$$K_a = \frac{[H_3O^+][A^-]}{[HA]} = \frac{(2.1 \times 10^{-3})^2}{(0.015 - 0.0021)} = 3.6 \times 10^{-4}$$

50.   Calculate the initial concentration of butyric acid:

$$[C_3H_7COOH] = \frac{0.20 \text{ mol}}{0.500 \text{ L}} = 0.40 \text{ M butyric acid}$$

pH = 2.60 and therefore $[H_3O^+] = 10^{-2.60} = 2.5 \times 10^{-3}$ M

So $[H_3O^+]_{eq} = [C_3H_7CO_2^-]_{eq} = 2.5 \times 10^{-3}$ and $[C_3H_7COOH]_{eq} = (0.40 - 0.0025)$

$$K_a = \frac{[C_3H_7CO_2^-][H_3O^+]}{[C_3H_7COOH]} = \frac{(2.5 \times 10^{-3})^2}{(0.40 - 0.0025)} = 1.6 \times 10^{-5}$$

52.   With a pH = 9.11, the solution has a pOH of 4.89 and $[OH^-] = 10^{-4.89} = 1.3 \times 10^{-5}$ M.
At equilibrium, $[H_2NOH] = [H_2NOH] - [OH^-] = (0.025 - 1.3 \times 10^{-5})$ or approximately
0.025 M.

$$K_b = \frac{[H_3NOH^+][OH^-]}{[H_2NOH]} = \frac{(1.3 \times 10^{-5})^2}{0.025} = 6.6 \times 10^{-9}$$

54.   The equilibrium concentrations of the species may be found as follows:

If we assume that some small amount (x) of the acid HA dissociates, the
equilibrium concentration of the acid HA will be $[HA]_i - x$. The amount of H_3O^+ will be
equal to the amount of molecular acid which dissociates, x. This will also be equal to the
concentration of the anion, A^-, present at equilibrium. Substitution into the K_a expression
yields:

$$K_a = \frac{[A^-][H_3O^+]}{[HA]} = \frac{(x)^2}{(0.040 - x)} = 4.0 \times 10^{-9}$$

Since the K_a for the acid is small, assume that the denominator may be simplified to yield:

$$\frac{x^2}{0.040} = 4.0 \times 10^{-9}$$

and $x = 1.3 \times 10^{-5}$ (to two significant figures)

$[H_3O^+] = [A^-] = 1.3 \times 10^{-5}$ M and [HA] = 0.040 M.

56.   Using the same logic as in question 54 above, we can write:

	HCN (aq) + H_2O (l)	$\Leftrightarrow$	H_3O^+ (aq) +	CN^- (aq)
Initial concentration:	0.025		0	0
Change (going to eq.)	-x		+x	+x
Eq. concentrations	0.025 - x		x	x

Substituting these values into the K_a expression for HCN:

$$K_a = \frac{[CN^-][H_3O^+]}{[HCN]} = \frac{(x)^2}{(0.025 - x)} = 4.0 \times 10^{-10}$$

Assuming that the denominator may be approximated as 0.025 M, we obtain:

$$\frac{x^2}{0.025} = 4.0 \times 10^{-10}$$

and $x = 3.2 \times 10^{-6}$. Since x represents $[H_3O^+]$ the pH $= -\log(3.2 \times 10^{-6})$ or 5.50.

58.   The $[C_2H_5COOH]_{initial}$ is: $\dfrac{0.588 \text{ g acid}}{0.250 \text{ L}} \cdot \dfrac{1 \text{ mol acid}}{74.08 \text{ g acid}} = 3.17 \times 10^{-2} \text{ M}$

Substituting into the K_a expression:

$$K_a = \frac{[C_2H_5COO^-][H_3O^+]}{[C_2H_5COOH]} = \frac{(x)^2}{(0.0317 - x)} = 1.4 \times 10^{-5}$$

If we simplify the denominator to be approximately equal to 0.0317, we can write:

$$x^2 = 1.4 \times 10^{-5} \cdot 0.0317 = 4.44 \times 10^{-7} \quad \text{and} \quad x = 6.6 \times 10^{-4}.$$

[Using the quadratic equation to solve the equation exactly gives a value of 6.6×10^{-4}.]

Since x represents $[H_3O^+]$ the pH $= -\log(6.7 \times 10^{-4})$ or 3.18.

60.   The equilibrium expression would be, for both acids:

$$\frac{[H_3O^+][A^-]}{[HA]} = K_a$$

Since $[H_3O^+] = [A^-]$ we can write $\dfrac{[H_3O^+]^2}{[HA]} = K_a$ and $[H_3O^+]^2 = [HA] \cdot K_a$ or

$$[H_3O^+] = \sqrt{[HA] \cdot K_a}$$

For barbituric acid:

$$[H_3O^+] = \sqrt{1 \times 10^{-2} \cdot 9.9 \times 10^{-5}} = 9.9 \times 10^{-4} \qquad \text{and pH} = 3.00$$

For nicotinic acid:

$$[H_3O^+] = \sqrt{1 \times 10^{-2} \cdot 1.4 \times 10^{-5}} = 3.7 \times 10^{-4} \qquad \text{and pH} = 3.43$$

NOTE: In general, the relative pH's of **equimolar solutions of two monoprotic acids** may be estimated by inspecting their respective K_a values. The **smaller the value of K_a**, the lower the concentration of hydrogen ion and **the higher the pH**. Therefore, (a) barbituric acid is the stronger acid (larger K_a), and

(b) nicotinic acid solution has the higher pH.

62.   For this monoprotic acid $K_a = 1.3 \times 10^{-3}$ and $100 \cdot K_a = 1.3 \times 10^{-1}$ and
$[HA]_{initial} = 1.0 \times 10^{-2}$. Since $[HA]_{initial} < 100 \cdot K_a$, we'll need to use the quadratic
equation.

For a monoprotic acid the K_a expression is:

$$K_a = \frac{[H_3O^+][A^-]}{[HA]} = 1.3 \times 10^{-3} \text{ and substitution yields } \frac{x^2}{0.010 - x} = 1.3 \times 10^{-3}$$

or $x^2 + (1.3 \times 10^{-3})x - 1.30 \times 10^{-5} = 0$

Using the quadratic equation you obtain: $x = 3.0 \times 10^{-3} = [H_3O^+]$ and $pH = 2.52$

64.   For the equilibrium we can write: $NH_3 \text{ (aq)} + H_2O \text{ (l)} \Leftrightarrow NH_4^+ \text{ (aq)} + OH^- \text{ (aq)}$

	NH_3	NH_4^+	OH^-
Initial concentration:	0.15	0	0
Change (going to eq.)	-x	+x	+x
Eq. concentrations	0.15 - x	x	x

Substituting these values into the K_b expression for NH_3:

$$K_b = \frac{[OH^-][NH_4^+]}{[NH_3]} = \frac{(x)^2}{(0.15 - x)} = 1.8 \times 10^{-5}$$

$x = 1.6 \times 10^{-3} = [NH_4^+] = [OH^-]$ and $[NH_3] = (0.15 - 0.0016) \approx 0.15$

and $pOH = -\log(1.6 \times 10^{-3}) = 2.78$; $pH = (14.00 - 2.78) = 11.22$

66.   The pH of a 0.12 M solution of aniline ($K_b = 4.2 \times 10^{-10}$) :
Using the logic of question 64, we can write the K_b expression for $C_6H_5NH_2$

$$K_b = \frac{[OH^-][C_6H_5NH_3^+]}{[C_6H_5NH_2]} = \frac{(x)^2}{(0.12 - x)} = 4.2 \times 10^{-10}$$

$x = 7.1 \times 10^{-6} = [C_6H_5NH_3^+] = [OH^-]$
and $[C_6H_5NH_2] = (0.12 - 7.1 \times 10^{-6}) \approx 0.12$

and $pOH = -\log(7.1 \times 10^{-6}) = 5.15$ and $pH = (14.00 - 5.15) = 8.85$

68.   The initial concentration of saccharin is:

$$\frac{1.50 \text{ g saccharin}}{0.500 \text{ L}} \cdot \frac{1 \text{ mol saccharin}}{183.2 \text{ g saccharin}} = 1.64 \times 10^{-2}$$

Substituting into the K_a expression gives:

$$K_a = \frac{[H_3O^+][A^-]}{[HA]} = 2.1 \times 10^{-12} = \frac{(x)^2}{1.64 \times 10^{-2} - x}$$

The magnitude of K_a allows us to simplify the denominator to 1.64×10^{-2}.

Solving for x we obtain: $x = [H_3O^+] = 1.85 \times 10^{-7}$ and pH = 6.73

70. $HF \text{ (aq)} + H_2O \text{ (l)} \Leftrightarrow F^- \text{ (aq)} + H_3O^+ \text{ (aq)}$

If pH = 2.30 then $[H_3O^+]$eq = $10^{-2.30}$ = 5.0×10^{-3} and $[F^-]$eq = 5.0×10^{-3} (from the stoichiometry of the reaction)

$$K_a = \frac{[F^-][H_3O^+]}{[HF]} = \frac{(5.0 \times 10^{-3})^2}{[HF]} = 7.2 \times 10^{-4}$$

and solving for [HF]eq = 0.035 M

Since $[H_3O^+]$ = 0.005, [HF]initial = 0.035 + 0.005 = 0.040 M

72.   Since the salts listed are all soluble sodium salts of the anions, the pH will be affected by the anions present (Na^+ does not hydrolyze). The equation for the hydrolysis is written:

$$A^- + H_2O \Leftrightarrow OH^- + HA$$

The extent to which this occurs is controlled by the K_b of the anion (A^-). The K_b values for the anions (found in Table 17.4) are:

Anion:	S^{2-}	F^-	PO_4^{3-}	CH_3COO^-
K_b:	7.7×10^{-2}	1.4×10^{-11}	2.8×10^{-2}	5.6×10^{-10}

The larger the value of K_b , the farther the equilibrium shown above lies to the right--producing more and more hydroxide ion (higher pH). **S^{2-} gives the highest pH, and F^- gives the lowest.**

74.   pH of aqueous solutions of: pH

    a. $NaHSO_4$ hydrolysis of HSO_4^- occurs, producing H_3O^+ < 7

    b. NH_4Br hydrolysis of NH_4^+ occurs, producing H_3O^+ < 7

    c. $NaNO_3$ no hydrolysis occurs from either cation or anion   = 7

    d. Na_2HPO_4 hydrolysis of HPO_4^{2-} occurs, producing OH^- > 7

76.   pH of aqueous solutions of: pH

    a. NH_4NO_3 hydrolysis of NH_4^+ occurs, producing H_3O^+ < 7

    b. Na_2CO_3 hydrolysis of CO_3^{2-} occurs, producing OH^- > 7

    c. KCH_3COO hydrolysis of CH_3COO^- occurs, producing OH^- > 7

    d. $LiBr$ no hydrolysis occurs   = 7

78.   Hydrolysis of the NH_4^+ produces H_3O^+ according to the equilibrium:

$$NH_4^+ \text{ (aq)} + H_2O \text{ (l)} \Leftrightarrow H_3O^+ \text{ (aq)} + NH_3 \text{ (aq)}$$

With the ammonium ion acting as an acid, to donate a proton, we can write the K_a expression:

$$K_a = \frac{[NH_3][H_3O^+]}{[NH_4^+]} = \frac{K_w}{K_b} = \frac{1.0 \times 10^{-14}}{1.8 \times 10^{-5}} = 5.6 \times 10^{-10}$$

The concentrations of both terms in the numerator are equal, and the concentration of ammonium ion is 0.20 M. (Note the approximation for the **equilibrium** concentration of NH_4^+ to be equal to the **initial** concentration.)

Substituting and rearranging we get

$$[H_3O^+] = \sqrt{0.20 \cdot 5.6 \times 10^{-10}} = 1.1 \times 10^{-5}$$

and the pH = 4.98.

80.   The hydrolysis of CN^- produces OH^- according to the equilibrium:

$$CN^- \text{ (aq)} + H_2O \text{ (l)} \Leftrightarrow HCN \text{ (aq)} + OH^- \text{ (aq)}$$

Calculating the concentrations of Na^+ and CN^- :

$$[Na^+]_i = [CN^-]_i = \frac{10.8 \text{ g NaCN}}{0.500 \text{ L}} \cdot \frac{1 \text{ mol NaCN}}{49.01 \text{ g NaCN}} = 4.41 \times 10^{-1} \text{ M}$$

$$\text{Then } K_b = \frac{K_w}{K_a} = \frac{1.0 \times 10^{-14}}{4.0 \times 10^{-10}} = 2.5 \times 10^{-5} = \frac{[HCN][OH^-]}{[CN^-]}$$

Substituting the [CN⁻] concentration into the K_b expression and noting that:

$$[OH^-]_e = [HCN]_e \quad \text{we may write}$$

$$[OH^-]_e = [(2.5 \times 10^{-5})(4.41 \times 10^{-1})]^{0.5} = 3.3 \times 10^{-3}$$

$$[H_3O^+] = \frac{1.0 \times 10^{-14}}{3.3 \times 10^{-3}} = 3.0 \times 10^{-12}$$

82. $HC_9H_7O_4$ (aq) + H_2O (aq) $\Leftrightarrow$ $C_9H_7O_4^-$ (aq) + H_3O^+ (aq)

$$K_a = \frac{[C_9H_7O_4^-][H_3O^+]}{[HC_9H_7O_4]} = 3.27 \times 10^{-4}$$

The initial concentration of aspirin is:

$$2 \text{ tablets} \cdot \frac{0.325 \text{ g}}{1 \text{ tablet}} \cdot \frac{1 \text{ mol } HC_9H_7O_4}{180.2 \text{ g}HC_9H_7O_4} \cdot \frac{1}{0.200 \text{ L}} = 1.80 \times 10^{-2} \text{ M } HC_9H_7O_4$$

Substituting into the K_a expression:

$$K_a = \frac{[H_3O^+]^2}{1.80 \times 10^{-2}} = 3.27 \times 10^{-4}$$

Since the $100 \cdot K_a \approx$ [aspirin], the quadratic equation is necessary to obtain a "good" value. Using the quadratic equation, $[H_3O^+] = 2.43 \times 10^{-3}$ and pH $= 2.61$.

84.   a. pH of 0.45 M H_2SO_3:

The equilibria for the diprotic acid are:

$$K_{a1} = \frac{[HSO_3^-][H_3O^+]}{[H_2SO_3]} = 1.2 \times 10^{-2} \quad \text{and } K_{a2} = \frac{[SO_3^{2-}][H_3O^+]}{[HSO_3^-]} = 6.2 \times 10^{-8}$$

For the first step of dissociation:

	H_2SO_3	HSO_3^-	H_3O^+
Initial concentration	0.45		
Change	-x	+x	+x
Equilibrium concentration	0.45 - x	+x	+x

$$K_{a1} = \frac{(x)(x)}{(0.45-x)} = 1.2 \times 10^{-2}$$

We must solve this expression with the quadratic equation since $(0.45 < 100 \cdot K_{a1})$.

The equilibrium concentrations for HSO_3^- and H_3O^+ ions are found to be 0.0677 M. The further dissociation is indicated by K_{a2}. Using the equilibrium concentrations obtained in the first step, we substitute into the K_{a2} expression.

	HSO_3^-	SO_3^{2-}	H_3O^+
Initial concentration	0.0677	0	0.0677
Change	-x	+x	+x
Equilibrium concentration	0.0677 - x	+x	0.0677 + x

$$K_{a2} = \frac{[SO_3^{2-}][H_3O^+]}{[HSO_3^-]} = \frac{(+x)(0.0677 + x)}{(0.0677 - x)} = 6.2 \times 10^{-8}$$

We note that x will be small in comparison to 0.0677, and we simplify the expression:

$$K_{a2} = \frac{(+x)(0.0677)}{(0.0677)} = 6.2 \times 10^{-8}$$

In summary the **concentrations of HSO_3^- and H_3O^+ ions have been virtually unaffected** by the second dissociation.

Then $[H_3O^+] = 0.0677$ M and pH = 1.17

b. The equilibrium concentration of SO_3^{2-} :

From the K_{a2} expression above: $[SO_3^{2-}] = 6.2 \times 10^{-8}$

86.   a. Concentrations of OH^-, $N_2H_5^+$, and $N_2H_6^{2+}$ in 0.010 M N_2H_4:

The K_{b1} equilibrium allows us to calculate $N_2H_5^+$ and OH^- formed by the reaction of N_2H_4 with H_2O. $K_{b1} = \frac{[N_2H_5^+][OH^-]}{[N_2H_4]} = 8.5 \times 10^{-7}$

	N_2H_4	$N_2H_5^+$	OH^-
Initial concentration	0.010		
Change	-x	+x	+x
Equilibrium concentration	0.010 - x	+x	+x

Substituting into the K_{b1} expression : $\frac{(x)(x)}{0.010 - x} = 8.5 \times 10^{-7}$

We can simplify the denominator $(0.010 > 100 \cdot K_{b1})$.

$$\frac{(x)(x)}{0.010} = 8.5 \times 10^{-7} \text{ and } x = 9.2 \times 10^{-5} = [N_2H_5^+] = [OH^-]$$

The second equilibrium (K_{b2}) indicates further reaction of the $N_2H_5^+$ ion with water. The step should consume some $N_2H_5^+$ and produce more OH^-. The magnitude of K_{b2} indicates that the equilibrium "lies to the left" and we anticipate that not much $N_2H_6^{2+}$ (or additional OH^-) will be formed by this interaction.

	$N_2H_5^+$	$N_2H_6^{2+}$	OH^-
Initial concentration	9.2×10^{-5}		9.2×10^{-5}
Change	$-x$	$+x$	$+x$
Equilibrium concentration	$9.2 \times 10^{-5} - x$	x	$9.2 \times 10^{-5} + x$

$$K_{b2} = \frac{[N_2H_6^{2+}][OH^-]}{[N_2H_5^+]} = 8.9 \times 10^{-16} = \frac{x \cdot (9.2 \times 10^{-5} + x)}{(9.2 \times 10^{-5} - x)}$$

Simplifying yields $\dfrac{x \cdot (9.2 \times 10^{-5})}{(9.2 \times 10^{-5})} = 8.9 \times 10^{-16}$; $x = 8.9 \times 10^{-16} = [N_2H_6^{2+}]$

In summary we see that the second stage produces a negligible amount of OH^- and consumes very little $N_2H_5^+$ ion. The equilibrium concentrations are:

$[N_2H_5^+]$: 9.2×10^{-5} M; $[N_2H_6^{2+}]$: 8.9×10^{-16} M; $[OH^-]$: 9.2×10^{-5} M

b.   The pH of the 0.010 M solution: $[OH^-] = 9.2 \times 10^{-5}$ M   pOH = 4.04 and pH = 14.0 - 4.04 = 9.96

88.   a. Mn^{2+} electron deficient   Lewis acid
      b. $:NH_2(CH_3)$ electron rich   Lewis base
      c. H_2NOH electron rich   Lewis base
      d. SO_2 donates an electron pair   Lewis base
      e. $Zn(OH)_2$ accepts electron pairs   Lewis acid

90.   **BH_3 is a Lewis acid** in that it seeks to complete the electron octet at B by forming a new bond with the lone pair at the nitrogen atom of NH_3.

92. ICl_3 accepts an electron pair from the chloride ion in forming the ICl_4^- ion. This behavior is that of a **Lewis acid**. The Lewis dot structure for ICl_3 is:

ICl_3 will have a T-shape and

ICl_4^- will be a square planar complex.

(6 electron pairs around the I atom, four of which are bonding pairs to Cl atoms.)

94.   Write the general equation for the formation of a ligand with the Cd ion:

$$Cd^{2+} + x L \Leftrightarrow [CdL_x]^{2+}$$

The equilibrium expression would take the form:

$$K_f = \frac{\{[CdL_x]^{2+}\}}{[Cd^{2+}][L]^x}$$

Note the use of {} to indicate the molar concentrations of the complex ion which is already enclosed in brackets. An inspection of the general equation reveals that the larger the value of K_f (formation or stability constant), the greater the amount of Cd^{2+} ion which is taken up as the complex. Hence the **cyanide ion (CN^-) would achieve the greater lowering of the Cd^{+2} ion concentration.**

96.   The equilibrium may be written: $Cu^{2+}(aq) + 4 NH_3 (aq) \Leftrightarrow [Cu(NH_3)_4]^{2+} (aq)$

If we assume that Cu^{2+} and NH_3 initially react completely to form the complex ion, we have the following initial concentrations:

$[Cu^{2+}]$ = 0

$[NH_3]$ = 3.0 - 4(0.050) = 2.8 M

$\{[Cu(NH_3)_4]^{2+}\}$ = 0.05 M

$$K_f = \frac{\{[Cu(NH_3)_4]^{2+} \}}{[Cu^{2+}][NH_3]^4} = 6.8 \times 10^{12}$$

Assume that x mol/L of the complex dissociates, the equilibrium concentrations resulting can then be substituted into the K_f expression:

$$K_f = \frac{0.05 - x}{(x)(2.8 + 4x)^4} = 6.8 \times 10^{12}$$

The magnitude of K_f for this complex indicates that the amount of complex which dissociates (x) will be **very small.**

So to simplify the mathematics of this last expression we may write:

$$K_f = \frac{0.05}{(x)(2.8)^4} = 6.8 \times 10^{12} \quad \text{and solving for x yields } 1.2 \times 10^{-16}$$

98. The equation for complex formation is: $Ag^+ (aq) + 2\,CN^- (aq) \Leftrightarrow [Ag(CN)_2]^- (aq)$

and the K_f expression is: $K_f = \dfrac{\{[Ag(CN)_2]^-\}}{[Ag^+][CN^-]^2} = 5.6 \times 10^{18}$

or, seen another way the instability constant: $K_i = \dfrac{[Ag^+][CN^-]^2}{\{[Ag(CN)_2]^-\}} = 1.8 \times 10^{-19}$

where the instability constant K_i is the reciprocal of the formation constant, K_f.

Assume that a small amount of the complex (C mol/L) dissociates, forming C mol/L of Ag^+ and 2 C mol/L of CN^-. Substituting into the K_i expression we have:

$$K_i = \frac{[C][2C]^2}{[0.010 - C]} = 1.8 \times 10^{-19}$$

Assuming that $(0.010 - C \approx 0.010)$ we may simplify the expression to:

$$\frac{4C^3}{0.010} = 1.8 \times 10^{-19} \quad \text{and } C = 7.6 \times 10^{-8} \text{ M} = [Ag^+]$$

100. $NaH (s) + H_2O (l) \rightarrow NaOH (aq) + H_2 (g)$

The resulting solution of sodium hydroxide would be basic.

102.

Specie	Behavior	Conjugate Partner
a. Br^-	base	HBr
b. $[Al(H_2O)_6]^{3+}$	acid	$[Al(H_2O)_5(OH)]^{2+}$
c. H_3PO_4	acid	$H_2PO_4^-$
d. CH_3COO^-	base	CH_3COOH

104. Predict whether the equilibrium lies predominantly to the left or to the right:
   a. right   HCl is a stronger acid than H_2CO_3.
   b. right NH_2^- is a stronger base than OH^-.
   c. left   phenoxide is a stronger base than SO_4^{2-}.
   d. left CO_3^{2-} is a stronger base than NH_3.

106.   The equation for the autoionization of water may be written:

$$2\,H_2O(l) \Leftrightarrow H_3O^+(aq) + OH^-\,(aq)$$

If K for this equilibrium at 37 °C is 2.5×10^{-14}, then we can write

$$[H_3O^+][OH^-] = 2.5 \times 10^{-14}$$

and since $[H_3O^+] = [OH^-]$ for a neutral solution then $[H_3O^+]^2 = [OH^-]^2 = 2.5 \times 10^{-14}$

or $[H_3O^+] = [OH^-] = 1.6 \times 10^{-7}$ and pH = 6.80

108.   The concentration of chloroacetic acid is:

$$\frac{0.0945\ g\ ClCH_2COOH}{0.100\ L} \cdot \frac{1\ mol\ ClCH_2COOH}{94.50\ g\ ClCH_2COOH} = 0.0100\ M\ ClCH_2COOH$$

Substituting into the K_a expression yields:

$$K_a = \frac{[H_3O^+][A^-]}{[HA]} = \frac{(x)^2}{(0.0100 - x)} = 1.40 \times 10^{-3}$$

Since $K_a \cdot 100 \approx [HA]$, it will be necessary to use the quadratic equation to solve for x.
The equation is: $x^2 + 1.40 \times 10^{-3}\,x - 1.40 \times 10^{-5} = 0$
and x = $[H_3O^+] = 3.10 \times 10^{-3}$ and pH = 2.508

110.   Of the solutions listed:

a. Acidic

0.1 M CH_3COOH	a weak acid
0.1 M NH_4Cl	a salt of a strong acid and weak base

b. Basic

0.1 M NH_3	a weak base
0.1 M Na_2CO_3	a salt of a strong base and a weak acid
0.1 M $NaCH_3COO$	a salt of a strong base and a weak acid

NaCl   neutral--NaCl is the salt of a strong acid and a strong base.

c. The more acidic:

CH3COOH K_a for acetic acid is greater than the corresponding K_a of the ammonium ion.

112. The solutions may be arranged in order of increasing pH by examining K_a's and K_b's in Table 17.4.

$$HCl < CH_3COOH < NaCl < NH_3 < NaCN < NaOH$$

114.   1 g nicotinic acid $\cdot$ $\dfrac{1 \text{ mol nicotinic acid}}{123.11 \text{ g nicotinic acid}}$ $\cdot$ $\dfrac{1}{0.060 \text{ L}}$ = $\dfrac{0.14 \text{ mol nicotinic acid}}{L}$

The pH of the solution is known to be 2.70 hence:

$$pH = 2.70 \Rightarrow [H_3O^+] = 10^{-2.70} = 2.0 \times 10^{-3} \text{ M}$$

If we represent nicotinic acid as HA, we may write the equilibrium of the acid in water as:

$$HA \text{ (aq)} + H_2O \text{ (l)} \Leftrightarrow H_3O^+ \text{ (aq)} + A^- \text{ (aq)}$$

and $K_a = \dfrac{[H_3O^+][A^-]}{[HA]} = \dfrac{(2.0 \times 10^{-3})(2.0 \times 10^{-3})}{(0.14 - 0.0020)} = 3.0 \times 10^{-5}$

116. As is the case with H_3O^+ in diprotic acids, the amount of OH^- ion furnished in the second equilibrium reaction is--for most purposes--negligible. Hence we may write:

$$K_{b1} = \frac{[OH^-]^2}{0.020} = 7.0 \times 10^{-7} \text{ and}$$

$$[OH^-] = (7.0 \times 10^{-7} \cdot 0.020)^{0.5} = 1.2 \times 10^{-4} \text{ M}$$

or a pOH = 3.93   and pH = 10.07

118.   a. The original concentrations of Ag^+ and $S_2O_3{}^{2-}$:

The "final" solution is prepared by mixing equal volumes of two solutions, hence the original concentrations of the two ions is exactly one-half of the ion concentrations prior to mixing. (Volume is doubled while the number of moles remains constant !)

     $[Ag^+] = 5.0 \times 10^{-4} \text{ M}$ $[S_2O_3{}^{2-}] = 2.5 \text{ M}$

b. Equilibrium concentrations:

Perhaps the easiest way to think about this is to calculate the concentrations of all species as a result of the dissociation of the complex. The maximum amount of complex ion which could be formed is 5.0×10^{-4} M (controlled by the limiting reagent, Ag). The concentration of thiosulfate (at equilibrium) would be [2.5 - 2(5.0 x 10^{-4})]-- closely approximated by the value 2.5. Substituting into the equilibrium expression we find:

$$K_f = \frac{\{[Ag(S_2O_3)_2]^{3-}\}}{[Ag^+][S_2O_3{}^{2-}]^2} = 2.0 \times 10^{13} = \frac{[5.0 \times 10^{-4} - x]}{[x][2.5 + 2x]^2}$$

If we ASSUME that the numerator will be approximated by 5.0×10^{-4}, and that the square term in the denominator will be approximated by 2.5, we can solve for x.

$$x = [Ag^+] = 1.0 \times 10^{-17} \text{ M and } \{[Ag(S_2O_3)_2]^{3-}\} = 5.0 \times 10^{-4} \text{ M}$$

Checking our assumptions, we find that x is indeed small with respect to 5.0×10^{-4} and 2x is negligible with respect to 2.5.

120. As hydrogen atoms are successively replaced by the very electronegative chlorine atoms, an increase in acidity is seen. This is quite understandable if you remember that an increasing "pull" on the electrons in the "carboxylate" end of the molecule--brought about by an **increasing number of chlorine atoms**--will weaken the O-H bond and **increase the acidity of the specie.**

122. a. Dimethyl ether is donating an electron pair to BF_3, and serves as the Lewis base. BF_3 accepts the electron pair, and serves as the Lewis acid.

b. BF_3 is a trigonal planar molecule, so the F-B-F bond angle is 120°. The complex has four groups around the B atom, so a bond angle of approximately 109° is anticipated.

c. $1.00 \text{ g complex} \cdot \dfrac{1 \text{ mol complex}}{113.9 \text{ g complex}} = 8.78 \times 10^{-3} \text{ mol complex}$

The pressure of the complex in the flask is then:

$$P_{complex} = \frac{nRT}{V} = \frac{8.78 \times 10^{-3} \text{ mol} \cdot 0.082057 \frac{L \cdot atm}{K \cdot mol} \cdot 298 \text{ K}}{0.565 \text{ L}} = 0.38 \text{ atm}$$

Now $K_f = \dfrac{P_{complex}}{P_{BF_3} \cdot P_{ether}} = 5.8 \text{ atm}^{-1}$

Now let us assume that x atm of the complex dissociate to produce BF_3 and ether.

Substituting we get $K_f = \dfrac{0.38 - x}{x \cdot x} = 5.8 \text{ atm}^{-1}$

Since $100 \cdot 5.8 > 0.38$ we must use the quadratic equation to obtain a solution.
$0.38 - x = 5.8 \text{ atm}^{-1} \cdot x^2$ or $5.8 x^2 + x - 0.38 = 0$ and the quadratic equation gives 0.18 atm. So the partial pressure of BF_3 = partial pressure of ether = 0.18 atm and the partial pressure of the complex at equilibrium is then $(0.38 - 0.18) = 0.20$ atm
The total pressure in the flask is :

$$P_t = P_{complex} + P_{BF3} + P_{ether}$$
$$= (0.20 + 0.18 + 0.18) = 0.56 \text{ atm.}$$

CHAPTER 18: Reactions Between Acids and Bases

8.   For the reaction of a strong base with a weak acid:

$$K_{net} = \frac{K_a}{K_w}$$

For $C_6H_5CO_2H$ reaction with NaOH: $K_{net} = \dfrac{6.3 \times 10^{-5}}{1.0 \times 10^{-14}}$ or 6.3×10^9

10.   The net reaction is:

$$CH_3COOH \text{ (aq)} + NaOH \text{ (aq)} \rightarrow CH_3COO^- \text{ (aq)} + Na^+ \text{ (aq)} + H_2O \text{ (l)}$$

The addition of 22.0 mL of 0.10 M NaOH (2.2 mmol NaOH) to 22.0 mL of 0.10 M CH_3COOH (2.2 mmol CH_3COOH) produces water and the soluble salt, sodium acetate (2.2 mmol $CH_3COO^- Na^+$). The acetate ion is the anion of a weak acid and reacts with water according to the equation:

$$CH_3COO^- \text{ (aq)} + H_2O \text{ (l)} \rightarrow CH_3COOH \text{ (aq)} + OH^- \text{ (aq)}$$

The equilibrium constant expression may be written:

$$K_b = \frac{[CH_3COOH][OH^-]}{[CH_3COO^-]} = 5.6 \times 10^{-10}$$

The concentration of acetate ion is: $\dfrac{2.2 \text{ mmol}}{(22.0 + 22.0) \text{ ml}} = 0.050 \text{ M}$

Equilibrium concentrations:

	$CH_3CO_2^-$	CH_3COOH	OH^-
Initial concentration	0.050		
Change (going to equilibrium)	-x	+x	+x
Equilibrium concentration	0.050 - x	+x	+x

$$K_b = \frac{[CH_3COOH][OH^-]}{[CH_3COO^-]} = \frac{x^2}{0.050 - x} = 5.6 \times 10^{-10}$$

Simplifying $(100 \cdot K_b \ll 0.050)$ we get $\dfrac{x^2}{0.050} = 5.6 \times 10^{-10}$

$$x = 5.3 \times 10^{-6} = [OH^-]$$

The hydrogen ion concentration is related to the hydroxyl ion concentration by the equation:

$$K_w = [H_3O^+][OH^-] = 1.0 \times 10^{-14}$$

$$[H_3O^+] = \frac{1.0 \times 10^{-14}}{[OH^-]} = \frac{1.0 \times 10^{-14}}{5.3 \times 10^{-6}} = 1.9 \times 10^{-9}$$

$$pH = 8.72$$

The pH is greater than 7, as we expect for a salt of a strong base and weak acid.

12.

	Equivalence point pH	Reacting Species	Reaction Controlling pH
a.	Greater than 7	CH_3COOH/KOH	Hydrolysis of CH_3COO^-
b.	Less than 7	NH_3/HCl	Hydrolysis of NH_4^+
c.	Equal to 7	$HNO_3/NaOH$	No Hydrolysis

14.   a. pH when 20.0 mL of 0.15 M HCl is mixed with 20.0 mL of 0.15 M NH_3

$(0.020 \cdot 0.15) = 3.0 \times 10^{-3}$ mol of HCl react with an equal number of mol of NH_3 to produce 3.0×10^{-3} mol of NH_4^+ in (20.0 + 20.0) 40.0 mL of solution.

$$[NH_4^+] = \frac{3.0 \times 10^{-3}\,mol}{0.040\ L} = 7.5 \times 10^{-2}\ M$$

Substituting into the K_a expression for NH_4^+:

$$K_a = \frac{[H_3O^+][NH_3]}{[NH_4^+]} = \frac{[H_3O^+]^2}{7.5 \times 10^{-2}} = 5.6 \times 10^{-10}$$

and $[H_3O^+] = 6.4 \times 10^{-6}\ M$; pH = 5.19

b. pH when 20.0 mL of 0.15 M HCl is mixed with 30.0 mL of 0.15 M NaOH:

The amount of strong base added exceeds the amount of acid present.
$[(0.030 \cdot 0.15) - (0.020 \cdot 0.15)]$ or 1.5×10^{-3} mol OH^- in excess.
This amount of excess base is present in (20.0 + 30.0) 50.0 mL solution.

$$[OH^-] = \frac{1.5 \times 10^{-3}\ mol}{0.050\ L} = 3.0 \times 10^{-2}\ M \quad \text{and pOH = 1.52 \quad and pH = 12.48.}$$

16. $$\frac{0.515\ g\ C_6H_5OH}{0.100\ L} \cdot \frac{1\ mol\ C_6H_5OH}{94.11\ g\ C_6H_5OH} = 5.47 \times 10^{-2}\ M\ C_6H_5OH$$

At the equivalence point 5.47×10^{-3} mol of NaOH will have been added. Phenol is a monoprotic acid. One mol of phenol reacts with one mol of sodium hydroxide.

The volume of 0.123 NaOH needed to provide this amount of base is:

$$\text{moles} = M \times V$$

$$5.47 \times 10^{-3} \text{ mol NaOH} = 0.123 \text{ mol NaOH} \times V$$

$$\text{or } 44.5 \text{ mL of the NaOH solution.}$$

The total volume would be (100. + 44.5) or 145 mL solution.

Sodium phenoxide is a soluble salt hence the initial concentration of both sodium and phenoxide ions will be equal to:

$$\frac{5.47 \times 10^{-3} \text{ mol}}{0.145 \text{ L}} = 3.77 \times 10^{-2} \text{ M}$$

The phenoxide ion however is the conjugate ion of a weak acid and undergoes hydrolysis.

$$C_6H_5O^- \text{ (aq)} + H_2O \text{ (l)} \Leftrightarrow C_6H_5OH \text{ (aq)} + OH^- \text{ (aq)}$$

	$C_6H_5O^-$	C_6H_5OH	OH^-
Initial	3.77×10^{-2} M		
Change	$-x$	$+x$	$+x$
Equilibrium	3.77×10^{-2} M $- x$	x	x

$$K_b = \frac{[C_6H_5OH][OH^-]}{[C_6H_5O^-]} = \frac{1.0 \times 10^{-14}}{1.3 \times 10^{-10}} = \frac{(x)(x)}{3.77 \times 10^{-2} - x} = 7.7 \times 10^{-5}$$

Since $7.7 \times 10^{-5} \cdot 100 < 3.77 \times 10^{-2}$ we simplify: $\dfrac{(x)(x)}{3.77 \times 10^{-2}} = 7.7 \times 10^{-5}$

$$x = 1.7 \times 10^{-3}$$

At the equivalence point: $[OH^-] = 1.7 \times 10^{-3}$

$$\text{and } [H_3O^+] = \frac{1.0 \times 10^{-14}}{1.7 \times 10^{-3}} = 5.9 \times 10^{-12}$$

$$\text{and pH} = 11.23$$

18.   At the equivalence point the moles of acid = moles of base.

$$(0.03546 \text{ L}) (0.0104 \text{ M HCl}) = 3.69 \times 10^{-4} \text{ mol HCl}$$

If this amount of base were contained in 25.0 mL of solution, **the concentration of NH₃ in the original solution was 0.0148M.**

At the equivalence point NH_4Cl will hydrolyze according to the equation:

$$NH_4^+ \text{ (aq)} + H_2O \text{ (l)} \Leftrightarrow NH_3 \text{ (aq)} + H_3O^+ \text{ (aq)}$$

$$K_a = \frac{[NH_3][H_3O^+]}{[NH_4^+]} = \frac{1.0 \times 10^{-14}}{1.8 \times 10^{-5}} = 5.6 \times 10^{-10}$$

The salt (3.69×10^{-4} mol) is contained in ($25.0 + 35.46$) 60.46 mL. Its concentration will

be $\frac{3.69 \times 10^{-4} \text{ mol}}{0.06046 \text{ L}}$ or 6.10×10^{-3}. Substituting into the K_a expression :

$$\frac{[H_3O^+]^2}{6.10 \times 10^{-3}} = 5.6 \times 10^{-10} \quad \text{and} : [H_3O^+] = 1.9 \times 10^{-6} \text{ and pH} = 5.73.$$

Since $[H_3O^+][OH^-] = 1.0 \times 10^{-14}$ then $[OH^-] = \frac{1.0 \times 10^{-14}}{1.9 \times 10^{-6}} = 5.4 \times 10^{-9}$

and the $[NH_4^+] = 6.1 \times 10^{-3}$ M

20. Examine the equilibria in each case:

a. NH_3 (aq) + H_2O (l) $\Leftrightarrow$ NH_4^+ (aq) + OH^- (aq)

As the NH_4Cl dissolves, ammonium ion are liberated, shifting the position of equilibrium
to the left--reducing OH^-, and decreasing the pH.

b. CH_3COOH + H_2O $\Leftrightarrow$ CH_3COO^- + H_3O^+

As sodium acetate dissolves, the additional acetate ion will shift the position of equilibrium
to the left--reducing H_3O^+, and increasing the pH.

c. $NaOH$ $\rightarrow$ Na^+ + OH^-

$NaOH$ is a strong base, and as such is totally dissociated. Since the added $NaCl$ does not
hydrolyze to any appreciable extent--no change in pH occurs.

22. $[NH_4^+] = \frac{2.20 \text{ g } NH_4Cl}{0.250 \text{ L}} \cdot \frac{1 \text{ mol } NH_4Cl}{53.49 \text{ g } NH_4Cl} = 0.165$ M

The equilibrium involved may be expressed as: NH_3 + H_2O $\Leftrightarrow$ NH_4^+ + OH^-

	NH_3	NH_4^+	OH^-
Initial concentration	0.12 M	0.165	0
Change	-x	+x	+x
Equilibrium concentration	0.12 - x	0.165 + x	x

Substitution into the K_a expression for the equilibrium :

$$K_a = \frac{(0.165 + x)(x)}{(0.12 - x)} = 1.8 \times 10^{-5}$$

If x is small with respect to both 0.12 and 0.165, simplify the expression to:

$$\frac{0.165x}{0.12} = 1.8 \times 10^{-5} \text{ and } x = [OH^-] = 1.3 \times 10^{-5}; \text{ pOH} = 4.88 \text{ and pH} = 9.12.$$

24.   pH of 0.15 M CH_3COOH:

The equilibrium may be written: $CH_3COOH + H_2O \rightarrow CH_3COO^- + H_3O^+$

$$K_a = \frac{[CH_3COO^-][H_3O^+]}{[CH_3COOH]} = 1.8 \times 10^{-5}$$

$$= \frac{[H_3O^+]^2}{(0.15)} = 1.8 \times 10^{-5} \quad \text{and } [H_3O^+] = 1.64 \times 10^{-3} \text{ and pH} = 2.78$$

pH after 83 g of $NaCH_3COO$ is added:

The concentration of sodium acetate is :

$$\frac{83 \text{ g NaCH}_3\text{COO}}{1.50 \text{ L}} \cdot \frac{1 \text{ mol NaCH}_3\text{COO}}{82.03 \text{ g NaCH}_3\text{COO}} = 0.675 \text{ M NaCH}_3\text{COO}$$

Assuming that there is no volume change when the sodium acetate is added, the concentrations can be substituted into the K_a expression:

$$K_a = \frac{(x + 0.675)(H_3O^+)}{0.15 - x} = 1.8 \times 10^{-5}$$

Simplifying: $\dfrac{0.675[H_3O^+]}{0.15} = 1.8 \times 10^{-5}$

$$[H_3O^+] = 4.0 \times 10^{-6} \text{ and pH} = 5.40$$

26.   As in study question 20, examine the equilibria involved:

a. Adding HCl will consume NH_3, lowering the pH.

b. Adding NaOH will consume acetic acid, raising the pH.

28.   Mass of NH_4Cl to be added to 5.0×10^2 mL of 0.10 M NH_3 to give pH = 9.00

1. The equilibrium for the NH_3 solution may be written:

$$NH_3 \text{ (aq)} + H_2O \text{ (l)} \Leftrightarrow NH_4^+ \text{ (aq)} + OH^- \text{ (aq)}$$

and $K_b = \dfrac{[NH_4^+][OH^-]}{[NH_3]} = 1.8 \times 10^{-5}$

With the desired pH (9.00) the pOH = 5.00, and $[OH^-] = 1.0 \times 10^{-5}$.
The $[NH_3]$ will be almost 0.10. (Some forms NH_4^+ and OH^- --but not much).

Substitution into the K_b expression yields: $\dfrac{[NH_4^+] \cdot 1 \times 10^{-5}}{0.10} = 1.8 \times 10^{-5}$

and $[NH_4^+] = 0.18$ M

2. Calculate the mass of NH_4Cl to prepare 5.0×10^2 mL of 0.18 M solution.

$\dfrac{0.18 \text{ mol } NH_4Cl}{1 \text{ L}} \cdot \dfrac{0.500 \text{ L}}{1} \cdot \dfrac{53.49 \text{ g } NH_4Cl}{1 \text{ mol } NH_4Cl} = 4.8 \text{ g } NH_4Cl$

30. The equilibrium of acetic acid and sodium acetate in water is:

$K_a = \dfrac{[H_3O^+][CH_3COO^-]}{[CH_3COOH]} = 1.8 \times 10^{-5}$

with a pH = 4.57, $[H_3O^+] = 2.7 \times 10^{-5}$. With the $[CH_3COOH] = 0.15$ M, we substitute :

$\dfrac{[2.7 \times 10^{-5}][CH_3COO^-]}{[0.15]} = 1.8 \times 10^{-5}$

and $[CH_3COO^-] = 0.10$ M

This concentration in 0.500 L contains $\dfrac{0.10 \text{ M}}{1 \text{ L}} \cdot \dfrac{0.500 \text{ L}}{1} \cdot \dfrac{82.0 \text{ g}}{1 \text{ mol } NaCH_3COO}$ or

4.1 g of sodium acetate (to two significant figures).

32. pH of buffer containing 12.2 g C_6H_5COOH and 7.20 g C_6H_5COONa in 250. mL of solution:

1. Calculate concentrations for the acid and its salt:

$12.2 \text{ g } C_6H_5COOH \cdot \dfrac{1 \text{ mol } C_6H_5COOH}{122.1 \text{ g } C_6H_5COOH} \cdot \dfrac{1}{250 \text{ L}} = 0.400 \text{ M } C_6H_5COOH$

$7.20 \text{ g } C_6H_5COONa \cdot \dfrac{1 \text{ mol } C_6H_5COONa}{144.1 \text{ g} C_6H_5COONa} \cdot \dfrac{1}{250 \text{ L}} = 0.200 \text{ M } C_6H_5COONa$

2. Substitute equilibrium concentrations into the equilibrium constant expression:

$C_6H_5COOH \text{ (aq)} + H_2O(l) \Leftrightarrow C_6H_5COO^- \text{ (aq)} + H_3O^+ \text{ (aq)}$

	C_6H_5COOH	$C_6H_5COO^-$	H_3O^+
Initial concentration	0.400	0.200	
Change (going to equilibrium)	-x	+x	+x
Equilibrium concentration	0.400 - x	0.200 + x	+x

$$K_a = \frac{[C_6H_5COO^-][H_3O^+]}{[C_6H_5COOH]} = \frac{(0.200 + x)(x)}{0.400 - x} = 6.3 \times 10^{-5}$$

Simplifying the equation: $\quad \dfrac{(0.200)(x)}{(0.400)} = 6.3 \times 10^{-5}$

$$x = 1.26 \times 10^{-4} = [H_3O^+] \quad \text{and} \quad pH = 3.90$$

Diluting this solution to 0.500 L will not change the pH. Notice that the concentrations of both acid and salt are changed, leaving the ratio of salt : acid and the pH unchanged.

34. The best combination to provide a buffer solution of pH 9 is (c) NH_3/NH_4NO_3. Combinations of acids and their salts provide buffer systems with pH < 7. HCl and NaCl don't provide a buffer at all.

36. a. The pH of the buffer solution is:

$$K_b = \frac{[NH_4^+][OH^-]}{[NH_3]} = \frac{(0.250)[OH^-]}{(0.500)} = 1.8 \times 10^{-5}$$

and solving for hydroxyl ion yields: $[OH^-] = 3.6 \times 10^{-5}$
$$pOH = 4.44 \quad pH = 9.56$$

b. pH after addition of 0.0100 mol HCl:

The basic component of the buffer (NH_3) will react with the HCl, producing more ammonium ion. The composition of the solution is then:

	NH_3	NH_4Cl
Moles present (before HCl is added)	0.250	0.125
Change (reaction)	- 0.0100	+ 0.0100
Following reaction	- 0.240	+ 0.135

The amounts of NH_3 and NH_4Cl following the reaction with HCl are only slightly different from those amounts prior to reaction. Converting these numbers into molar concentrations and substitution of the concentrations into the K_b expression yields:

$$K_b = \frac{[NH_4^+][OH^-]}{[NH_3]} = \frac{(0.270)[OH^-]}{(0.480)} = 1.8 \times 10^{-5}$$

$[OH^-] = 3.2 \times 10^{-5}$; pOH = 4.49, and the new pH = 9.51.

38.   The equilibrium under consideration may be written:

$$H_2PO_4^- + H_2O \Leftrightarrow HPO_4^{2-} + H_3O^+$$

The NaOH added corresponds to 5.31×10^{-3} M NaOH.

Calculating the concentrations of species present:

	$H_2PO_4^-$	HPO_4^{2-}	H_3O^+
Concentrations(before NaOH):	0.132	0.132	
Change on addition of NaOH:	$- 5.31 \times 10^{-3}$	$+ 5.31 \times 10^{-3}$	$+ 5.31 \times 10^{-3}$
Resulting concentrations:	0.127	0.137	
Change on going to equilibrium	0.127 - x	0.137 + x	x

$$K_a \text{ (for } [H_2PO_4^-]) = \frac{[H_3O^+][HPO_4^{2-}]}{[H_2PO_4^-]} = 6.2 \times 10^{-8}$$

$$= \frac{[x][0.137 + x]}{[0.127 - x]} = 6.2 \times 10^{-8}$$

and $x = [H_3O^+] = 5.7 \times 10^{-8}$ and pH = 7.24

40.   The titration of 0.10 M NaOH with 0.10 M HCl (a strong base vs a strong acid)

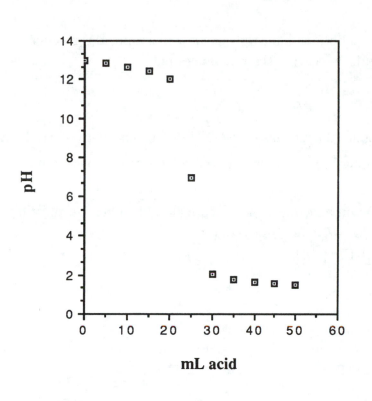

mL acid

The initial pH of a 0.10 M NaOH would be pOH = -log[0.1]
pOH = 1 and pH = 13.

When 12.5 mL of 0.10 M HCl have been added, one-half of the NaOH initially present will be consumed, leaving 0.5 (0.025 L • 0.10 mol/L) or 1.25×10^{-3} mol NaOH in 37.5 mL--therefore a concentration of 0.0333 M NaOH. pOH = 1.48 and pH = 12.52.

At the equivalence point (25.0 mL of the 0.10 M acid are added) there is only NaCl present. Since this salt does not hydrolyze, the pH at that point is exactly 7.0.

Once a total of 50.0 mL of acid are added, there is an excess of 1.25×10^{-3} mol of HCl. Contained in a total volume of 75.0 mL of solution, the [HCl] = 0.0167 and the pH =1.8.

42.   a. pH of 20.0 mL of 0.11 M NH_3:

For the weak base, NH_3, the equilibrium in water is represented as:

$$NH_3 \text{ (aq)} + H_2O \text{ (l)} \Leftrightarrow NH_4^+ \text{ (aq)} + OH^- \text{ (aq)}$$

The slight dissociation of NH_3 would form equimolar amounts of NH_4^+ and OH^- ions.

$$K_b = \frac{[NH_4^+][OH^-]}{[NH_3]} = \frac{(x)(x)}{0.11 - x} = 1.8 \times 10^{-5}$$

Simplifying, we get : $\frac{x^2}{0.11} = 1.8 \times 10^{-5}$ $x = 1.4 \times 10^{-3} = [OH^-]$

pOH = 2.85 and pH = 11.15

Addition of HCl will consume NH_3 and produce NH_4^+ (the conjugate) according to the net equation: NH_3 (aq) + H^+ (l) $\Leftrightarrow$ NH_4^+ (aq)

The strong acid will drive this equilibrium to the right so we will assume this reaction to be complete. Let us first calculate the moles of NH_3 initially present:

$$(0.0200 \text{ L}) (0.11 \frac{\text{mol } NH_3}{\text{L}}) = 0.0022 \text{ mol } NH_3$$

Reaction with the HCl will produce the conjugate acid, NH_4^+. The task is two-fold. First calculate the amounts of the conjugate pair present. Second substitute the concentrations into the K_b expression. [One time-saving hint: **The ratio of concentrations** and the **ratio of the amounts (moles)** will have the same numerical value. One can substitute the amounts of the conjugate pair into the K_b expression.]

$$K_b = \frac{[NH_4^+][OH^-]}{[NH_3]} = 1.8 \times 10^{-5}$$

Present initially: 2.2 millimol NH_3

mL of 0.10 M HCl added	millimol HCl added	millimol NH_3 after reaction	millimol NH_4^+ after reaction	[OH$^-$] after reaction	pH
5.00	0.50	1.7	0.50	6.1×10^{-5}	9.79
11.0	1.1	1.1	1.1	1.8×10^{-5}	9.26
15.0	1.5	0.70	1.5	8.40×10^{-6}	8.92
20.0	2.0	0.20	2.0	1.8×10^{-6}	8.26

b. When 22.0 mL of the 0.10 M HCl has been added (total solution volume = 42.0 mL), the reaction is at the equivalence point. All the NH_3 will be consumed, leaving the salt, NH_4Cl. The NH_4Cl (2.2 millimol) has a concentration of 5.2×10^{-2} M. This salt, being formed from a weak base and strong acid, undergoes hydrolysis.

$$NH_4^+(aq) + H_2O \text{ (l)}   \Leftrightarrow NH_3 \text{ (aq)} + H_3O^+ \text{ (aq)}$$

	NH_4^+	NH_3	H_3O^+
Initial concentration	5.2×10^{-2}		
Change (going to equilibrium)	-x	+x	+x
Equilibrium concentration	5.2×10^{-2} - x	+x	+x

$$K_a = \frac{[NH_3][H_3O^+]}{[NH_4^+]} = 5.6 \times 10^{-10} = \frac{x^2}{5.2 \times 10^{-2} - x}$$

$$\approx \frac{x^2}{5.2 \times 10^{-2}} = 5.6 \times 10^{-10} \quad \text{and} \quad x = 5.4 \times 10^{-6} = [H_3O^+]$$

and pH = 5.27 (equivalence point)

c. The midpoint of the titration occurs when 11.0 mL of the acid have been added. At that point the amount of base and salt present are equal. An examination of the K_b expression will show that under these conditions the $[OH^-] = K_b$. Hence a pOH of 4.74 and a pH = 9.26.

d. From Table 17.3 we see that the best indicator to use is Methyl Red. This indicator would be yellow prior to the equivalence point and red past that point.

e. All points except the one for pH after the addition of 25.0 mL have been listed above. Addition of acid in excess of 22.0 mL will result in a solution which is essentially a strong acid.

After the addition of 25.0 mL substances present are:

millimol HCl added: 2.50
millimol NH_3 present: <u>2.20</u>
excess HCl present :   0.30 millimol

This HCl is present in a total volume of 45.0 mL of solution, hence the calculation for a strong acid proceeds as follows:

$$[H_3O^+] = \frac{0.30 \text{ millimol HCl}}{45.0 \text{ milliliters}} = 6.7 \times 10^{-3} \text{ M} \quad \text{and pH} = 2.18.$$

The graph for these data is:

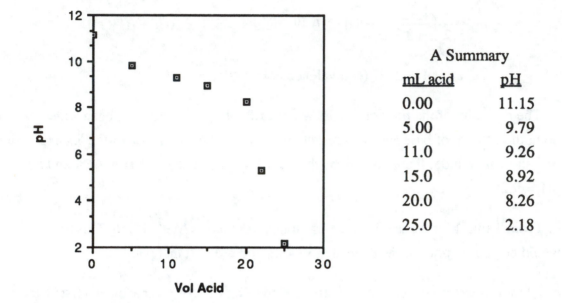

A Summary

mL acid	pH
0.00	11.15
5.00	9.79
11.0	9.26
15.0	8.92
20.0	8.26
25.0	2.18

44.   Suitable indicators for titrations:

a. HCl with pyridine: A solution of pyridinium chloride would have a pH of approximately 3. A suitable indicator would be thymol blue.

b. NaOH with formic acid: The salt formed at the equivalence point is sodium formate. Hydrolysis of the formate ion would give rise to a basic solution (pH ≈ 8.5). Phenolphthalein would be a suitable indicator.

c. Hydrazine and HCl: The salt, hydrazinium hydrochloride, hydrolyzes. The hydrazinium ion would produce an acidic solution (pH ≈ 4.5). A suitable indicator would be bromcresol green.

46.   The addition of NaOH to a solution of acetic acid will (1) consume acetic acid and (2) produce sodium acetate. This solution--providing that there is **both** acetic acid and sodium acetate--is a buffer.

$$CH_3COOH \text{ (before reaction)} \quad 5.19 \times 10^{-2} \text{ moles}$$
$$NaOH \text{ (before reaction):} \quad 2.50 \times 10^{-2} \text{ moles}$$

NaOH reacts on a one-to-one basis with acetic acid. Following the reaction, the amounts of material present are:

$$CH_3COOH \quad (5.19 \times 10^{-2} - 2.50 \times 10^{-2}) \quad = 2.69 \times 10^{-2}$$
$$NaCH_3COO \qquad\qquad\qquad\qquad\qquad = 2.50 \times 10^{-2}$$

Substituting into the K_a expression gives:

$$K_a = \frac{[CH_3COO^-][H_3O^+]}{[CH_3COOH]} = \frac{(2.50 \times 10^{-2})[H_3O^+]}{(2.69 \times 10^{-2})} = 1.8 \times 10^{-5}$$

$[H_3O^+] = 1.9 \times 10^{-5}$ M and pH = 4.71.

48. The addition of NaOH to a solution of HF will (1) consume HF and (2) produce NaF. This solution--providing that there is **both** HF and NaF -- is a buffer.

HF (before reaction) 2.00×10^{-2} moles

NaOH (before reaction): 0.400 g NaOH $\cdot \dfrac{1 \text{ mol NaOH}}{40.00 \text{ g NaOH}} = 0.0100$ mol

NaOH reacts on a one-to-one basis with HF. Following the reaction the amounts of materials present are:

$$CH_3COOH \quad (2.00 \times 10^{-2} - 1.00 \times 10^{-2}) \quad = 1.00 \times 10^{-2} \text{ moles}$$
$$NaF \qquad\qquad\qquad\qquad\qquad\qquad = 1.00 \times 10^{-2} \text{ moles}$$

Substituting into the K_a expression gives:

$$K_a = \frac{[F^-][H_3O^+]}{[HF]} = \frac{(1.00 \times 10^{-2})[H_3O^+]}{(1.00 \times 10^{-2})} = 7.2 \times 10^{-4}$$

$[H_3O^+] = 7.2 \times 10^{-4}$ and pH = 3.14

[Note that you can use the **ratio** of moles in lieu of the molar concentrations in this situation]

50. At the equivalence point, the number of moles of acid and base are equal. 25.0 mL of 0.120 M HCO_2H contain 0.00300 moles. The volume of 0.105 M NaOH which contains

that amount of HCO_2H is $\dfrac{(0.025)(0.120)}{0.105} = 0.0286$ L.

The total volume of solution is $(0.025 \text{ L} + 0.0286) \, 0.0536$ L. The concentration of the salt formed when the NaOH reacts with HCO_2H is then $\dfrac{0.00300 \text{ mol}}{0.0536 \text{ L}}$ or 0.0560 M.

Substituting into the K_b expression for the formate ion gives:

$$K_b = \frac{[HCO_2H][OH^-]}{[HCO_2^-]} = \frac{[OH^-]^2}{0.0560} = \frac{1.0 \times 10^{-14}}{1.8 \times 10^{-4}}$$

$[OH^-] = 1.7 \times 10^{-6}$ and $pH = 8.25$

52.   The equilibrium for the $HNO_2/NaNO_2$ conjugate pair may be written:

$$HNO_2 + H_2O \Leftrightarrow H_3O^+ + NO_2^-$$

and the K_a expression is then: $K_a = \dfrac{[H_3O^+][NO_2^-]}{[HNO_2]} = 4.5 \times 10^{-4}$

The concentrations of the acid and salt are:

$$\frac{0.150 \text{ mol NaNO}_2}{0.500 \text{ L}} = 0.300 \text{ M NaNO}_2 \qquad \frac{0.200 \text{ mol HNO}_2}{0.500 \text{ L}} = 0.400 \text{ M HNO}_2$$

If the pH is to be 4.00 then $[H_3O^+] = 1.00 \times 10^{-4}$. The solution also contains 0.400 mol HNO_2 /L.

The $[NO_2^-]$ necessary is then: $\dfrac{(1.00 \times 10^{-4})[NO_2^-]}{(0.400)} = 4.5 \times 10^{-4}$ and $[NO_2^-] = 1.8$ M

The concentration of NO_2 needed is 1.8 mol/L. The amount initially present is 0.300 mol/L, so we must add 1.5 mol of $NaNO_2$/L.

$$\frac{1.5 \text{ mol NaNO}_2}{L} \cdot \frac{0.500 \text{ L}}{1} \cdot \frac{69.00 \text{ g NaNO}_2}{1 \text{ mol NaNO}_2} = 52 \text{ g NaNO}_2$$

54.   Listed in order of increasing pH the solutions are:

     $HCl < CH_3COOH < CH_3COOH/CH_3COONa < NaCl < NH_3/NH_4Cl < NH_3$

The two acids are of lowest pH with the stronger acid--at equal molarities--having the lower pH. The acetic acid/acetate buffer will be of slightly higher pH, but less than 7. NaCl, being the salt of a strong acid and strong base will have a pH of 7. The ammonia/ammonium chloride buffer will be basic, but the salt will keep the pH lower than that of the remaining base, NH_3.

56.   a. At the equivalence point in an acid/base titration, only salt remains. The titration of a weak acid with a strong base forms a salt. The conjugate anion of the acid hydrolyzes. For an example, let's use the formate anion.

$$HCOO^- \text{ (aq)} + H_2O \text{ (l)} \Leftrightarrow HCOOH \text{ (aq)} + OH^- \text{ (aq)}$$

The position of the equilibrium is determined by the relative strengths of the acids and bases. An acid stronger than HCOOH would shift the equilibrium to the left to a greater extent than HCOOH, reducing the OH^- concentration to a greater extent. The reduced OH^- concentration results in a lower pH. In summary the stronger the acid, the lower the pH of the equivalence point. HB is therefore a stronger acid than HA.

   b. Since HB is a stronger acid than HA, the conjugate base B^- is a weaker base than A^-.

58.   a. The pH of 0.0256 M $C_6H_5NH_2$ is:

$$K_b = \frac{[C_6H_5NH_3^+][OH^-]}{[C_6H_5NH_2]} = 4.2 \times 10^{-10}$$

and since $[C_6H_5NH_3^+] = [OH^-]$ then $\frac{[OH^-]^2}{0.0256} = 4.2 \times 10^{-10}$

and $[OH^-] = 3.3 \times 10^{-6}$ pOH = 5.48 and pH = 8.52

   b. At the equivalence point, moles acid = moles base--and the amount of base present is:

(0.025 L)(0.0256 M $C_6H_5NH_2$) = 6.4×10^{-4} mol base

To furnish that amount of HCl from a solution that is 0.0195 M HCl requires 0.0328 L

At the equivalence point there are $\dfrac{6.4 \times 10^{-4} \text{ mol salt}}{0.0578 \text{ L}}$ = 0.0111 M salt

The pH of this salt solution is controlled by the equilibrium:

$$C_6H_5NH_3^+ \text{ (aq)} + H_2O \text{ (l)} \Leftrightarrow C_6H_5NH_2 \text{ (aq)} + H_3O^+ \text{ (aq)}$$

$$K_a = \frac{[C_6H_5NH_2][H_3O^+]}{[C_6H_5NH_3^+]} = \frac{1.0 \times 10^{-14}}{4.2 \times 10^{-10}} = \frac{[H_3O^+]^2}{0.0111} = 2.38 \times 10^{-5}$$

$[H_3O^+] = 5.1 \times 10^{-4}$ and pH = 3.29

   c. At the midpoint of the titration, half the base will be consumed--producing an equal number of moles of salt.

moles of base consumed = 1/2(6.4×10^{-4} mol) = 3.2×10^{-4} mol

moles of salt produced = 3.2×10^{-4} mol

Substituting into the K_b expression as in part (a) gives:

$$\frac{[3.2 \times 10^{-4}][OH^-]}{(3.2 \times 10^{-4})} = 4.2 \times 10^{-10} \qquad pOH = 9.38$$

$$\text{and } [H_3O^+] = 2.38 \times 10^{-5} \qquad pH = 4.62$$

Note that in the K_b expression, you can substitute the **number of moles of each specie** for the **concentration of each specie** since the moles and their concentration will be in the same proportion.

d. With an equivalence point of 3.29, this titration could be followed with bromphenol blue, which would be yellow at a pH below the equivalence point and blue above the equivalence point.

e. The addition of acid results in (1) consumption of aniline and (2) production of the salt, anilinium chloride. The solution containing both the weak base and its conjugate salt is a buffer. The pH can be calculated:

$$K_b = \frac{[C_6H_5NH_3^+][OH^-]}{[C_6H_5NH_2]} = 4.2 \times 10^{-10}$$

rearranging : $[OH^-] = 4.2 \times 10^{-10} \cdot \dfrac{[C_6H_5NH_2]}{[C_6H_5NH_3^+]}$

Note that as in part (c) the mathematical expression has the **ratio** of base to acid--allowing you to substitute the number of moles of base and salt instead of their actual concentrations.

	vol acid	moles acid added	moles aniline	moles salt produced
1.	5.00	9.75×10^{-5}	5.43×10^{-4}	9.75×10^{-5}
2.	10.0	1.95×10^{-4}	4.45×10^{-4}	1.95×10^{-4}
3.	15.0	2.93×10^{-4}	3.47×10^{-4}	2.93×10^{-4}
4.	20.0	3.90×10^{-4}	2.50×10^{-4}	3.90×10^{-4}
5.	24.0	4.68×10^{-4}	1.72×10^{-4}	4.68×10^{-4}
6.	30.0	5.85×10^{-4}	5.50×10^{-5}	5.85×10^{-4}

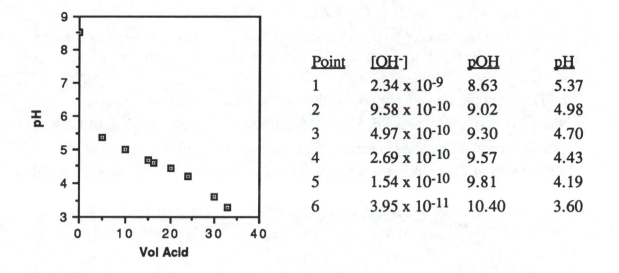

Point	$[OH^-]$	pOH	pH
1	2.34×10^{-9}	8.63	5.37
2	9.58×10^{-10}	9.02	4.98
3	4.97×10^{-10}	9.30	4.70
4	2.69×10^{-10}	9.57	4.43
5	1.54×10^{-10}	9.81	4.19
6	3.95×10^{-11}	10.40	3.60

60. The ratio of the two buffer components can be calculated from the equilibrium expression:

$$K_a = \frac{[H_3O^+][HPO_4^{2-}]}{[H_2PO_4^-]} = 6.2 \times 10^{-8}$$

With pH = 7.40, $[H_3O^+] = 4.0 \times 10^{-8}$. Substituting this value into the K_a expression and rearranging gives:

$$\frac{[HPO_4^{2-}]}{[H_2PO_4^-]} = \frac{6.2 \times 10^{-8}}{4.0 \times 10^{-8}} = 1.6$$

The requested ratio, however, is the reciprocal of $\frac{[HPO_4^{2-}]}{[H_2PO_4^-]}$. Taking the reciprocal gives a ratio of 0.65.

62. According to the equation on p 751 of the text :

$$[H_3O^+] = \left(\frac{K_w \cdot K_a}{K_b}\right)^{0.5}$$

$$[H_3O^+] = \left(\frac{(1.0 \times 10^{-14})(6.3 \times 10^{-5})}{1.8 \times 10^{-5}}\right)^{0.5} = 1.9 \times 10^{-7} \, M$$

and pH = 6.73

64.   a. pH of 0.010 M solution of picric acid (HA)

$K_a = 5.1 \times 10^{-1}$ and [HA] = 0.010 M

The magnitude of K_a is such that we can treat picric acid as a strong acid. Or

[HA] = 0.010 M $\Rightarrow$ [H_3O^+] = 0.010 M and pH = 2.00.

b. The pH at the equivalence point is that of a solution of sodium picrate.

Since [NaOH] = [picric acid], 50.0 mL of the acid will require 50.0 mL of the NaOH to exactly neutralize it. The salt concentration will be 1/2(0.010 M) or 0.0050 M.

The hydrolysis of the picrate anion (A^-) may be written: $A^- + H_2O \Leftrightarrow HA + OH^-$

$$K_b = \frac{[HA][OH^-]}{[A^-]} = \frac{[OH^-]^2}{5.0 \times 10^{-3}} = \frac{1.0 \times 10^{-14}}{5.1 \times 10^{-1}} \begin{array}{l} \leftarrow K_w \\ \leftarrow K_a \end{array}$$

and [OH^-] = 9.90×10^{-9} ; pOH = 8.00 and pH = 6.00

c. The pH at the mid-point of the titration:

At the mid-point of a titration of this acid with the NaOH, one-half of the acid has been consumed. This will be 1/2(5.0×10^{-4} mol) contained in (50.0 + 25.0) 75.0 mL. The concentration of remaining acid is then $\frac{2.5 \times 10^{-4} \text{ mol}}{0.075 \text{ L}}$ or 3.33×10^{-3} M. Picric acid is a strong acid with a concentration of 3.33×10^{-3} M.

The [H_3O^+] = [picric acid] = 3.3×10^{-3} M so pH = 2.48.

66.   Since the pH = 9.50 at the midpoint of the titration, we know that pOH = 4.50,

[OH^-] = 3.16×10^{-5} , and K_b for ethanolamine = 3.2×10^{-5}.

The amount of base in the solution is: (0.0250 L • 1.0×10^{-2}) = 2.50×10^{-4} mol

The amount of 0.0095 M HCl which contains that amount is: 26.3 mL

The concentration of salt at this point is : $\frac{2.50 \times 10^{-4} \text{ mol}}{(0.0250 \text{ L} + 0.0263 \text{ L})}$ = 4.87×10^{-3} M

Substitution into the K_a expression for the salt yields:

$$K_a = \frac{[HOCH_2CH_2NH_2][H_3O^+]}{[HOCH_2CH_2NH_3^+]} = \frac{[H_3O^+]^2}{4.87 \times 10^{-3}} = \frac{1.0 \times 10^{-14}}{3.16 \times 10^{-5}} \begin{array}{l} \leftarrow K_w \\ \leftarrow K_b \end{array}$$

So [H_3O^+] = 1.2×10^{-6} and pH = 5.91

67.   a. Approximate bond angles for:

   (a) C=C-C in the ring : $120°$ for sp^2 hybridized C
   (b) C-C=O: $120°$ for sp^2 hybridized C
   (c) C-O-H: $109°$ for sp^3 hybridized C
   (d) C-C-H: $120°$ for sp^2 hybridized C

   b. Number of π bonds : 4   c. C atoms in the ring are sp^2 hybridized
      Number of σ bonds : 16   C atom in -COOH group: sp^2 hybridized

   d. pH of the solution containing 1.00 g of salicylic acid in 460 mL of water

   The concentration of the salicylic acid is:

   $$\frac{1.00 \text{ g salicylic acid}}{0.46 \text{ L}} \cdot \frac{1 \text{ mol salicylic acid}}{138.1 \text{ g salicylic acid}} = 1.6 \times 10^{-2} \text{ mol}$$

   $$K_a = \frac{[A^-][H_3O^+]}{[HA]} = \frac{[x][x]}{1.6 \times 10^{-2} - x} = 1.1 \times 10^{-3}$$

   Since $100 \cdot K_a \approx [HA]$, the quadratic equation must be used to find an exact solution.

   $$x^2 = 1.73 \times 10^{-5} - 1.1 \times 10^{-3}(x)$$
   $$x^2 + 1.1 \times 10^{-3}(x) - 1.73 \times 10^{-5} = 0$$
   $$x = [H_3O^+] = 3.65 \times 10^{-3} \text{ and the pH} = 2.44$$

   e. The percentage of the acid in the form of salicylate ion :

   From the equilibrium expression:
   $$\frac{[H_3O^+][salicylate]}{[acid]} = 1.1 \times 10^{-3}$$

   At a pH = 2.0, $[H_3O^+] = 1 \times 10^{-2}$

   Substituting into the equilibrium expression:

   $$\frac{[1 \times 10^{-2}][salicylate]}{[acid]} = 1.1 \times 10^{-3} \text{ or}$$

   $$\frac{[salicylate]}{[acid]} = \frac{1.1 \times 10^{-3}}{1 \times 10^{-2}} = 0.11$$

   or 11 % of the acid as the salicylate ion.

f. The pH at the midpoint of the titration :

At the midpoint of the titration, the concentration of the acid and its conjugate ion are equal. Therefore $[H_3O^+] = K_a = 1.1 \times 10^{-3}$ and pH $= 2.96$

The pH at the equivalence point can be calculated after the concentration of the sodium salicylate is calculated.

$$0.0250 \text{ mL} \cdot 0.014 \text{ M salicylic acid} = 3.5 \times 10^{-4} \text{ moles}$$

The volume of 0.010 M NaOH needed to neutralize this amount is 35.0 mL.

The concentration of the salt is: $\dfrac{3.5 \times 10^{-4}}{(0.0250 + 0.0350)} = 5.8 \times 10^{-3} \text{ M}$

Substituting into the K_b expression for the salicylate ion:

$$K_b = \frac{[\text{salicylic acid}][OH^-]}{[\text{salicylate}]} = \frac{[OH^-]^2}{5.83 \times 10^{-3}} = \frac{1.0 \times 10^{-14}}{1.1 \times 10^{-3}} \begin{matrix} \leftarrow K_w \\ \leftarrow K_a \end{matrix}$$

and $[OH^-] = 2.3 \times 10^{-7}$; pOH $= 6.64$ and pH $= 7.36$

9.   Two insoluble salts of:

        a. Cl^- $AgCl$ and Hg_2Cl_2

        b. S^{2-} CuS and FeS

        c. Zn^{2+} ZnS and $Zn_3(PO_4)_2$

        d. Fe^{2+} FeS and $Fe(OH)_2$

11.   Using the table of solubility guidelines, predict water solubility for the following:

        a. $(NH_4)_2S$ Ammonium salts are **soluble**.

        b. $ZnCO_3$ Carbonates are generally **insoluble**.

        c. $Mg(OH)_2$ Hydroxides are generally **insoluble**.

        d. FeS Sulfides are generally **insoluble**.

        e. $BaSO_4$ Sr^{2+}, Ba^{2+}, and Pb^{2+} form **insoluble** sulfates.

        f. Hg_2Cl_2 <u>Most chlorides</u> are <u>soluble</u>. Ag^+, Pb^{2+}, Hg_2^{2+}

                              chlorides are **insoluble**.

13.   When Na_2CO_3 is added to an aqueous solution of $AgNO_3$, the reaction is:

$$2\,Ag^+\,(aq)\ +\ 2\,NO_3^-\,(aq)\ +\ 2\,Na^+\,(aq)\ +\ CO_3^{2-}\,(aq) \rightarrow$$
$$Ag_2CO_3\,(s)\ +\ 2\,Na^+\,(aq)\ +\ 2\,NO_3^-\,(aq)$$

Ag_2CO_3 precipitates.

15.   Precipitate expected:

        a. $Ag^+\,(aq)\ +\ Cl^-\,(aq)\ \rightarrow\ AgCl\,(s)$

        b. $Mg^{2+}\,(aq)\ +\ 2\,OH^-\,(aq)\ \rightarrow\ Mg(OH)_2\,(s)$

        c. No precipitate is expected since combinations of the ions give only soluble salts.

        d. $Pb^{2+}\,(aq)\ +\ 2\,Cl^-\,(aq)\ \rightarrow\ PbCl_2\,(s)$

17.   a. $ZnS\,(s)\ \Leftrightarrow\ Zn^{2+}\,(aq)\ +\ S^{2-}\,(aq)$; $K_{sp} = [Zn^{2+}][S^{2-}]$ $= 1.1 \times 10^{-21}$

      b. $SnI_2\,(s)\ \Leftrightarrow\ Sn^{2+}\,(aq) + 2\,I^-\,(aq)$; $K_{sp} = [Sn^{2+}][\,I^-]^2$ $= 1.0 \times 10^{-4}$

      c. $NiCO_3\,(s)\ \Leftrightarrow\ Ni^{2+}\,(aq) + CO_3^{2-}\,(aq)$; $K_{sp} = [Ni^{2+}][CO_3^{2-}] = 6.6 \times 10^{-9}$

      d. $Ag_2SO_4\,(s)\ \Leftrightarrow\ 2\,Ag^+\,(aq) + SO_4^{2-}\,(aq)$; $K_{sp} = [Ag^+]^2[SO_4^{2-}] = 1.7 \times 10^{-5}$

19.   If $K_{sp} = [Mg^{2+}]^3[AsO_4^{3-}]^2$, the formula for magnesium arsenate is $Mg_3(AsO_4)_2$.

21.   The equilibrium for the solid/ion equilibria is:

$$CoS\ (s) \Leftrightarrow Co^{2+}\ (aq) + S^{2-}\ (aq) \qquad K_{sp} = [Co^{2+}][S^{2-}].$$

As the solid CoS is added to the water, it dissolves--liberating Co^{2+} and S^{2-} ions. The concentration of Co^{2+} (and also S^{2-}) is reported as 7.7×10^{-11} M.
The K_{sp} for CoS is:

$$K_{sp} = [Co^{2+}][S^{2-}] = (7.7 \times 10^{-11})(7.7 \times 10^{-11}) = 5.9 \times 10^{-21}$$

23.   The K_{sp} expression for silver acetate would be: $K_{sp} = [Ag^+][CH_3COO^-]$

The molar concentration corresponding to 1.0 g in 100.0 mL is:

$$\frac{10.\ g\ AgCH_3COO}{1.00\ L} \cdot \frac{1\ mol\ AgCH_3COO}{166.9\ g\ AgCH_3COO} = 6.0 \times 10^{-2}\ M$$

and the $K_{sp} = [Ag^+][CH_3COO^-] = (6.0 \times 10^{-2})^2 = 3.6 \times 10^{-3}$

25.   The K_{sp} expression for BaF_2 is : $K_{sp} = [Ba^{2+}][F^-]^2$. The stoichiometry for the solid indicates that, as one formula unit of the solid dissolves, two fluoride ions and one barium ion are formed. In the saturated solution $[Ba^{2+}] = 0.5\ [F^-]$. We can write the K_{sp} expression: $K_{sp} = [0.5\ F^-][F^-]^2$ or $0.5 \cdot [F^-]^3$.
Substituting the concentration of fluoride ion:

$$K_{sp} = 0.5 \cdot (1.50 \times 10^{-2})^3 = 1.69 \times 10^{-6}$$

27. 7.4×10^{-3} g of $Zn(CN)_2$ in 0.50 L corresponds to:

$$\frac{7.4 \times 10^{-3}\ g\ Zn(CN)_2}{0.50\ L} \cdot \frac{1\ mol\ Zn(CN)_2}{117\ g\ Zn(CN)_2} = 1.3 \times 10^{-4}\ M$$

The concentrations--in the saturated solution--would be:

$$[Zn^{2+}] = 1.3 \times 10^{-4}\ M \quad and \quad [CN^-] = 2 \cdot [Zn^{2+}]$$

The K_{sp} expression is:

$$K_{sp} = [Zn^{2+}][CN^-]^2 = (Zn^{2+})(2 \cdot [Zn^{2+}])^2\ or\ 4 \cdot [Zn^{2+}]^3$$
$$K_{sp} = 4 \cdot [1.3 \times 10^{-4}]^3 = 8.1 \times 10^{-12}$$

29.   With pH = 12.40, pOH = 1.60 and $[OH^-]$ = 2.51 x 10^{-2} M

The stoichiometry of $Ca(OH)_2$ indicates that $[Ca^{2+}]$ is one-half $[OH^-]$, so

$[Ca^{2+}]$ = 1.26 x 10^{-2} M; K_{sp} = $[Ca^{2+}][OH^-]^2$ = (1.26 x 10^{-2})(2.51 x 10^{-2})^2 = 7.9 x 10^{-6}

31.   a. The K_{sp} for AgCN is 1.2 x 10^{-16}. The equation for AgCN dissolving is:

$$AgCN\ (s)\ \Leftrightarrow\ Ag^+\ (aq)\ +\ CN^-\ (aq)$$

From the equation we see that $[Ag^+]$ = $[CN^-]$.

The K_{sp} expression for AgCN is: K_{sp} = $[Ag^+][CN^-]$ = 1.2 x 10^{-16}

$$K_{sp} = [Ag^+]^2 = [CN^-]^2 = 1.2\ x\ 10^{-16}$$

$$[Ag^+] = [CN^-] = 1.1\ x\ 10^{-8}\ M$$

These concentrations tell us that 1.1 x 10^{-8} mol/L of AgCN dissolve.

b. Solubility of AgCN in g/L:

$$1.1\ x\ 10^{-8}\ \frac{mol\ AgCN}{L}\ \cdot\ \frac{134\ g\ AgCN}{1\ mol\ AgCN}\ =\ 1.5\ x\ 10^{-6}\ \frac{g\ AgCN}{L}$$

33.   Milligrams of gold(III)/L in saturated gold(III) iodide solution:

The equilibrium equation is: $AuI_3\ (s)\ \Leftrightarrow\ Au^{3+}\ (aq) + 3\ I^-\ (aq)$

and the K_{sp} expression is: $K_{sp} = [Au^{3+}][I^-]^3 = 1.0\ x\ 10^{-46}$. The equation shows that for each mol of the solid that dissolves/L <u>one</u> mol of Au^{3+} and <u>three</u> mol of I^- are formed. If we let x = mol/L of AuI_3 that dissolve, then $[Au^{3+}]$ = x and $[I^-]$ = 3x. Substitution into the K_{sp} expression yields:

$$K_{sp} = (x)(3x)^3 = 1.0\ x\ 10^{-46}$$
$$27\ x^4 = 1.0\ x\ 10^{-46}$$
$$x = 1.4\ x\ 10^{-12} = [Au^{3+}]\ \ and$$

$$1.4\ x\ 10^{-12}\ \frac{mol\ Au^{3+}}{L}\ \cdot\ \frac{197\ g\ Au^{3+}}{1\ mol\ Au^{3+}}\ \cdot\ \frac{1000\ mg\ Au^{3+}}{1.0\ g\ Au^{3+}}\ =\ 2.7\ x\ 10^{-7}\frac{mg\ Au^{3+}}{L}$$

35.   Milligrams of Ag^+ in 1.00 x 10^2 mL of saturated Ag_2CO_3 solution.

The equilibrium equation is: $Ag_2CO_3\ (s) \Leftrightarrow 2\ Ag^+\ (aq) + CO_3^{2-}\ (aq)$

if x mol/L of Ag_2CO_3 dissolve, $[Ag^+]$ = 2x and $[CO_3^{2-}]$ = x

$$K_{sp} = [Ag^+]^2[CO_3^{2-}] = (2x)^2(x) = 8.1\ x\ 10^{-12}$$

$$4x^3 = 8.1 \times 10^{-12} \quad \text{and} \quad x = 1.27 \times 10^{-4}$$

$$[Ag^+] = 2x = 2.54 \times 10^{-4} \frac{\text{mol } Ag^+}{L}$$

and $2.54 \times 10^{-4} \dfrac{\text{mol } Ag^+}{L} \cdot \dfrac{107.868 \text{ g } Ag^+}{1 \text{ mol } Ag^+} \cdot \dfrac{1000 \text{ mg } Ag^+}{1.0 \text{ g } Ag^+} = 27 \text{ mg } Ag^+$

and the amount in 1.00×10^2 mL is: $\dfrac{27 \text{ mg } Ag^+}{1 \text{ L}} \cdot \dfrac{0.1 \text{ L}}{1} = 2.7 \text{ mg } Ag^+$

37. The more soluble of the substances would--in each case--have the greater K_{sp}.

a. AgSCN   b. $SrSO_4$ c. MgF_2 d. ZnS

39. Compounds in order of increasing solubility in H_2O:

Compound	K_{sp}
BaF_2	1.7×10^{-6}
$BaCO_3$	8.1×10^{-9}
Ag_2CO_3	8.1×10^{-12}

To determine the relative solubilities find the molar solubilities. The equation representing BaF_2 dissolving in H_2O is: $BaF_2(s) \Leftrightarrow Ba^{2+}$ (aq) $+ 2 F^-$ (aq)

For each mol of BaF_2 dissolving, one mol of Ba^{2+} and two mol of F^- are produced. If x mol/L of BaF_2 dissolve, the equilibrium concentrations of Ba^{2+} and F^- are x and 2x respectively.

$$K_{sp} = [Ba^{2+}][F^-]^2 = (x)(2x)^2 = 1.7 \times 10^{-6}$$
$$4x^3 = 1.7 \times 10^{-6}$$
$$x = 7.5 \times 10^{-3}$$

The molar solubility of BaF_2 is then 7.5×10^{-3} M.

Similarly for $BaCO_3$:

$$K_{sp} = [Ba^{2+}][CO_3^{2-}] = (x)(x) = 8.1 \times 10^{-9} \quad \text{and} \quad x = 9.0 \times 10^{-5}$$

The molar solubility of $BaCO_3$ is then 9.0×10^{-5} M.

and for Ag_2CO_3:

$$K_{sp} = [Ag^+]^2[CO_3^{2-}] = (2x)^2(x) = 8.1 \times 10^{-12}$$
$$4x^3 = 8.1 \times 10^{-12}$$
$$\text{and} \quad x = 1.3 \times 10^{-4}$$

The molar solubility of Ag_2CO_3 is 1.3×10^{-4}.

In order of increasing solubility: $BaCO_3 < Ag_2CO_3 < BaF_2$

41.   a. The concentration corresponding to 5.0×10^{-3} g AgCl in 1.0 L of water:

$$\frac{5.0 \times 10^{-3} \text{ g AgCl}}{1} \cdot \frac{1 \text{ mol AgCl}}{143 \text{ g AgCl}} = 3.5 \times 10^{-5} \text{ M}$$

Q for this solution would be $[Ag^+][Cl^-] = (3.5 \times 10^{-5})^2 = 1.2 \times 10^{-9}$

Since $Q > K_{sp}$ for AgCl (1.8×10^{-10}), not all the solid dissolves.

b. The concentration corresponding to 5.0×10^{-3} g $NiCO_3$ in 1.0 L of water:

$$\frac{5.0 \times 10^{-3} \text{ g NiCO}_3}{1} \cdot \frac{1 \text{ mol NiCO}_3}{119 \text{ g NiCO}_3} = 4.2 \times 10^{-5} \text{ M}$$

Q for this solution would be $[Ni^{2+}][CO_3^{2-}] = (4.2 \times 10^{-5})^2 = 1.8 \times 10^{-9}$

Since $Q < K_{sp}$ for $NiCO_3$ (6.6×10^{-9}), all the solid dissolves.

43.   Precipitation of a substance will occur if the reaction quotient exceeds the K_{sp} for the substance.

For $NiCO_3$: $K_{sp} = [Ni^{2+}][CO_3^{2-}] = 6.6 \times 10^{-9}$

a. If $[Ni^{2+}] = 2.4 \times 10^{-3}$ M and $[CO_3^{2-}] = 1.0 \times 10^{-6}$ M

The reaction quotient is $[Ni^{2+}][CO_3^{2-}] = (2.4 \times 10^{-3})(1.0 \times 10^{-6}) = 2.4 \times 10^{-9}$

and precipitation does not occur.

b. If $[Ni^{2+}] = 2.4 \times 10^{-3}$ M and $[CO_3^{2-}] = 1.0 \times 10^{-4}$ M

The reaction quotient is $[Ni^{2+}][CO_3^{2-}] = (2.4 \times 10^{-3})(1.0 \times 10^{-4}) = 2.4 \times 10^{-7}$

and precipitation occurs.

45.   If $Zn(OH)_2$ is to precipitate, the reaction quotient (Q) must exceed the K_{sp} for the salt.

4.0 mg of NaOH in 10. mL corresponds to a concentration of:

$$[OH^-] = \frac{4.0 \times 10^{-3} \text{ g NaOH}}{0.0100 \text{ L}} \cdot \frac{1 \text{ mol NaOH}}{40.0 \text{ g NaOH}} = 0.01 \text{ M}$$

The value of Q is :

$$[Zn^{2+}][OH^-]^2 = (1.6 \times 10^{-4})(1.0 \times 10^{-2})^2 = 1.6 \times 10^{-8}$$

The value of Q is greater than the K_{sp} for the salt (4.5×10^{-17}) , so $Zn(OH)_2$ precipitates.

261

47.   1350 mg Mg^{2+}/L corresponds to a concentration of :

$$[Mg^{2+}] = \frac{1.350 \text{ g } Mg^{2+}}{1 \text{ L}} \cdot \frac{1 \text{ mol } Mg^{2+}}{24.305 \text{ g } Mg^{2+}} = 5.55 \times 10^{-2} \text{ M}$$

For $Mg(OH)_2$ to begin precipitation, Q must exceed K_{sp} for $Mg(OH)_2$ (1.5×10^{-11}).

$$Q = (5.55 \times 10^{-2})(OH^-)^2 = 1.5 \times 10^{-11} \text{ and } [OH^-] = 1.6 \times 10^{-5}$$

49. K_{sp} for CaF_2: $K_{sp} = [Ca^{2+}][F^-]^2 = 3.9 \times 10^{-11}$

If $[Ca^{2+}] = 2.1 \times 10^{-3}$ M, the $[F^-]$ which will form a saturated solution is:

$$[Ca^{2+}][F^-]^2 = 3.9 \times 10^{-11}$$

$$[F^-]^2 = \frac{3.9 \times 10^{-11}}{2.1 \times 10^{-3}} = 1.9 \times 10^{-8}$$

$$[F^-] = 1.4 \times 10^{-4} \text{ M}$$

The mass of KF to produce 500. mL of a 1.4×10^{-4} M F^- is:

$$\frac{1.4 \times 10^{-4} \text{ mol } F^-}{1 \text{ L}} \cdot \frac{1 \text{ mol KF}}{1 \text{ mol } F^-} \cdot \frac{58.1 \text{ g KF}}{1 \text{ mol KF}} \cdot \frac{1.0 \times 10^3 \text{ mg KF}}{1.0 \text{ g KF}} \cdot \frac{0.500 \text{ L}}{1}$$

$$= 4.0 \text{ mg KF}$$

The addition of a mass of KF greater than 4.0 mg would initiate precipitation.

51.   Calculate the concentration of barium ion and sulfate ion after mixing--but before reaction.

$$0.0012 \text{ M } Ba^{2+} \cdot \frac{50. \text{ mL}}{(50. + 25 \text{ mL})} = 8.0 \times 10^{-4} \text{ M } Ba^{2+}$$

$$1.0 \times 10^{-6} \text{ M } SO_4^{2-} \cdot \frac{25 \text{ mL}}{(50. + 25 \text{ mL})} = 3.3 \times 10^{-7} \text{ M } SO_4^{2-}$$

$$Q = [Ba^{2+}][SO_4^{2-}] = (8.0 \times 10^{-4})(3.3 \times 10^{-7}) = 2.7 \times 10^{-10}$$

and since $Q > K_{sp}$ (1.1×10^{-10}) $BaSO_4$ precipitates.

53.   Calculate the concentration of lead ion and chloride ion after mixing--but before reaction.

$$0.0010 \text{ M Pb}^{2+} \cdot \frac{10.\text{ mL}}{(10. + 5.0 \text{ mL})} = 6.7 \times 10^{-4} \text{ M Pb}^{2+}$$

$$0.015 \text{ M Cl}^- \cdot \frac{5.0 \text{ mL}}{(10. + 5.0 \text{ mL})} = 5.0 \times 10^{-3} \text{ M Cl}^-$$

$$Q = [\text{Pb}^{2+}][\text{Cl}^-]^2 = (6.7 \times 10^{-4})(5.0 \times 10^{-3})^2 = 1.7 \times 10^{-8}$$

and since $Q < K_{sp}$ (1.7×10^{-5}) PbCl_2 does not precipitate.

55.   These metal sulfides are 1:1 salts. The K_{sp} expression has the general form:

$$K_{sp} = [\text{M}^{2+}][\text{S}^{2-}]$$

To determine the $[\text{S}^{2-}]$ necessary to begin precipitation, we can divide the equation by the metal ion concentration to obtain:

$$\frac{K_{sp}}{[\text{M}^{2+}]} = [\text{S}^{2-}]$$

The concentration of the three metal ions under consideration are each 0.10 M. Substitution of the appropriate K_{sp} for the sulfides and 0.10 M for the metal ion concentration yields the sulfide ion concentrations in the table below. As the soluble sulfide is added to the metal ion solution, the sulfide ion concentration increases from zero molarity. The lowest sulfide ion concentration is reached first, with higher concentrations reached later. The order of precipitation is listed in the last column of the table below.

Compound	K_{sp}	Maximum $[\text{S}^{2-}]$	Order of Precipitation
CoS	5.9×10^{-21}	5.9×10^{-20}	2
MnS	5.1×10^{-15}	5.1×10^{-14}	3
NiS	3.0×10^{-21}	3.0×10^{-20}	1

57.   The K_{sp} expressions for the hydroxides of Fe^{3+}, Pb^{2+}, and Al^{3+} are:

$$[\text{Fe}^{3+}][\text{OH}^-]^3 = 6.3 \times 10^{-38} \qquad [\text{Pb}^{2+}][\text{OH}^-]^2 = 2.8 \times 10^{-16}$$

$$[\text{Al}^{3+}][\text{OH}^-]^3 = 1.9 \times 10^{-33}$$

The solution in question contains the cations each in 0.1 M concentration.

Substituting this value for the metal ion concentrations, and solving for the [OH⁻] yields:

Fe(OH)₃: $[OH^-] = (6.3 \times 10^{-37})^{1/3} = 8.6 \times 10^{-13}$ M

Pb(OH)₂: $[OH^-] = (2.8 \times 10^{-15})^{1/2} = 5.3 \times 10^{-8}$ M

Al(OH)₃: $[OH^-] = (1.9 \times 10^{-32})^{1/3} = 2.7 \times 10^{-11}$ M

The salts would precipitate in the order: Fe(OH)₃, Al(OH)₃, Pb(OH)₂.

59.   Following the same process as in question 55, we calculate the [F⁻] necessary to begin precipitation of the three salts; $K_{sp} = [M^{2+}][F^-]^2$ and $[M^{2+}] = 1.5 \times 10^{-2}$ M.

$$\text{For MgF}_2 \ [F^-] = \sqrt{\frac{6.4 \times 10^{-9}}{1.5 \times 10^{-2}}} = 6.5 \times 10^{-4} \text{ M}$$

$$\text{For CaF}_2 \ [F^-] = \sqrt{\frac{3.9 \times 10^{-11}}{1.5 \times 10^{-2}}} = 5.1 \times 10^{-5} \text{ M}$$

$$\text{For BaF}_2 \ [F^-] = \sqrt{\frac{1.7 \times 10^{-6}}{1.5 \times 10^{-2}}} = 1.1 \times 10^{-2} \text{ M}$$

The salts precipitate in the order: CaF₂, MgF₂, BaF₂.

61.   The equilibrium for AgSCN dissolving is: AgSCN (s) $\Leftrightarrow$ Ag⁺ (aq) + SCN⁻ (aq).
As x mol/L of AgSCN dissolve in pure water, x mol/L of Ag⁺ and x mol/L of SCN⁻ are produced.

$$K_{sp} = [Ag^+][SCN^-] = x^2 = 1.0 \times 10^{-12} \qquad \text{and } x = 1.0 \times 10^{-6} \text{ M}$$

So 1.0×10^{-6} mol AgSCN/L dissolve in pure water.

The equilibrium for AgSCN dissolving in NaSCN (0.010 M) is like that above. Equimolar amounts of Ag⁺ and SCN⁻ ions are produced as the solid dissolves. However the [SCN⁻] is augmented by the soluble NaSCN.

$$K_{sp} = [Ag^+][SCN^-] = (x)(x + 0.010) = 1.0 \times 10^{-12}$$

We can simplify the expression by assuming that x + 0.010 ≈ 0.010.

$$(x)(0.010) = 1.0 \times 10^{-12} \quad \text{and } x = 1.0 \times 10^{-10} \text{ M}$$

The solubility of AgSCN in 0.010 M NaSCN is 1.0×10^{-10} M -- reduced by four orders of magnitude from its solubility in pure water. LeChatelier strikes again!

63.   The K_{sp} expression for Ag_3PO_4 is: $K_{sp} = [Ag^+]^3[PO_4^{3-}] = 1.3 \times 10^{-20}$

a. The stoichiometry of the compound is such that when the solid dissolves,
$$[Ag^+] = 3 \cdot [PO_4^{3-}]$$
Substituting into the K_{sp} expression gives $(3 \cdot [PO_4^{3-}])^3[PO_4^{3-}] = 1.3 \times 10^{-20}$

or $27 \cdot [PO_4^{3-}]^4 = 1.3 \times 10^{-20}$ and $[PO_4^{3-}] = 4.7 \times 10^{-6}$

The molar amount of solid that dissolves will be equal to the concentration of phosphate ion. Expressing this concentration in mg/mL:

$$\frac{4.7 \times 10^{-6} \text{ mol}}{1 \text{ L}} \cdot \frac{419 \text{ g } Ag_3PO_4}{1 \text{ mol}} \cdot \frac{1000 \text{ mg}}{1 \text{ g}} \cdot \frac{1 \text{ L}}{1000 \text{ mL}} = 2.0 \times 10^{-3} \text{ mg/mL}$$

b. The solubility of the salt in 0.020 M $AgNO_3$:

While the ratio of silver and phosphate ions from the solid dissolving remains the same, the equilibrium concentration of Ag^+ will be increased by the 0.020 M Ag^+ ion. If we represent that molar solubility of Ag_3PO_4 as x, the concentrations at equilibrium will be:
$$[PO_4^{3-}] = x \quad \text{and} [Ag^+] = 3x + 0.020$$
Substituting the values into the K_{sp} expression:
$$[Ag^+]^3[PO_4^{3-}] = 1.3 \times 10^{-20} = (3x + 0.020)^3 x = 1.3 \times 10^{-20}$$

The amount of silver phosphate which dissolves is small, hence let's simplify the expression to: $(0.020)^3 x = 1.3 \times 10^{-20}$ and solving for x:

$x = 1.6 \times 10^{-15}$ M, and expressing this concentration in mg/mL:

$$\frac{1.6 \times 10^{-15} \text{ mol}}{1 \text{ L}} \cdot \frac{419 \text{ g } Ag_3PO_4}{1 \text{ mol}} \cdot \frac{1000 \text{ mg}}{1 \text{ g}} \cdot \frac{1 \text{ L}}{1000 \text{ mL}} = 6.8 \times 10^{-13} \text{ mg/mL}$$

65.   The K_{sp} expression for $AgIO_3$ is: $[Ag^+][IO_3^-] = 1.0 \times 10^{-8}$.
Calculate the solubility of $AgIO_3$ in 0.020 M KIO_3:

$K_{sp} = [Ag^+][IO_3^-] = 1.0 \times 10^{-8}$ and if x mol/L of the solid dissolves we obtain: $x \cdot (x + 0.020) = 1.0 \times 10^{-8}$

If we assume that x is negligible with respect to 0.020, we can simplify the expression to:
$$x \cdot 0.020 = 1.0 \times 10^{-8} \quad \text{and } x = 5.0 \times 10^{-7} \text{ M}.$$

At equilibrium: $[Ag^+] = 5.0 \times 10^{-7}$ M

$[IO_3^-] = 5.0 \times 10^{-7} + 0.020 = 0.020$ M and

$[K^+] = 0.020$ M

The amount of solid $AgIO_3$ added to the 100. mL of solution is:

$$\frac{0.10 \text{ g } AgIO_3}{0.100 \text{ L}} \cdot \frac{1 \text{ mol } AgIO_3}{283 \text{ g } AgIO_3} = 3.5 \times 10^{-3} \text{ M}$$

So $AgIO_3$ will dissolve until the concentrations listed above are reached. Note that not all the solid will dissolve.

67.   Mixing equal volumes of two solutions halves the original concentration of each solution

So $[Ag^+] = 0.0075$ M   and $[Cl^-] = 0.00060$ M

$Q = [Ag^+][Cl^-] = (7.5 \times 10^{-3})(6.0 \times 10^{-4}) = 4.5 \times 10^{-6}$

Since K_{sp} for AgCl $= 1.8 \times 10^{-10}$, **AgCl precipitates** and continues to precipitate until $[Ag^+][Cl^-] = K_{sp}$.

Mass of AgCl precipitated:

Since AgCl is a 1:1 salt, and Cl^- is present is the smaller amount, the maximum amount of AgCl is no greater than the amount of Cl^- present.

Cl^- present: $(0.0250 \text{ L})(1.2 \times 10^{-3} \text{ M}) = 3.0 \times 10^{-5}$ mol Cl^-

3.0×10^{-5} mol $Cl^- \cdot \dfrac{1 \text{ mol AgCl}}{1 \text{ mol } Cl^-} \cdot \dfrac{143.4 \text{ g AgCl}}{1 \text{ mol AgCl}} = 4.3 \times 10^{-3}$ g AgCl precipitates

Concentration remaining:

$[Ag^+] = 7.5 \times 10^{-3} - 6.0 \times 10^{-4} = 6.9 \times 10^{-3}$

In a saturated solution: $[Ag^+][Cl^-] = 1.8 \times 10^{-10}$. If the $[Ag^+] = 6.9 \times 10^{-3}$,

then $[Cl^-] = \dfrac{1.8 \times 10^{-10}}{6.9 \times 10^{-3}} = 2.6 \times 10^{-8}$ M

$[NO_3^-]$: Since the nitrate ion doesn't participate in any of the precipitations, the $[NO_3^-]$ is simply 1/2 its original value (The volume of the final solution is twice that of the original.) $= 7.5 \times 10^{-3}$ M

$[Na^+]$: Like the nitrate ion, the concentration of sodium is one-half its original value or 6.0×10^{-4} M.

69.   a. To determine the maximum concentration of oxalate ion before the Mg^{2+} salt begins to precipitate, substitute the concentration of Mg^{2+} into the K_{sp} expression:

$$K_{sp} = [Mg^{2+}][C_2O_4^{2-}] = 8.6 \times 10^{-5} \text{ and for a solution in which}$$

$$[Mg^{2+}] = 0.020 \text{ the maximum } [C_2O_4^{2-}] = \frac{8.6 \times 10^{-5}}{0.020} = 4.3 \times 10^{-3} \text{ M}.$$

b. When the magnesium salt just begins to precipitate, the $[C_2O_4^{2-}] = 4.3 \times 10^{-3}$.
    At that point the $[Ca^{2+}]$ would be:

$$K_{sp} = [Ca^{2+}][C_2O_4^{2-}] = 2.3 \times 10^{-9}$$

and the $[Ca^{2+}]$ would be $\dfrac{2.3 \times 10^{-9}}{4.3 \times 10^{-3}} = 5.3 \times 10^{-7} \text{ M}.$

71.   The more soluble salt will have the greater K_{sp}. The respective K_{sp}'s for the compounds
        are: $PbI_2 = 8.7 \times 10^{-9}$; $PbCO_3 = 1.5 \times 10^{-13}$

        a. The K_{sp} expressions for these are:

$$K_{sp} = [Pb^{2+}][I^-]^2 = 8.7 \times 10^{-9} \qquad \text{and } K_{sp} = [Pb^{2+}][CO_3^{2-}] = 1.5 \times 10^{-13}$$

Substituting the appropriate concentrations and solving for $[Pb^{2+}]$:

$$[Pb^{2+}] = \frac{8.7 \times 10^{-9}}{(0.10)^2} \qquad\qquad [Pb^{2+}] = \frac{1.5 \times 10^{-13}}{0.10}$$

$$[Pb^{2+}] = 8.7 \times 10^{-7} \text{ M} \qquad\qquad [Pb^{2+}] = 1.5 \times 10^{-12} \text{ M}$$

So **$PbCO_3$ will begin to precipitate first**.

b. When PbI_2 begins to precipitate, $[Pb^{2+}] = 8.7 \times 10^{-7}$ and for a solution that is saturated in $PbCO_3$ we can write: $[Pb^{2+}][CO_3^{2-}] = 1.5 \times 10^{-13}$

Solving for $[CO_3^{2-}] = \dfrac{1.5 \times 10^{-13}}{8.7 \times 10^{-7}} = 1.7 \times 10^{-7} \text{ M}$

73.   Separate the following pairs of ions:
        a. **Ba^{2+} and Na^+** : Since most sodium salts are soluble, it is simple to find a barium salt which is not soluble--e.g. the sulfate. Addition of dilute sulfuric acid should provide a source of SO_4^{2-} ions in sufficient quantity to precipitate the barium ions.

b. **Ca^{2+} and Zn^{2+}:** Zn has a relatively insoluble hydroxide ($K_{sp} = 4.5 \times 10^{-17}$). Addition of NaOH should precipitate $Zn(OH)_2$, while leaving most of the calcium ion in solution.

c. **Bi^{3+} and Cd^{2+}:** As in the case of part b, the hydroxide ion will serve as an effective reagent for selective precipitation of the two ions with the less soluble $Bi(OH)_3$ ($K_{sp} = 3.2 \times 10^{-40}$) precipitating well before the cadmium salt ($K_{sp} = 1.2 \times 10^{-14}$).

75. The equation for AgI dissolving is:

$$AgI \text{ (s)} \Leftrightarrow Ag^+ \text{ (aq)} + I^- \text{ (aq)}. \qquad K_{sp} = [Ag^+][I^-] = 1.5 \times 10^{-16}$$

The formation of the silver ammine complex may be written:

$$Ag^+ \text{ (aq)} + 2\, NH_3 \text{ (aq)} \Leftrightarrow [Ag(NH_3)_2]^+ \text{ (aq)}$$

$$K_f = \frac{[Ag(NH_3)_2]^+}{[Ag^+][NH_3]^2} = 1.6 \times 10^7$$

Adding the two equations, Ag^+ (aq) on the "left" side of the second equation cancels with the same substance on the "right" side of the first equation giving an overall equation of:

$$AgI \text{ (s)} + 2\, NH_3 \text{ (aq)} \Leftrightarrow [Ag(NH_3)_2]^+ \text{ (aq)} + I^- \text{ (aq)}$$

$$K_{overall} = K_{sp} \cdot K_f = [Ag^+][I^-] \cdot \frac{\{[Ag(NH_3)_2]^+\}}{[Ag^+][NH_3]^2} = \frac{\{[Ag(NH_3)_2]^+\}[I^-]}{[NH_3]^2}$$

$$= (1.5 \times 10^{-16})(1.6 \times 10^7) = 2.4 \times 10^{-9}$$

77. For AgI to form, the K_{sp} must be exceeded. In a saturated solution of AgCl the operative equilibrium is:

$$[Ag^+][Cl^-] = 1.8 \times 10^{-10} \text{ and the } [Ag^+] = (1.8 \times 10^{-10})^{1/2} \text{ or } 1.3 \times 10^{-5}.$$

For AgI the $K_{sp} = [Ag^+][I^-] = 1.5 \times 10^{-16}$. Dividing the K_{sp} by the silver ion concentration tells us that once the iodide ion exceeds $\left(\dfrac{1.5 \times 10^{-16}}{1.3 \times 10^{-5}}\right)$ or 1.2×10^{-11} M AgI will begin to precipitate.

The equilibrium constant expression for this process is:

$$K_{overall} = \frac{K_{sp}\,(AgCl)}{K_{sp}\,(AgI)} = \frac{1.8 \times 10^{-10}}{1.5 \times 10^{-16}} = 1.2 \times 10^6 = \frac{[Cl^-]}{[I^-]}$$

79.   Will 5.0 mL of 2.5 M NH_3 dissolve 1.0×10^{-4} mol of $AgCl$?

The reaction that must occur for $AgCl$ to dissolve is:

$$AgCl \text{ (s)} + 2 NH_3 \text{ (aq)} \Leftrightarrow [Ag(NH_3)_2]^+ \text{ (aq)} + Cl^- \text{ (aq)}$$

The reaction is the sum of (1) $AgCl$ dissolving and (2) $[Ag(NH_3)_2]^+$ forming from Ag^+ and NH_3. Accordingly,

$$K_{overall} = K_{sp} \cdot K_f$$
$$K_{overall} = (1.8 \times 10^{-10})(1.6 \times 10^7) = 2.9 \times 10^{-3}$$

$$K_{overall} = \frac{\{[Ag(NH_3)_2]^+\}[Cl^-]}{[NH_3]^2} = 2.9 \times 10^{-3}$$

Calculate the $[NH_3]$ necessary to dissolve the $AgCl$.

1. If the $AgCl$ dissolves, the equilibrium amount of Cl^- will be 1.0×10^{-4} moles, and the amount of the complex will be 1.0×10^{-4} moles. In 5.0 mL, the concentrations will be:

$$\frac{1.0 \times 10^{-4} \text{ mol}}{5.0 \times 10^{-3} \text{ L}} = 0.020 \text{ M. Substituting into the } K_{overall} \text{ expression:}$$

$$K_{overall} = \frac{\{[Ag(NH_3)_2]^+\}[Cl^-]}{[NH_3]^2} = 2.9 \times 10^{-3}$$

$$\frac{(0.020)(0.020)}{[NH_3]^2} = 2.9 \times 10^{-3}$$

$$[NH_3] = 0.37 \text{ M}$$

2. The total ammonia necessary is the ammonia to form the complex $(2 \cdot 0.020) = 4.0 \times 10^{-2}$ M NH_3 and the ammonia to increase the NH_3 concentration to 0.37 M.

$$[NH_3] = 4.0 \times 10^{-2} + 0.37 = 0.41 \text{ M.}$$

The concentration of NH_3 is 2.5 M. We see that this concentration of NH_3 would be sufficient to dissolve the $AgCl$.

81. $CaCO_3$ dissolves to furnish Ca^{2+} and CO_3^{2-} ions. The carbonate ions react with added HCl to form carbonic acid, H_2CO_3. This acid decomposes to form water and carbon dioxide. The reduction in carbonate ion concentration (Remember LeChatelier !) causes more $CaCO_3$ to dissociate---giving the overall effect of "greater" solubility. When $CaSO_4$ dissolves, any sulfate ion can react with the HCl forming the acid, H_2SO_4. This acid does not undergo decomposition, and the resulting reaction of the HCl with the SO_4^{2-} is minimal.

83.   Preparation of $CaSO_4$ from $CaCO_3$:
$$CaCO_3 \text{ (s)} + H_2SO_4 \text{ (aq)} \rightarrow CaSO_4 \text{ (s)} + H_2O \text{ (l)} + CO_2 \text{ (g)}$$

85.   Prepare an aqueous solution of magnesium chloride. Addition of $AgNO_3$ in sufficient
        quantity will precipitate AgCl, leaving magnesium ions and nitrate ions in solution.
        Remove the solid AgCl by filtration, and reduce the remaining aqueous solution to dryness.

87.   If a solution contains CO_3^{2-}, addition of aqueous $AgNO_3$ will produce a white precipitate
        (Ag_2CO_3 or AgCl). Addition of excess HCl will dissolve Ag_2CO_3 but not AgCl.
        Alternatively addition of the acid HCl to a carbonate solution will result in the formation of
        carbonic acid, H_2CO_3, which decomposes quickly to water and carbon dioxide. Gas
        bubbles forming in the solution would be a good indication of the presence of carbonate
        ions.

89.   a. $K_{dissolve}$ for ZnS: $K_{dissolve} = K_{sp} \cdot \dfrac{1}{K_a}$

$$= 1.1 \times 10^{-21} \cdot \frac{1}{1.3 \times 10^{-20}} = 0.085$$

        b. Calculate $[Zn^{2+}]$ if $[H_3O^+] = 0.30$ M and $[H_2S] = 0.10$ M.

$$\text{Since } K_{dissolve} = \frac{[Zn^{2+}][H_2S]}{[H_3O^+]^2} = 0.085$$

$$[Zn^{2+}] = \frac{(0.085)(0.30)^2}{(0.10)} = 7.61 \times 10^{-2} \text{ or } 7.6 \times 10^{-2} \text{ (to 2 sig. figs.)}$$

91.   Determining the more soluble of two substances with the same general composition
        (dipositive cation and dipositive anion) is done by noting that the substance with the greater
        K_{sp} is the more soluble. Aragonite ($K_{sp} = 6.0 \times 10^{-9}$) is slightly more soluble than calcite
        ($K_{sp} = 3.8 \times 10^{-9}$)

93.   Determine if $Q > K_{sp}$ (for AgCl):
$$K_{sp} = [Ag^+][Cl^-] = 1.8 \times 10^{-10}$$
$$Q = (1.0 \times 10^{-5})(2.0 \times 10^{-4}) = 2.0 \times 10^{-9}$$
        Since Q exceeds K_{sp}, AgCl will precipitate. (You are found out!)

95.   a. The general K_{sp} expression for an $M(OH)_2$ compound is: $K_{sp} = [M^{2+}][OH^-]^2$
       If the concentration of both M^{2+} and OH^- are 1.0×10^{-6}
$$Q = (1.0 \times 10^{-6})(1.0 \times 10^{-6})^2 = 1.0 \times 10^{-18}$$
K_{sp} for $Pb(OH)_2 = 2.8 \times 10^{-16}$
K_{sp} for $Zn(OH)_2 = 4.5 \times 10^{-11}$
Since $Q < K_{sp}$ for both salts, neither substance precipitates.

b. The maximum $[OH^-]$ before $Pb(OH)_2$ precipitates:

Immediately before $Pb(OH)_2$--the more soluble salt--precipitates, the reaction quotient, Q, will be exactly equal to K_{sp}.

$$K_{sp} = [Pb^{2+}][OH^-]^2 = 2.8 \times 10^{-16} \qquad \text{and} \quad [Pb^{2+}] = 1.0 \times 10^{-6}$$

$$[OH^-]^2 = \frac{2.8 \times 10^{-16}}{1.0 \times 10^{-6}} \text{ and } [OH^-] = 1.7 \times 10^{-5} \text{ M}$$

97.   Calculate the concentrations of Ca^{2+} and OH^- after mixing (but before reaction)

$$0.010 \text{ M Ca}^{2+} \cdot \frac{15.0 \text{ mL}}{(15.0 + 25.0 \text{ mL})} = 3.75 \times 10^{-3} \text{ M Ca}^{2+}$$

$$0.0010 \text{ M OH}^- \cdot \frac{25.0 \text{ mL}}{(15.0 + 25.0 \text{ mL})} = 6.25 \times 10^{-4} \text{ M OH}^-$$

$$Q = [Ca^{2+}][OH^-]^2 = (3.75 \times 10^{-3})(6.25 \times 10^{-4})^2 = 1.46 \times 10^{-9}$$

Since K_{sp} (7.9×10^{-6}) $> Q$ so no $Ca(OH)_2$ precipitates.

Concentration of the ions:
$[Ca^{2+}] = 3.8 \times 10^{-3}$ M (to two significant figures)
$[Cl^-] = 2(3.75 \times 10^{-3}) = 7.5 \times 10^{-3}$ M
$[Na^+] = 6.3 \times 10^{-4}$ M
The pH of the solution:
$[OH^-] = 6.25 \times 10^{-4}$ M $\rightarrow$ pOH = 3.20 and pH = 10.80.

99. $AgCl (s) \Leftrightarrow Ag^+ (aq) + Cl^- (aq) \qquad K_{sp} = 1.8 \times 10^{-10}$

At equilibrium $[Ag^+] = [Cl^-] = (1.8 \times 10^{-10})^{1/2}$ so $[Ag^+] = [Cl^-] = 1.3 \times 10^{-5}$ M

$$\frac{1.0 \text{ mg NaCl}}{L} \cdot \frac{1.0 \text{ g NaCl}}{1000 \text{ mg NaCl}} \cdot \frac{1 \text{ mol NaCl}}{58.4 \text{ g NaCl}} = 1.7 \times 10^{-5} \text{ M NaCl}$$

To calculate the equilibrium concentration after the NaCl is added:

$$K_{sp} = [Ag^+][Cl^-] = (x)(x + 1.7 \times 10^{-5}) = 1.8 \times 10^{-10}$$

and solving via the quadratic equation yields: $x = 7.4 \times 10^{-6}$ M

So $[Ag^+] = 7.4 \times 10^{-6}$ and $[Cl^-] = (1.7 \times 10^{-5} + 7.4 \times 10^{-6}) = 2.4 \times 10^{-5}$ M

If $[Ag^+] = 1.3 \times 10^{-5}$ M before the NaCl was added and 7.4×10^{-6} M after the NaCl was added, the difference (5.6×10^{-6} mol/L) was precipitated. On a mass basis this is:

$$\frac{5.6 \times 10^{-6} \text{ mol}}{L} \cdot \frac{143 \text{ g AgCl}}{1 \text{ mol AgCl}} \cdot \frac{1000 \text{ mg AgCl}}{1.0 \text{ g AgCl}} \cdot \frac{1.0 \text{ L}}{1} = 0.80 \text{ mg AgCl}$$

101.   The equation for $Zn(OH)_2$ dissolving is : $Zn(OH)_2$ (s) $\Leftrightarrow$ Zn^{2+} (aq) $+$ 2 OH^- (aq)

If a saturated solution of the base has a pH = 8.65 then pOH = 5.35 and
$[OH^-] = 4.5 \times 10^{-6}$ M. The stoichiometry of the solid says that for two hydroxyl ions, one zinc ion is produced. So the concentration of $Zn^{2+} = 1/2 \cdot (4.5 \times 10^{-6}) = 2.2 \times 10^{-6}$.

The K_{sp} is then: $K_{sp} = [Zn^{2+}][OH^-]^2 = (2.2 \times 10^{-6})(4.5 \times 10^{-6})^2 = 4.5 \times 10^{-17}$.

103.   Each of the salts can be prepared by the addition of the appropriate acid to solid $MgCO_3$.
The acid will react, forming $CO_2 + H_2O$. Heating the resulting solution to dryness will
result in the collection of the appropriate salt. The acids are:
a. magnesium sulfate $\leftarrow$ sulfuric acid   b. magnesium bromide $\leftarrow$ hydrobromic acid
c. magnesium oxalate $\leftarrow$ oxalic acid   d. magnesium perchlorate $\leftarrow$ perchloric acid
e. magnesium fluoride $\leftarrow$ hydrofluoric acid

105.   If 0.581 g of $Mg(OH)_2$ is dissolved in 1.00 L of water, the concentration is:

$$\frac{0.581 \text{ g Mg(OH)}_2}{1.00 \text{ L}} \cdot \frac{1 \text{ mol Mg(OH)}_2}{58.32 \text{ g Mg(OH)}_2} = 9.96 \times 10^{-3} \text{ M}$$

To determine if all the $Mg(OH)_2$ dissolves, calculate the molar solubility of $Mg(OH)_2$.

$$K_{sp} = [Mg^{2+}][OH^-]^2 = 1.5 \times 10^{-11}$$

If x mol/L of $Mg(OH)_2$ dissolves, $[Mg^{2+}] = x$ and $[OH^-] = 2x$.

and $\quad K_{sp} = [x][2x]^2 = 1.5 \times 10^{-11}$

$$4x^3 = 1.5 \times 10^{-11} \quad \text{and} \quad x = 1.6 \times 10^{-4} \text{ M}$$

In water, only 1.6×10^{-4} mol $Mg(OH)_2$ will dissolve, and since the 0.581 g of $Mg(OH)_2$ corresponds to a concentration of 9.96×10^{-3} M, **not all the solid dissolves.**

If the solution is buffered at pH = 5.00 then $[OH^-] = 1.0 \times 10^{-9}$. The solubility of $Mg(OH)_2$ in this solution may be calculated as above. The hydroxide ion concentration will be the contribution from the dissolution of the solid and from the buffer. Given these conditions, calculate the $[Mg^{2+}]$ which the solution could tolerate:

$$K_{sp} = [Mg^{2+}][OH^-]^2 = 1.5 \times 10^{-11}$$
$$[Mg^{2+}](1.0 \times 10^{-9})^2 = 1.5 \times 10^{-11}$$
$$[Mg^{2+}] = 1.5 \times 10^7 \text{ M}$$

Note that this concentration greatly exceeds the concentration of magnesium ion that could be furnished by the $Mg(OH)_2$ (9.96×10^{-3} M). **So at pH = 5.00 all of the solid will dissolve.**

107. a. The equilibrium expression for the equation may be written:

$$K = \frac{\{[Ag(S_2O_3)_2]^{3-}\}[Br^-]}{[S_2O_3^{2-}]^2}$$

   NOTE: We have used { } to indicate concentration when complex ions are present.

The K_{sp} expression for AgBr is written: $K_{sp} = [Ag^+][Br^-] = 3.3 \times 10^{-13}$

The K_f for $[Ag(S_2O_3)_2]^{3-}$ may be written: $\quad K_f = \dfrac{\{[Ag(S_2O_3)_2]^{3-}\}}{[Ag^+][S_2O_3^{2-}]^2} = 2.0 \times 10^{13}$

$K_{overall} = K_{sp} \cdot K_f = \dfrac{\{[Ag(S_2O_3)_2]^{3-}\}[Br^-]}{[S_2O_3^{2-}]^2} = (3.3 \times 10^{-13})(2.0 \times 10^{13}) = 6.6$

b. 1.0 g AgBr in 1.0 L corresponds to $\dfrac{1.0 \text{ g AgBr}}{1 \text{ L}} \cdot \dfrac{1 \text{ mol AgBr}}{188 \text{ g AgBr}} = 5.3 \times 10^{-3}$ M

Note that the concentration, 5.3×10^{-3} M, corresponds to the concentration of both silver complex and bromide ion. Note also that for each mole of silver complex, two moles of thiosulfate ion are consumed.

The amount of $S_2O_3^{2-}$ consumed is then $(2 \cdot 5.3 \times 10^{-3})$ or 1.1×10^{-2} M.

Substituting into the $K_{overall}$ expression gives:

$$\frac{\{[Ag(S_2O_3)_2]^{3-}\}[Br^-]}{[S_2O_3^{2-}]^2} = \frac{(5.3 \times 10^{-3})(5.3 \times 10^{-3})}{[S_2O_3^{2-}]^2} = 6.6$$

and $[S_2O_3^{2-}]^2 = 4.3 \times 10^{-6}$ and $[S_2O_3^{2-}] = 2.1 \times 10^{-3}$

Note that this concentration is the equilibrium concentration for $S_2O_3^{2-}$.

(the result of $[S_2O_3^{2-}]_{initial} - [S_2O_3^{2-}]_{change}$)

$$2.1 \times 10^{-3} \text{ M} = [S_2O_3^{2-}]_{initial} - 1.1 \times 10^{-2} \text{ M}$$
$$1.3 \times 10^{-2} \text{ M} = [S_2O_3^{2-}]_{initial}.$$

This corresponds to 1.3×10^{-2} moles of $Na_2S_2O_3$.

The mass is: 1.3×10^{-2} mol $Na_2S_2O_3 \cdot \dfrac{158 \text{ g } Na_2S_2O_3}{1 \text{ mol } Na_2S_2O_3} = 2.1$ g $Na_2S_2O_3$

109.   a. $AlCl_3 + H_3PO_4 \rightarrow AlPO_4 + 3 \text{ HCl}$

     b. 152 g $AlCl_3 \cdot \dfrac{1 \text{ mol } AlCl_3}{133.3 \text{ g} AlCl_3} = 1.14$ mol $AlCl_3$

       0.750 M $H_3PO_4 \cdot 3.0$ L $= 2.3$ mol H_3PO_4

Examining the moles-available and moles-required ratios, we note that $AlCl_3$ is the limiting reagent, so the amount of $AlPO_4$ obtainable is:

$$1.14 \text{ mol } AlCl_3 \cdot \frac{1 \text{ mol } AlPO_4}{1 \text{ mol } AlCl_3} \cdot \frac{122.0 \text{ } AlPO_4}{1 \text{ mol } AlPO_4} = 139 \text{ g } AlPO_4$$

     c. $\dfrac{25.0 \text{ g } AlPO_4}{1 \text{ L}} \cdot \dfrac{1 \text{ mol } AlPO_4}{122.0 \text{ g } AlPO_4} = 0.205$ M $AlPO_4$

Determine if this is a saturated solution:

$$K_{sp} = [Al^{3+}][PO_4^{3-}] = 1.3 \times 10^{-20}$$
$$[Al^{3+}] = [PO_4^{3-}] = 1.1 \times 10^{-10} \text{ M}$$

Since 25.0 g of $AlPO_4$ corresponds to a greater concentration, the solution is saturated--only 1.1×10^{-10} mol/L $AlPO_4$ will dissolve.

d. Since the phosphate ion is capable of reacting with H_3O^+ to form other ions (HPO_4^{2-}, $H_2PO_4^-$), addition of HCl will shift the phosphate equilibrium toward formation of these protonated anions--reducing $[PO_4^{3-}]$.

The reduction of $[PO_4^{3-}]$ will **increase the amount of** $AlPO_4$ which dissolves.

e. Concentrations after mixing:

$$[Al^{3+}] = 2.5 \times 10^{-3} \text{ M } Al^{3+} \cdot \frac{1.50 \text{ L}}{(1.50 + 2.50)} = 9.4 \times 10^{-4} \text{ M}$$

$$[PO_4^{3-}] = 3.5 \times 10^{-2} \text{ M } PO_4^{3-} \cdot \frac{2.50 \text{ L}}{(1.50 + 2.50 \text{ L})} = 2.2 \times 10^{-2} \text{ M}$$

$$Q = [Al^{3+}][PO_4^{3-}] = (9.4 \times 10^{-4})(2.2 \times 10^{-2}) = 2.1 \times 10^{-5}$$

Since $Q > K_{sp}$ (1.3×10^{-20}), **$AlPO_4$ precipitates.**

The Al^{3+} is the limiting reagent and the maximum amount of $AlPO_4$ that can form is :

$$\frac{9.4 \times 10^{-4} \text{ mol } Al^{3+}}{L} \cdot \frac{4.0 \text{ L}}{1} \cdot \frac{1 \text{ mol } AlPO_4}{1 \text{ mol } Al^{3+}} \cdot \frac{122 \text{ g } AlPO_4}{1 \text{ mol } AlPO_4} = 0.46 \text{ g } AlPO_4$$

CHAPTER 20: The Spontaneity of Chemical Reactions: Entropy and Free Energy

6. The sample with the higher entropy:

    a. CO_2 vapor at 0 °C -- The entropy of a substance is greatest in its gaseous form.

    b. Dissolved sugar   -- The entropy of a dissolved substance is greater than the entropy of the pure solid.

    c. Mixture   -- The entropy of each pure substance contained in separate beakers will be lower than the entropy of the solution formed by mixing the two liquids.

8. Compound with the higher entropy:

    a. $AlCl_3$ - Entropy increases with molecular complexity.

    b. CH_3CH_2I - Entropy increases with molecular complexity.

    c. NH_4Cl (aq) - Entropy of solutions is greater than that of solids.

10. Entropy changes:

    a. C (graphite) $\rightarrow$ C (diamond)

    ΔS^o = 2.311 J/K • mol - 5.140 J/K • mol

    = - 3.363 J/K • mol

    The decrease in entropy reflects the greater order of diamond.

    b. Na (g) $\rightarrow$ Na (s)

    ΔS^o =   51.21 J/K • mol - 153.112 J/K • mol

    = - 102.50 J/K • mol

    The lower entropy of the solid state is evidenced by the negative sign.

    c. Hg (l) $\rightarrow$ Hg (g)

    ΔS^o = 115 J/K • mol - 16.02 J/K • mol

    = 99 J/K • mol

    The increase in entropy is expected with the transition to the disordered state of a gas.

12. a. ΔS^o for the transition of $(C_2H_5)_2O$ (l) $\rightarrow$ $(C_2H_5)_2O$ (g)

    $$\Delta S^o = \frac{\Delta H_{vap}}{T} = \frac{26.0 \times 10^3 \text{ J/mol}}{308 \text{ K}} = 84.4 \text{ J/K} \cdot \text{mol}$$

    b. ΔS^o for the transition of $(C_2H_5)_2O$ (g) $\rightarrow$ $(C_2H_5)_2O$ (l) is -84.4 J/K • mol.

    Note the reduction in entropy in the phase change from gas to liquid.

14.   For the reaction: $3\,C$ (graphite) $+ 4\,H_2$ (g) $\rightarrow C_3H_8$ (g)

$\Delta S^o = 1 \cdot S^o\,C_3H_8 - [3 \cdot S^o\,C$ (graphite) $+ 4 \cdot S^o\,H_2$ (g)$]$

$= (1\,mol)(269.9\,J/K \cdot mol) -$

$[(3\,mol)(5.740\,J/K \cdot mol) + (4\,mol)(130.684\,J/K \cdot mol)]$

$= - 270.1\,J/K \cdot mol$

16.   Calculate the standard molar entropy change for each substance from its elements:

a. H_2 (g) $+ \frac{1}{2}\,O_2$ (g) $\rightarrow H_2O$ (l)

$\Delta S^o = (1\,mol)(69.91\,J/K \cdot mol)$

$- [(1\,mol)(130.684\,J/K \cdot mol) + (\frac{1}{2}\,mol)(205.138\,J/K \cdot mol)] = - 163.34\,J/K$

b. Mg (s) $+ \frac{1}{2}\,O_2$ (g) $\rightarrow MgO$ (s)

$\Delta S^o = (1\,mol)(26.94\,J/K \cdot mol)$

$- [(1\,mol)(32.68\,J/K \cdot mol) + (\frac{1}{2}\,mol)(205.138\,J/K \cdot mol)] = - 108.31\,J/K$

c. Ca (s) $+ S$ (s,rhombic) $\rightarrow CaS$ (s)

$\Delta S^o = (1\,mol)(56.5\,J/K \cdot mol)$

$- [(1\,mol)(41.42\,J/K \cdot mol) + (1\,mol)(31.80\,J/K \cdot mol)] = - 16.7\,J/K$

d. $2\,Al$ (s) $+ \frac{3}{2}\,O_2$ (g) $\rightarrow Al_2O_3$ (s)

$\Delta S^o = (1\,mol)(50.92\,J/K \cdot mol)$

$- [(2\,mol)(28.3\,J/K \cdot mol) + (\frac{3}{2}\,mol)(205.138\,J/K \cdot mol)] = - 313.4\,J/K$

18.   Calculate ΔS^o for the following reactions:

a. Pb (s) $+ Cl_2$ (g) $\rightarrow PbCl_2$ (s)

$\Delta S^o = 1 \cdot S^o\,PbCl_2$ (s) $- [1 \cdot S^o\,Pb$ (s) $+ 1 \cdot S^o\,Cl_2$ (g)$]$

$\Delta S^o = (1\,mol)(136.0\,J/K \cdot mol) -$

$[(1\,mol)(64.8\,J/K \cdot mol) + (1\,mol)(223.1\,J/K \cdot mol)] = - 151.9\,J/K$

b. C_2H_5OH (l) $+ 3\,O_2$ (g) $\rightarrow 2\,CO_2$ (g) $+ 3\,H_2O$ (l)

$\Delta S^o = [2 \cdot S^o\,CO_2$ (g) $+ 3 \cdot S^o\,H_2O$ (l)$] - [1 \cdot S^o\,C_2H_5OH$ (l) $+ 3 \cdot S^o\,O_2$ (g)$]$

$\Delta S^o = [(2\,mol)(213.14\,J/K \cdot mol) + (3\,mol)(69.91\,J/K \cdot mol)] -$

$[(1\,mol)(160.7\,J/K \cdot mol) + (3\,mol)(205.14\,J/K \cdot mol)]$

$= - 138.9\,J/K$

c. $4 \, Cr \, (s) \; + \; 3 \, O_2 \, (g) \; \rightarrow \; 2 \, Cr_2O_3 \, (s)$

$\Delta S^o \; = \; [(2 \, mol)(81.2 \, J/K \cdot mol)] -$

$\qquad\qquad\qquad [(4 \, mol)(23.77 \, J/K \cdot mol) + (3 \, mol)(205.138 \, J/K \cdot mol)]$

$\qquad = \; - 548.1 \, J/K$

20.   Calculate ΔG^o for the following reactions:

a. $Pb \, (s) + \; Cl_2 \, (g) \; \rightarrow \; PbCl_2 \, (s)$

$\quad$ 64.8 $\qquad$ 223.1 $\qquad\quad$ 136.0 $\qquad\quad$ $S^o \, (J/K \cdot mol)$

$\qquad$ 0 $\qquad\qquad$ 0 $\qquad\quad$ - 359.41 $\qquad$ $\Delta H^o_f \, (kJ/mol)$

In study question 18.a we calculated ΔS^o for this process (- 151.9 J/K).

$\qquad \Delta H^o_{rxn} \; = \; \Delta H^o_f \, PbCl_2 \, (s) - [\Delta H^o_f \, Pb \, (s) + \Delta H^o_f \, Cl_2 \, (g)]$

$\qquad\qquad\qquad = \; (- 359.41 \, kJ/mol)(1 \, mol) - 0$

$\qquad\qquad\qquad = \; - 359.41 \, kJ$

$\qquad \Delta G^o_{rxn} = \Delta H^o_f - T \, \Delta S^o_{rxn} \; = \; - 359.41 \, kJ - (298 \, K)(- 151.9 \, J/K)(\frac{1.000 \, kJ}{1000 \, J})$

$\qquad\qquad\qquad = \; - 314.1 \, kJ$

b. $Mg \, (s) + \frac{1}{2} O_2 \, (g) \; \rightarrow \; MgO \, (s)$

$\quad$ 32.68 $\qquad$ 205.14 $\qquad\qquad$ 26.94 $\qquad$ $S^o \, (J/K \cdot mol)$

$\qquad$ 0 $\qquad\qquad$ 0 $\qquad\qquad$ - 601.10 $\quad$ $\Delta H^o_f \, (kJ/mol)$

In study question 16.b we calculated ΔS^o for this process (- 108.31 J/K).

$\qquad \Delta H^o_{rxn} \; = \Delta H^o_f \, MgO \, (s) - [\Delta H^o_f \, Mg \, (s) + \frac{1}{2} \Delta H^o_f \, O_2 \, (g)]$

$\qquad\qquad\qquad = \; (- 601.70 \, kJ/mol)(1 \, mol) - 0$

$\qquad\qquad\qquad = \; - 601.70 \, kJ$

$\qquad \Delta G^o_{rxn} = \Delta H^o_f - T \, \Delta S^o_{rxn} \; = \; - 601.70 \, kJ - (298 \, K)(- 108.31 \, J/K)(\frac{1.000 \, kJ}{1000 \, J})$

$\qquad\qquad\qquad = \; - 601.70 \, kJ + 32.28 \, kJ$

$\qquad\qquad\qquad = \; - 569.42 \, kJ$

c. NH_3 (g) + HCl (g)   → NH_4Cl (s)

192.45   186.91   94.6 S^o (J/K•mol)

- 46.11   - 92.31   - 314.43 ΔH^o_f (kJ/mol)

ΔS^o_{rxn} = 1 • S^o NH_4Cl (s) - [1 • S^o NH_3 (g) + 1 • S^o HCl (g)]

         = (1 mol)(94.6 J/K • mol) - [(1 mol)(192.45 J/K • mol) +

                                                        (1 mol)(186.91 J/K • mol)]

         = - 284.8 J/K

ΔH^o_{rxn} = 1 • ΔH^o_f NH_4Cl (s) - [1 • ΔH^o_f NH_3 (g) + 1 • ΔH^o_f HCl (g)

         = (1 mol)(- 314.43 kJ/mol) -

                     [(1 mol)(- 46.11 kJ/mol) + (1 mol)(- 92.31 kJ/mol)]

         = - 176.01 kJ

ΔG^o_{rxn} = ΔH^o_f - T ΔS^o_{rxn}

         = - 176.01 kJ - (298 K)(- 284.76 J/K)($\frac{1.000 \text{ kJ}}{1000 \text{ J}}$)

         = - 176.01 kJ + 84.86 kJ

         = - 91.15 kJ

Parts a and b correspond to formation of one mole of a substance from its elements, each in their standard state--ΔG^o_f. The values obtained for ΔG^o agree with those in Appendix J. The values for ΔG^o for **all three equations** are negative, indicating that they **are spontaneous**.

22.   Calculate the molar free energies of formation for:

a. CS_2 (g)   The reaction is: C (graphite) + 2 S (s,rhombic) → CS_2 (g)

ΔH^o_f = (1 mol)(117.36 kJ/mol) - [0 + 0] = 117.36 kJ/mol

ΔS^o = (1 mol)(237.84 J/K • mol)

         - [(1 mol)(5.740 J/K • mol) + (2 mol)(31.80 J/K • mol)] = 168.50 J/K

ΔG^o_f = ΔH^o_f - TΔS^o

         = (1 mol)(117.36 kJ/mol) - (298 K)(168.50 J/K)($\frac{1.000 \text{ kJ}}{1000. \text{ J}}$)

         = 67.15 kJ   Appendix value: 67.12 kJ/mol

b. N_2H_4 (l)   The reaction is: $N_2(g) + 2 H_2(g) \rightarrow N_2H_4$ (l)

$\Delta H^o_f = (1 \text{ mol})(50.63 \text{ kJ/mol}) - [0 + 0] = 50.63 \text{ kJ/mol}$

$\Delta S^o = (1 \text{ mol})(121.21 \text{ J/K} \cdot \text{mol}) -$
$[(1 \text{ mol})(191.61 \text{ J/K} \cdot \text{mol}) + (2 \text{ mol})(130.684 \text{ J/K} \cdot \text{mol})] = -331.77 \text{ J/K}$

$\Delta G^o_f = \Delta H^o_f - T\Delta S^o$
$= (1 \text{ mol})(50.63 \text{ kJ/mol}) - (298 \text{ K})(-331.77 \text{ J/K})(\frac{1.000 \text{ kJ}}{1000. \text{ J}})$
$= 149.50 \text{ kJ}$ Appendix value: 149.34 kJ/mol

c. $COCl_2$ (g)   The reaction is: C (graphite) $+ \frac{1}{2} O_2$ (g) $+ Cl_2$ (g) $\rightarrow COCl_2$ (g)

$\Delta H^o_f = (1 \text{ mol})(-218.8 \text{ kJ/mol}) - [0 + 0 + 0] = -218.8 \text{ kJ/mol}$

$\Delta S^o = (1 \text{ mol})(283.53 \text{ J/K} \cdot \text{mol}) -$
$[(1 \text{ mol})(5.740 \text{ J/K} \cdot \text{mol}) + (\frac{1}{2} \text{ mol})(205.138 \text{ J/K} \cdot \text{mol}) +$
$(1 \text{ mol})(223.066 \text{ J/K} \cdot \text{mol})] = -47.85 \text{ J/K}$

$\Delta G^o_f = \Delta H^o_f - T\Delta S^o$
$= (1 \text{ mol})(-218.8 \text{ kJ/mol}) - (298 \text{ K})(-47.85 \text{ J/K})(\frac{1.000 \text{ kJ}}{1000. \text{ J}})$
$= -204.5 \text{ kJ}$ Appendix value: -204.6 kJ/mol

The formation of $COCl_2$ is predicted to be spontaneous.

24.   Formation of Fe_2O_3 (s): $2 Fe$ (s) $+ 3/2 O_2$ (g) $\rightarrow Fe_2O_3$ (s)

ΔG^o_{rxn} for formation of 1.00 mole of Fe_2O_3 (s):

$\Delta H^o_{rxn} = [1 \cdot \Delta H^o_f Fe_2O_3 \text{ (s)}] - [2 \cdot \Delta H^o_f Fe \text{ (s)} + \frac{3}{2} \cdot \Delta H^o_f O_2 \text{ (g)}]$
$= (1 \text{ mol})(-824.2 \text{ kJ/mol})] - 0 = -824.2 \text{ kJ}$

$\Delta S^o_{rxn} = [1 \cdot S^o Fe_2O_3 \text{ (s)}] - [2 \cdot S^o Fe \text{ (s)} + \frac{3}{2} \cdot S^o O_2 \text{ (g)}]$
$= [(1 \text{ mol})(81.40 \text{ J/K} \cdot \text{mol})] -$
$[(2 \text{ mol})(21.18 \text{ J/K} \cdot \text{mol}) + (\frac{3}{2} \text{ mol})(205.14 \text{ J/K} \cdot \text{mol})$
$= -275.87 \text{ J/K}$

$$\Delta G^o{}_{rxn} = \Delta H^o{}_f - T \, \Delta S^o{}_{rxn}$$

$$= (1 \text{ mol})(-824.2 \text{ kJ/mol}) - (298 \text{ K})(- 275.87 \text{ J/K})(\frac{1.000 \text{ kJ}}{1000. \text{ J}})$$

$$= - 742.0 \text{ kJ (in good agreement with } \Delta G^o{}_{rxn} \text{ for } Fe_2O_3 \text{ in Appendix J)}$$

Free energy change for formation of 454 g of Fe_2O_3 (s):

$$454 \text{ g of } Fe_2O_3 \cdot \frac{1 \text{ mol } Fe_2O_3}{160 \text{ g } Fe_2O_3} \cdot \frac{-742.0 \text{ kJ}}{1 \text{ mol } Fe_2O_3} = -2.11 \times 10^3 \text{ kJ}$$

26.   Calculate $\Delta G^o{}_{rxn}$ for the following reactions:

a. Ca (s) + Cl_2 (g) → $CaCl_2$ (s)

$$\Delta G^o{}_{rxn} = (1 \text{ mol})(- 748.1 \text{ kJ/mol}) - [0 + 0]$$

$$= - 748.1 \text{ kJ}$$

b. HgO (s) → Hg (l) + $\frac{1}{2}$ O_2(g)

$$\Delta G^o{}_{rxn} = [0 + 0] - (1 \text{ mol})(-58.539 \text{ kJ/mol})$$

$$= 58.539 \text{ kJ}$$

c. NH_3 (g) + 2 O_2 (g) → HNO_3 (l) + H_2O (l)

$$\Delta G^o{}_{rxn} = [(1 \text{ mol})(- 80.71 \text{ kJ/mol}) + (1 \text{ mol})(- 237.129 \text{ kJ/mol})$$

$$- [(1 \text{ mol})(- 16.45 \text{ kJ/mol}) + 0]$$

$$= - 301.39 \text{ J/K}$$

Reactions in parts **a** and **c** are predicted to be spontaneous:

28.   Reaction:

	$C_6H_{12}O_6$ (s) →	2 C_2H_5OH (l) +	2 CO_2 (g)
$\Delta H^o{}_f$ (kJ/mol)	- 1260.0	- 277.7	- 393.51
$\Delta S^o{}_f$ (J/K • mol)	289	160.7	213.74
$\Delta G^o{}_{rxn}$ (kJ/mol)	- 918.8	- 174.8	- 394.4

$$\Delta H^o{}_{rxn} = [2 \cdot \Delta H^o{}_f \, C_2H_5OH \text{ (l)} + 2 \cdot \Delta H^o{}_f \, CO_2 \text{ (g)}] - [1 \cdot \Delta H^o{}_f \, C_6H_{12}O_6 \text{ (s)}]$$

$$= [(2 \text{ mol})(- 277.7 \text{ kJ/mol}) + (2 \text{ mol})(- 393.51 \text{ kJ/mol})] -$$

$$[(1 \text{ mol})(- 1260.0 \text{ kJ/mol})]$$

$$= - 82.4 \text{ kJ}$$

$$\Delta S^o_{rxn} = [2 \cdot S^o\ C_2H_5OH\ (l) + 2 \cdot S^o\ CO_2\ (g)] - [1 \cdot S^o\ C_6H_{12}O_6\ (aq)]$$
$$= [(2\ mol)(160.7\ J/K \cdot mol) + (2\ mol)(213.74\ J/K \cdot mol)] -$$
$$[(1\ mol)(289\ J/K \cdot mol)]$$
$$= 460.\ J/K$$

$$\Delta G^o_{rxn} = [2 \cdot \Delta G^o\ C_2H_5OH\ (l) + 2 \cdot \Delta G^o\ CO_2\ (g)] - [1 \cdot \Delta G^o\ C_6H_{12}O_6\ (aq)]$$
$$= [(2\ mol)(-174.8\ kJ/mol) + (2\ mol)(-394.359\ kJ/mol)] -$$
$$[(1\ mol)(-918.8\ kJ/mol)]$$
$$= -219.6\ kJ$$

30.   a. $\Delta G^o_{rxn} = +48.2\ kJ$ for the reaction: $MgCO_3(s) \rightarrow MgO\ (s) + CO_2\ (g)$

$+48.2\ kJ = [(1mol)(-569.43\ kJ/mol) + (1\ mol)(-394.359\ kJ/mol)] - \Delta G^o_f\ MgCO_3\ (s)$

$+48.2\ kJ = (-963.789\ kJ) - \Delta G^o_f\ MgCO_3\ (s)$

$-1012.0\ kJ/mol = \Delta G^o_f\ MgCO_3\ (s)$

b. $\Delta G^o_{rxn} = +137.6\ kJ$ for the reaction: $NOF\ (g) \rightarrow NO\ (g) + \frac{1}{2}\ F_2\ (g)$

$+137.6\ kJ = [(1mol)(86.55\ kJ/mol) + (\frac{1}{2}\ mol)(0\ kJ/mol)] - \Delta G^o_f\ (NOF\ (g)\ )$

$+137.6\ kJ = (86.55\ kJ) - \Delta G^o_f\ NOF\ (g)$

$-51.1\ kJ = \Delta G^o_f\ NOF\ (g)$

32.   Determine ΔH^o, ΔS^o, and ΔG^o for: $C_8H_{16}\ (g) + H_2\ (g) \rightarrow C_8H_{18}\ (g)$

ΔH^o_f (kJ/mol)	-82.93	0	-208.45
S^o_f (J/K $\cdot$ mol)	462.8	130.7	463.6

$$\Delta H^o_{rxn} = [(1\ mol)(-208.45\ kJ/mol)] - [(1\ mol)(-82.93\ kJ/mol) + 0]$$
$$= -125.52\ kJ$$

$$\Delta S^o_{rxn} = [(1\ mol)(463.6\ J/K \cdot mol)] -$$
$$[(1\ mol)(462.8\ J/K \cdot mol) + (1\ mol)(130.7\ J/K \cdot mol)]$$
$$= -129.9\ J/K$$

$$\Delta G^o_{rxn} = \Delta H^o_f - T\ \Delta S^o_{rxn}$$
$$= -125.52\ kJ - (298\ K)(-129.9\ J/K)(\frac{1.000\ kJ}{1000\ J})$$
$$= -86.81\ kJ$$

The negative value for ΔG^o_{rxn} indicates that the reaction is spontaneous under standard conditions.

34.   Calculate ΔG^o for the reaction: O_3 (g) + H_2O (g) → H_2O_2 (l) + O_2 (g)

$$\Delta G^o_{rxn} = [\Delta G^o_f\, H_2O_2\, (l) + \Delta G^o_f\, O_2\, (g)] - [\Delta G^o_f\, O_3\, (g) + \Delta G^o_f\, H_2O\, (g)]$$
$$= [(-120.35\text{ kJ}) + (0)] - [(163.2\text{ kJ}) + (-228.572\text{ kJ})]$$
$$= -54.98\text{ kJ}$$

The reaction is thermodynamically feasible, and the sign of ΔG^o_{rxn} (< 0) indicates that it is spontaneous under standard conditions.

36.   Calculate K_p for the reaction:
$$\tfrac{1}{2} N_2(g) + \tfrac{1}{2} O_2(g) \to NO\,(g) \qquad \Delta G^o_f = +86.55\text{ kJ/mol NO}$$

$$\Delta G^o_{rxn} = -RT \ln K_p$$
$$86.55 \times 10^3\text{ J/mol} = -(8.314\text{ J/K·mol})(298\text{ K}) \ln K_p$$
$$-34.93 = \ln K_p$$
$$6.7 \times 10^{-16} = K_p$$

Note that the + value of ΔG^o_f results in a value of K_p which is small--reactants are favored. A negative value would result in a large K_p -- a process in which the products were favored.

38.   The ΔG^o_{rxn} is calculated by the equation: $\Delta G^o_{rxn} - RT \ln K$

$$\Delta G^o_{rxn} = -(8.314\text{ J/K·mol})(298\text{ K})(\frac{1.000\text{ kJ}}{1000.\text{ J}}) \ln (2.5)$$
$$= -2.27\text{ kJ/mol}$$

40.   a. For the reaction: $SiCl_4$ (g) + 2 H_2O (l) → SiO_2 (s) + 4 HCl (aq)
       ΔG^o_f (kJ/mol)   - 616.98   - 237.13   - 856.64   - 131.23

$$\Delta G^o_{rxn} = [(1\text{ mol})(-856.64\text{ kJ/mol}) + (4\text{ mol})(-131.23\text{ kJ/mol})] -$$
$$[(1\text{ mol})(-616.98\text{ kJ/mol}) + (2\text{ mol})(-237.13\text{ kJ/mol})]$$
$$= -290.32\text{ kJ}$$

The reaction is predicted to be spontaneous.

b. Calculate K_p:

$$\Delta G^{\circ}_{rxn} = -RT \ln K_p$$

$$-290.32 \times 10^3 \text{ J} = -(8.314 \text{ J/K} \cdot \text{mol})(298 \text{ K}) \ln K_p$$

$$117 = \ln K_p$$

$$7.77 \times 10^{50} = K_p$$

The negative value for ΔG° indicates that the products are favored at 25 °C.

42.   The estimated entropy of n-pentane should be approximately 40 J/K • mol greater than that of n-butane.

$$310.0 + 40 \approx 350 \text{ J/K} \cdot \text{mol}$$

44.   The standard state entropy of Br_2 is greater than that of I_2 in agreement with the fact that the **entropy of liquids is** usually **greater than that of solids**.

46.

Process	ΔH°	ΔS°	ΔG°
a. Electrolysis of H_2O (l) to form gaseous H_2, O_2	+	+	+

The process does not proceed spontaneously ($\Delta G^{\circ} > 0$). Two moles of water (l) produce three moles of gas ($\Delta S^{\circ} > 0$). Enthalpy of formation of a compound is greater than the enthalpy of formation for elements (in their standard states) ($\Delta H^{\circ} > 0$).

Process	ΔH°	ΔS°	ΔG°
b. Dissolving NH_4Cl in water	+	+	-

The dissolution of NH_4Cl proceeds spontaneously upon mixing the solid with water ($\Delta G^{\circ} < 0$). The entropy of a substance in solution is greater than the entropy of a solid ($\Delta S^{\circ} > 0$). The reduction in temperature is a result of energy <u>absorbed</u> by the dissolving process ($\Delta H^{\circ} > 0$).

Process	ΔH°	ΔS°	ΔG°
c. Explosion of dynamite	-	+	-

The explosion of dynamite releases much heat ($\Delta H^{\circ} < 0$). The explosion produces many moles of gas from a liquid ($\Delta S^{\circ} > 0$). Unfortunately for many miners, the reaction proceeds spontaneously (and often very rapidly--but that's for "kinetics" to explain).

	Process	ΔH^o	ΔS^o	ΔG^o
d.	Combustion of gasoline	-	+	-

The combustion of gasoline in an automobile releases heat ($\Delta H^o < 0$). There is an increase in the number of moles of gaseous products over moles of reactants ($\Delta S^o > 0$). With a negative ΔH^o and a positive ΔS^o, ΔG^o should also be negative.

48. The ΔG^o_{rxn} is calculated by the equation: $\Delta G^o_{rxn} - RT \ln K$

$$\Delta G^o_{rxn} = -(8.314 \text{ J/K} \cdot \text{mol})(298 \text{ K})(\frac{1.000 \text{ kJ}}{1000. \text{ J}}) \ln (0.15)$$

$$= 4.70 \text{ kJ/ mol N}_2\text{O}_4$$

We can compare the values from Appendix J.

$$\Delta G^o_{rxn} = [2 \cdot \Delta G^o_f \text{ NO}_2 \text{ (g)}] - \Delta G^o_f \text{ N}_2\text{O}_4 \text{ (g)}]$$

$$= (2 \text{ mol})(51.31 \text{ kJ/mol}) - (1 \text{ mol})(97.89 \text{ kJ/mol})$$

$$= 4.73 \text{ kJ/ mol N}_2\text{O}_4$$

50. a. For the reaction: $C_6H_6 \text{ (g)} + 3 H_2 \text{ (g)} \rightarrow C_6H_{12} \text{ (g)}$

$$\Delta G^o_{rxn} = \Delta H^o_f - T \Delta S^o_{rxn}$$

$$= -206.1 \text{ kJ} - (298 \text{ K})(-363.12 \text{ J/K} \cdot \text{mol})(\frac{1.000 \text{ kJ}}{1000 \text{ J}})$$

$$= -97.9 \text{ kJ}$$

The reaction is predicted to be spontaneous. The reaction is enthalpy driven.

b. ΔG^o_f for benzene

$$\Delta G^o_{rxn} = [\Delta G^o_f \text{ C}_6\text{H}_{12} \text{ (g)}] - [\Delta G^o_f \text{ C}_6\text{H}_6 \text{ (g)} + 3 \cdot \Delta G^o_f \text{ H}_2 \text{ (g)}]$$

$$-97.9 \text{ kJ} = (31.76 \text{ kJ}) - (\Delta G^o_f \text{ C}_6\text{H}_6 \text{ (g)} + 0)$$

$$-129.7 \text{ kJ} = -\Delta G^o_f \text{ C}_6\text{H}_6 \text{ (g)}$$

$$129.7 \text{ kJ} = \Delta G^o_f \text{ C}_6\text{H}_6 \text{ (g)}$$

52. Calculate ΔG^o_{rxn} for:

	$C_6H_{12}O_6 \text{ (s)} + 6 O_2 \text{ (g)} \rightarrow 6 CO_2 \text{ (g)} + 6 H_2O \text{ (l)}$			
ΔH^o_f (kJ/mol)	- 1260	0	- 393.51	- 285.830
S^o (J/K $\cdot$ mol)	288.9	205.138	213.74	69.91

$$\Delta H^o_{rxn} = [(6 \text{ mol})(-393.51 \text{ kJ/mol}) + (6 \text{ mol})(-285.830 \text{ kJ/mol})] -$$
$$[(1 \text{ mol})(-1260 \text{ kJ/mol}) + 0]$$

$$= -2816 \text{ kJ}$$

$$\Delta S^o_{rxn} = [(6 \text{ mol})(213.74 \text{ J/K} \cdot \text{mol}) + (6 \text{ mol})(69.91 \text{ J/K} \cdot \text{mol})] -$$
$$[(1 \text{ mol})(288.9 \text{ J/K} \cdot \text{mol}) + (6 \text{ mol})(205.138 \text{ J/K} \cdot \text{mol})]$$
$$= 182.5 \text{ J/K}$$

$$\Delta G^o_{rxn} = \Delta H^o_f - T \Delta S^o_{rxn}$$
$$= -2816 \text{ kJ} - (298 \text{ K})(182.5 \text{ J/K})(\frac{1.000 \text{ kJ}}{1000 \text{ J}})$$
$$= -2870 \text{ kJ}$$

54.   a. Reaction: $C(s) + H_2O(g) \rightarrow CO(g) + H_2(g)$

ΔH^o_f (kJ/mol) 0 -241.82 -110.52 0

S^o (J/K $\cdot$ mol) 5.74   188.83   197.67   130.68

$$\Delta H^o_{rxn} = [(1 \text{ mol})(-110.52 \text{ kJ/mol}) + 0] - [0 + (1 \text{ mol})(-241.82 \text{ kJ/mol})]$$
$$= 131.30 \text{ kJ}$$

$$\Delta S^o_{rxn} = [(1 \text{ mol})(197.67 \text{ J/K} \cdot \text{mol}) + (1 \text{ mol})(130.68 \text{ J/K} \cdot \text{mol})] -$$
$$[(1 \text{ mol})(5.74 \text{ J/K} \cdot \text{mol}) + (1 \text{ mol})(188.83 \text{ J/K} \cdot \text{mol})]$$
$$= 133.78 \text{ J/K}$$

$$\Delta G^o_{rxn} = \Delta H^o_f - T \Delta S^o_{rxn}$$
$$= 131.30 \text{ kJ} - (298 \text{ K})(133.78 \text{ J/K})(\frac{1.000 \text{ kJ}}{1000 \text{ J}})$$
$$= 91.4 \text{ kJ}$$

b. K_p for the reaction: $\Delta G^o = - RT \ln K_p$

$$91.4 \times 10^3 \text{ J} = -(8.314 \text{ J/K} \cdot \text{mol})(298 \text{ K}) \ln K_p$$
$$-36.9 = \ln K_p$$
$$9.5 \times 10^{-17} = K_p$$

c. The reaction is predicted to be nonspontaneous at 25 °C ($\Delta G^o > 0$).

Temperature at which reaction becomes spontaneous:

$$\Delta G^o_{rxn} = \Delta H^o_f - T \Delta S^o_{rxn}$$

with the ΔH^o and ΔS^o calculated for this equation (from part a), find the temperature at which $\Delta G^o = 0$.

$$0 = 131.30 \text{ kJ} - T \cdot (133.78 \times 10^{-3} \text{ kJ/K})$$
$$-131.30 \text{ kJ} = - T \cdot (133.78 \times 10^{-3} \text{ kJ/K})$$
$$981.5 \text{ K} = T$$

At any T greater than this (708.5 °C) ΔG^o will be negative--and the process spontaneous.

d. Temperature at which $K_p = 1.0 \times 10^{-4}$:

$$\Delta G^\circ_{rxn} = -RT \ln K_p$$
$$91.4 \times 10^3 \text{ J} = -(8.314 \text{ J/K} \cdot \text{mol})(T) \ln (1.0 \times 10^{-4})$$
$$1190 \text{ K} = T \text{ or approximately } 920 \text{ °C.}$$

56.   For the reaction: $CH_3OH \text{ (l)} \rightarrow CH_4 \text{ (g)} + \frac{1}{2} O_2 \text{ (g)}$

ΔH°_f (kJ/mol)	-238.7	-74.8	0
S° (J/K $\cdot$ mol)	126.8	186.264	205.138

a. Entropy change for the reaction:
$$\Delta S^\circ_{rxn} = [(1 \text{ mol})(186.264 \text{ J/K} \cdot \text{mol}) + (\frac{1}{2} \text{ mol})(205.138 \text{ J/K} \cdot \text{mol})] -$$

$$[(1 \text{ mol})(126.8 \text{ J/K} \cdot \text{mol})]$$

$$= 162.0 \text{ J/K}$$

The **increase** in entropy is **anticipated** with the production of gases from a liquid.

b. Is the reaction spontaneous?

Calculate ΔH°_{rxn} :
$$\Delta H^\circ_{rxn} = [(1 \text{ mol})(- 74.8 \text{ kJ/mol}) + 0] - [(1 \text{ mol})(- 238.7 \text{ kJ/mol})]$$
$$= 163.9 \text{ kJ}$$

Calculate ΔG°_{rxn} :
$$\Delta G^\circ_{rxn} = \Delta H^\circ - T\Delta S^\circ$$
$$= 163.9 \text{ kJ} - (298 \text{ K})(162.0 \text{ J/K})(\frac{1.000 \text{ kJ}}{1000 \text{ J}})$$
$$= 115.6 \text{ kJ}$$

The reaction is not spontaneous. This is not surprising especially when one considers the many industrial and commercial uses of methanol.

c. Temperature at which the reaction becomes spontaneous:

Calculate the temperature at which $\Delta G = 0$:
$$0 = 163.9 \text{ kJ} - T(0.1620 \text{ kJ/K})$$
$$- 163.9 \text{ kJ} = - T(0.1620 \text{ kJ/K})$$
$$1012 \text{ K} = T \text{ or } 739 \text{ °C}$$

Any temperature greater than 739 °C would be sufficient to make the reaction spontaneous.

58.   Calculate ΔG^o for the reaction: H_2S (g) $+ \frac{1}{2} O_2$ (g) $\rightarrow$ H_2O (l) $+$ S (s)

$$\Delta G^o_{rxn} = [\Delta G^o_f\, H_2O\ (l) + \Delta G^o_f\, S\ (s)] - [\Delta G^o_f\, H_2S\ (g) + \frac{1}{2} \cdot \Delta G^o_f\, O_2\ (g)]$$

$$= [-237.129\ kJ\ +\ 0] - [-33.56\ kJ + 0]$$

$$= -203.57\ kJ$$

Note that to obtain the desired reaction, we must multiply the H_2S equation by 24. The ΔG^o for that process is therefore $(24)(-203.57\ kJ) = -4885.7\ kJ$

Now add the two equations: ΔG^o_{rxn}

$$6\ CO_2\ (g) + 6\ H_2O\ (l) \rightarrow C_6H_{12}O_6\ (aq) + 6\ O_2\ (g) \qquad +\ 2870.\ \ kJ$$

$$24\ H_2S\ (g) + 12\ O_2\ (g) \rightarrow 24\ H_2O\ (l) + 24\ S\ (s) \qquad \underline{-\ 4885.7\ kJ}$$

$$-\ 2016\ \ \ kJ$$

For an overall equation:

$$24\ H_2S\ (g)\ + 6\ O_2\ (g) + 6\ CO_2\ (g) \rightarrow 18\ H_2O\ (l) + 24\ S\ (s) + C_6H_{12}O_6\ (aq)$$

With an overall ΔG^o_{rxn} which is less than zero, the reaction should be spontaneous.

60.   Predict changes in the equilibrium position for the reaction:

$$H_2\ (g) + ZnO\ (s) \Leftrightarrow H_2O\ (g) + Zn\ (s)$$

Change	Shift in equilibrium
a. Addition of solid ZnO	no change
b. Addition of H_2 (g)	shift to the right
c. Removal of Zn (s)	no change
d. Removal of H_2O (g)	shift to the right

Note that the addition or removal of a solid has no change on the position of equilibrium. Examination of the equilibrium constant will show that these terms do not appear. The concentrations of solids are a function of their respective densities, not the amount of solid present.

e. Effect on the equilibrium by increasing the temperature:
This effect can be predicted by examining the ΔH^o and ΔS^o for the reaction.

$$\Delta H^o_{rxn} = [(1\ mol)(-241.818\ kJ/mol)\ +\ 0] - [0\ +\ (1\ mol)(-348.28\ kJ/mol)]$$

$$= 106.46\ kJ$$

ΔS^o_{rxn} = [(1 mol)(188.825 J/K • mol) + (1 mol)(41.63 J/K • mol)] -

$\qquad\qquad$ [(1 mol)(130.684 J/K • mol) + (1 mol)(43.64 J/K • mol)]

$\qquad$ = 56.13 J/K

From the calculations above, we see that with both ΔH^o_{rxn} and ΔS^o_{rxn} having positive values, **an increase in temperature** will result in a less positive (or, said another way—more negative) value for ΔG^o, favoring the right side of the equilibrium.

f. $\qquad \Delta G^o_{rxn} = \Delta H^o - T\,\Delta S^o$

$$= 106.46 \text{ kJ} - (298 \text{ K})(56.13 \text{ J/K})(\frac{1.000 \text{ kJ}}{1000 \text{ J}}) = 89.73 \text{ kJ}$$

g. The positive value for ΔG^o indicates a process that is **nonspontaneous** at 25 ˚C.

h. The comparison of the equilibrium constant, K, with respect to one (1) can be decided by observing that $\Delta G^o > 0$.

$$\Delta G^o = - RT \ln K$$

Since R and T (at room temperature) are positive, a positive value for ΔG^o, means that K is less than 1.

CHAPTER 21: Electrochemistry: The Study of Oxidation-Reduction Reactions

12. For the reaction: $Cu\ (s)\ +\ 2\ Ag^+\ (aq)\ \rightarrow\ Cu^{2+}\ (aq)\ +\ 2\ Ag\ (s)$

    a. The half reactions are:

    b. | Processes: | Compartment |
    |---|---|

    $Cu\ (s)\ \rightarrow\ Cu^{2+}\ (aq)\ +\ 2\ e^-$ oxidation anode

    $Ag^+\ (aq)\ +\ 1\ e^-\ \rightarrow\ Ag\ (s)$ reduction cathode

14. The copper electrode is found to be the external anode (+) and the tin electrode the external cathode (-). The copper electrode is therefore the **internal cathode** and the tin electrode the **internal anode**.

    The half-reactions occurring in the half-cells are:

	Processes:	Compartment
$Cu^{2+}\ (aq)\ +\ 2\ e^-\ \rightarrow\ Cu\ (s)$	reduction	cathode
$Sn\ (s)\ \rightarrow\ Sn^{2+}\ (aq)\ +\ 2\ e^-$	oxidation	anode

    The abbreviated notation for the cell is: $Sn\ |\ Sn^{2+}\ (aq)\ \|\ Cu^{2+}\ (aq)\ |\ Cu$

16. For a cell with $E° = +0.46$, the $\Delta G°$ is:

$$\Delta G° = -nFE°$$

$$= -2\ mole\ \bullet\ 9.65\ x\ 10^4\ \frac{J}{volt\ \bullet\ mol}\ \bullet\ 0.46\ volt\ \bullet\ \frac{1.000\ kJ}{1000\ J}$$

$$= -89\ kJ$$

18. Calculate $E°$ and decide if each reaction is spontaneous:

    a. $2\ I^-\ (aq)\ +\ Zn^{2+}\ (aq)\ \rightarrow\ I_2\ (aq)\ +\ Zn\ (s)$

	process	potential
$2\ I^-\ (aq)\ \rightarrow\ I_2\ (aq)\ +\ 2\ e^-$	oxidation	- 0.535 V
$Zn^{2+}\ (aq)\ +\ 2\ e^-\ \rightarrow\ Zn\ (s)$	reduction	- 0.763 V

    Process is not spontaneous. $E°$ cell $= -1.298$ V

    b. $Zn^{2+}\ (aq)\ +\ Ni\ (s)\ \rightarrow\ Zn\ (s)\ +\ Ni^{2+}\ (aq)$

	process	potential
$Ni\ (s)\ \rightarrow\ Ni^{2+}\ (aq)\ +\ 2\ e^-$	oxidation	+ 0.25 V
$Zn^{2+}\ (aq)\ +\ 2\ e^-\ \rightarrow\ Zn\ (s)$	reduction	- 0.763 V

    Process is not spontaneous. $E°$ cell $= -0.51$ V

c. $2\,Cl^-(aq) + Cu^{2+}(aq) \rightarrow Cu(s) + Cl_2(g)$

	process	potential
$2\,Cl^-(aq) \rightarrow Cl_2(g) + 2\,e^-$	oxidation	-1.360 V
$Cu^{2+}(aq) + 2\,e^- \rightarrow Cu(s)$	reduction	$+\underline{0.337\text{ V}}$

Process is not spontaneous.　　$E°$ cell $= -1.023$ V

d. $Sn^{2+}(aq) + Br_2(l) \rightarrow Sn^{4+}(aq) + 2\,Br^-(aq)$

	process	potential
$Sn^{2+}(aq) \rightarrow Sn^{4+}(aq) + 2\,e^-$	oxidation	-0.15 V
$Br_2(l) + 2\,e^- \rightarrow 2\,Br^-(aq)$	reduction	$+\underline{1.066\text{ V}}$

Process is not spontaneous.　　$E°$ cell $= +0.92$ V

20.　a. $Sn^{2+}(aq) + 2\,Ag(s) \rightarrow Sn(s) + 2\,Ag^+(aq)$

Sn^{2+} is reduced (-0.14 V); Ag is oxidized (-0.7994 V)

$E° = (-0.14 - 0.7994) = -0.94$ V　　　　not spontaneous

b. $Zn(s) + Sn^{4+}(aq) \rightarrow Sn^{2+}(aq) + Zn^{2+}(aq)$

Sn^{4+} is reduced ($+0.15$ V); Zn is oxidized ($+0.763$ V)

$E° = (+0.15 + 0.763) = +0.91$ V　　　　spontaneous

c. $I_2(aq) + 2\,Br^-(aq) \rightarrow 2\,I^-(aq) + Br_2(l)$

I_2 is reduced ($+0.535$ V); Br^- is oxidized (-1.066 V)

$E° = (+0.535 - 1.066) = -0.531$ V　　　　not spontaneous

22.　For the half-reactions listed:

a. Weakest oxidizing agent: V^{2+}

The more positive the $E°$, the better the oxidizing ability of the specie.

b. Strongest oxidizing agent: Cl_2

c. Strongest reducing agent: V

The more negative the $E°$, the better the reducing ability of a substance.

d. Weakest reducing agent: Cl^-

e. Pb (s) **cannot** reduce V^{2+} (aq).

Since the reduction potential of lead is less negative than that of V^{2+}, lead cannot reduce V^{2+}.

f. I_2 cannot oxidize Cl^- to Cl_2

The greater the value of $E°$, the better the oxidizing ability of a substance.

The $E°$ for Cl_2 is greater than that for I_2, hence I_2 cannot oxidize Cl^- to Cl_2.

g. Pb(s) can reduce Cl_2, and I_2

The comment from part c applies. The reduction potential of Pb is more negative than that for Cl_2 or I_2, making Pb capable of reducing either of these substances.

24.   a. Maximum positive standard potential:

$$Cl_2 \text{ (g)} + V \text{ (s)} \rightarrow 2\,Cl^- \text{ (aq)} + V^{2+} \text{ (aq)} \qquad E° = +2.54 \text{ V}$$

potential

b. I_2 (s) + 2 e$^-$ $\rightarrow$ 2 I$^-$ (aq) $\qquad\qquad$ + 0.535 V

 V (s) $\rightarrow$ V^{2+} (aq) + 2 e$^-$ $\qquad\qquad$ + 1.18 V

$\qquad\qquad\qquad\qquad\qquad\qquad$ $E°_{cell}$ = + 1.72 V

The spontaneous reaction is: I_2 (s) + V (s) $\rightarrow$ 2 I$^-$ (aq) + V^{2+} (aq)

26.   The standard reduction potentials for the two species are:

$\qquad\qquad$ Zn^{2+} (aq) + 2 e$^-$ $\rightarrow$ Zn (s) $\qquad$ $E°$ = - 0.763 V

$\qquad\qquad$ Ag^+ (aq) + 1 e$^-$ $\rightarrow$ Ag (s) $\qquad$ $E°$ = + 0.7994 V

a. The spontaneous reaction would be:

$\qquad\qquad$ 2 Ag^+ (aq) + Zn (s) $\rightarrow$ 2 Ag (s) + Zn^{2+} (aq)

The potential for the cell would be: $\qquad$ $E°$ = (+ 0.7994 + 0.763) = + 1.56 V

b. Silver is the cathode and zinc is the anode. The abbreviated notation for the cell is:

$\qquad\qquad$ Zn | Zn^{2+} (aq) || Ag^+ (aq)| Ag

c. The diagram for the cell:

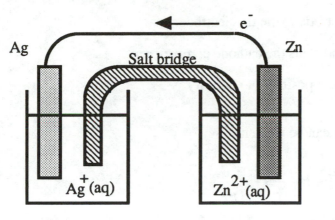

d. A strip of Zinc would serve as the anode.

e. Electrons would flow from the zinc electrode to the silver electrode.

f. Nitrate ions would flow from the Ag compartment (site of increasing electrons) to the Zn compartment (as electrons leave the external Zn

electrode, an increasing positive charge would develop).

28.   The potentials involved in the reaction are:

	process	potential
$Cr\ (s) \rightarrow Cr^{3+}\ (aq) + 2\ e^-$	oxidation	+ 0.74 V
$Fe^{2+}\ (aq) + 2\ e^- \rightarrow Fe\ (s)$	reduction	- 0.44 V

$$E°_{cell} = +0.30\ V$$

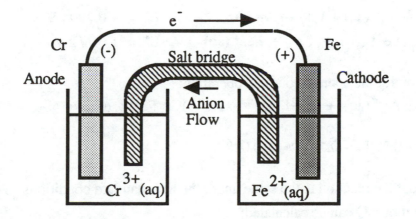

30.   The standard reduction potentials for the two species are:

$$Zn^{2+}\ (aq) + 2\ e^- \rightarrow Zn\ (s) \qquad E° = -0.763\ V$$
$$Sn^{2+}\ (aq) + 2\ e^- \rightarrow Sn\ (s) \qquad E° = -0.14\ \ V$$

a. The spontaneous reaction would be:

$$Sn^{2+}(aq) + Zn\ (s) \rightarrow Sn\ (s) + Zn^{2+}\ (aq)$$

b. The potential for the cell would be: $E^\circ = (-0.14 + 0.763) = +0.62$ V

c. Electrons move from the zinc electrode to the tin electrode.

d. Zinc is the anode compartment and tin is the cathode compartment.

e. Tin is the cathode and would have a positive (+) polarity.

32.   The Nernst equation for this reaction can be written:

$$E_{cell} = E^0_{cell} - \frac{0.0592}{n} \log \frac{[Fe^{2+}]^2}{[Fe^{3+}]^2[I^-]^2}$$

Substituting appropriate values:

$$E_{cell} = (0.771 - 0.535) - \frac{0.0592}{2} \log \frac{(0.1)^2}{(0.1)^2(0.1)^2}$$

$$= 0.236 - 0.0296 \cdot \log (100)$$

$$= 0.177 \text{ V}$$

Note that the potential of the cell, E_{cell}, is decreased from that of the standard potential of the cell.

34.   From Appendix I:

$$Ag^+ (aq) + 1\ e^- \rightarrow Ag\ (s) \qquad E^\circ = +0.7994 \text{ V}$$
$$Fe^{3+} (aq) + 1\ e^- \rightarrow Fe^{2+} (aq) \qquad E^\circ = +0.771 \text{ V}$$

a. The reaction for the spontaneous cell operation:
$$Ag^+ (aq) + Fe^{2+} (aq) \rightarrow Ag\ (s) + Fe^{3+} (aq)$$

b. $E^\circ_{cell} = (+0.7994 - 0.771) = +0.028$ V

c. **Assume** that the reaction is the same under the non-standard conditions as under the standard conditions. Q can be calculated:

$$Q = \frac{[Fe^{3+}]}{[Fe^{2+}][Ag^+]} = \frac{(1.0 \text{ M})}{(1.0 \text{ M})(0.1 \text{ M})} = 10$$

Substitution into the Nernst equation yields:

$$E_{cell} = E^\circ_{cell} - \frac{0.0592}{n} \log Q$$

$$= +0.03 \text{ V} - \frac{0.0592}{1} \log (10) = -0.03 \text{ V}$$

The cell reaction--contrary to our assumption--is not the same as under standard conditions. The net reaction is now:

$$Ag\ (s)\ +\ Fe^{3+}\ (aq) \rightarrow Ag^+\ (aq)\ +\ Fe^{2+}\ (aq)$$

36.   To calculate the equilibrium constants for the reactions, we use the Nernst equation expressed as follows:

$$E_{cell}\ =\ E°_{cell}\ -\ \frac{0.0592}{n}\ \log K$$

At equilibrium $E_{cell} = 0$. $E°_{cell}$ must be calculated.

a.
		potential
$2\ I^-\ (aq)\ \rightarrow\ I_2\ (aq) + 2\ e^-$		- 0.535 V
$Fe^{3+}\ (aq)\ +\ 1\ e^- \rightarrow Fe^{2+}\ (aq)$		+0.771 V
	$E°_{cell}\ = +\ 0.236\ V$	

The number of electrons in the balanced overall equation (n) is 2.
Substituting we get:

$$0\ =\ 0.236\ V\ -\ \frac{0.0592}{2}\ \log K$$

$$\frac{(2)(0.236)}{(0.0592)}\ =\ \log K\ =\ 7.97\ \text{and } K\ =\ 9.4 \times 10^7$$

b.
		potential
$I_2\ (aq) + 2\ e^-\ \rightarrow\ 2\ I^-\ (aq)$		+ 0.535 V
$2\ Br^-\ (aq)\ \rightarrow\ Br_2\ (l) + 2\ e^-$		- 1.066 V
	$E°_{cell} = -\ 0.531\ V$	

The number of electrons in the balanced overall equation (n) is 2.
Substituting we get:

$$0\ =\ -\ 0.531\ V\ -\ \frac{0.0592}{2}\ \log K$$

$$\frac{(2)(-\ 0.531)}{(0.0592)}\ =\ \log K\ =\ -17.94\ \text{and } K = 1.2 \times 10^{-18}$$

38.   The half-reaction for the reduction of silver is: $Ag^+\ (aq)\ +\ 1\ e^- \rightarrow Ag\ (s)$

$$\frac{107.9\ g}{1\ mol\ Ag} \cdot \frac{1\ mol\ Ag}{1\ mol\ e^-} \cdot \frac{1\ mol\ e^-}{9.65 \times 10^4\ C} \cdot \frac{1\ C}{1\ amp \cdot s} \cdot \frac{0.015\ amp}{1}$$

$$\cdot \frac{60\ s}{1\ min} \cdot \frac{155\ min}{1}\ =\ 0.16\ g\ Ag$$

40.   Solutions to problems of this sort are best solved by beginning with a factor containing the desired units. Connecting this factor to data provided usually gives a direct path to the answer.

units desired
↓

$$\frac{63.55 \text{ g Cu}}{1 \text{ mol Cu}} \cdot \frac{1 \text{ mol Cu}}{2 \text{ mol e}^-} \cdot \frac{1 \text{ mol e}^-}{9.65 \times 10^4 \text{ C}} \cdot \frac{1 \text{ C}}{1 \text{ amp} \cdot \text{s}}$$

$$\cdot \frac{2.50 \text{ amps}}{1} \cdot \frac{3600 \text{ s}}{\text{hr}} \cdot \frac{2.00 \text{ hr}}{1} = 5.93 \text{ g Cu}$$

The second factor $\left(\frac{1 \text{ mol Cu}}{2 \text{ mol e}^-}\right)$ is arrived at by looking at the reduction half-reaction:

$$Cu^{2+} (aq) + 2 e^- \rightarrow Cu (s)$$

All other factors are either data or common unity factors (e.g. $\frac{3600 \text{ s}}{1 \text{ hr}}$).

42.   Current flowing if 0.052 g Ag are deposited in 450 s:

$$\frac{0.052 \text{ g Ag}}{1} \cdot \frac{1 \text{ mol Ag}}{108 \text{ g Ag}} \cdot \frac{1 \text{ mol e}^-}{1 \text{ mol Ag}} \cdot \frac{9.65 \times 10^4 \text{ C}}{1 \text{ mol e}^-} = 46 \text{ C}$$

and with this charge flowing in 450 s, the current is: $\frac{46 \text{ C}}{450 \text{ s}} = 0.10$ amperes

44.   The mass of aluminum produced in 8.0 hr by a current of 1.0×10^5 amp:

$$\frac{26.98 \text{ g Al}}{1 \text{ mol Al}} \cdot \frac{1 \text{mol Al}}{3 \text{ mol e}^-} \cdot \frac{1 \text{ mol e}^-}{9.65 \times 10^4 \text{ C}} \cdot \frac{1 \text{ C}}{1 \text{ amp} \cdot \text{s}}$$

$$\cdot \frac{1.0 \times 10^5 \text{ amp}}{1} \cdot \frac{3600 \text{ s}}{1 \text{ hr}} \cdot \frac{8.0 \text{ hr}}{1} = 2.7 \times 10^5 \text{ g Al}$$

46.   The mass of lead consumed by a current of 1.0 amp for 50. hours:
During discharge of the lead storage battery, the anode reaction is Pb (s) → Pb^{2+} (aq) + 2e.

$$\frac{207 \text{ g Pb}}{1 \text{ mol Pb}} \cdot \frac{1 \text{ mol Pb}}{2 \text{ mol e}^-} \cdot \frac{1 \text{ mol e}^-}{9.65 \times 10^4 \text{ C}} \cdot \frac{1 \text{ C}}{1 \text{ amp} \cdot \text{s}}$$

$$\cdot \frac{1.0 \text{ amp}}{1} \cdot \frac{3600 \text{ s}}{1 \text{ hr}} \cdot \frac{50. \text{ hr}}{1} = 190 \text{ g Pb}$$

296

48.   Calculate the number of Joules (V • C) associated with 1.0 amp of current for 100. hours.

$$\frac{1.0 \text{ amp}}{1} \cdot \frac{1 \text{ C}}{1 \text{ amp} \cdot \text{s}} = \frac{1.0 \text{ C}}{1 \text{ s}}$$ With a voltage of 12.0 volts, the number of joules is:

$$\frac{12.0 \text{ V}}{1} \cdot \frac{1.0 \text{ C}}{1 \text{ s}} = \frac{12.0 \text{ V} \cdot \text{C}}{1 \text{ s}}$$ and since 1 V • C = 1 Joule we obtain

$$\frac{12 \text{ J}}{1 \text{ s}}$$ and since 1 watt $= \frac{1 \text{ J}}{\text{s}}$ then the power is 12 watts.

50. $22.66 \times 10^9 \text{ lbs Cl}_2 \cdot \dfrac{454 \text{ g Cl}_2}{1.0 \text{ lb Cl}_2} \cdot \dfrac{1 \text{ mol Cl}_2}{70.906 \text{ g Cl}_2} \cdot \dfrac{2 \text{ mol e}^-}{1 \text{ mol Cl}_2} \cdot \dfrac{9.65 \times 10^4 \text{ C}}{1 \text{ mol e}^-}$

$$= 2.800 \times 10^{16} \text{ C}$$

The power is then:

$$\frac{2.800 \times 10^{16} \text{ C}}{1} \cdot \frac{4.6 \text{ V}}{1} \cdot \frac{1 \text{ J}}{1 \text{ V} \cdot \text{C}} \cdot \frac{1 \text{ kwh}}{3.6 \times 10^6 \text{ J}} = 3.6 \times 10^{10} \text{ kwh.}$$

52.   a. At the anode, oxidation of I⁻ occurs: $2 \text{ I}^- \text{(aq)} \rightarrow \text{I}_2 \text{(aq)} + 2 \text{ e}^-$ Sign: **positive**

b. At the cathode, reduction of water occurs:
$$2 \text{ H}_2\text{O (l)} + 2 \text{ e}^- \rightarrow \text{H}_2 \text{ (g)} + 2 \text{ OH}^- \text{(aq)} \quad \text{Sign: \textbf{negative}}$$

c. $\dfrac{0.050 \text{ amp}}{1} \cdot \dfrac{5.0 \text{ hr}}{1} \cdot \dfrac{3600 \text{ s}}{1 \text{ hr}} \cdot \dfrac{1 \text{ C}}{1 \text{ amp} \cdot \text{s}} \cdot \dfrac{1 \text{ mol e}^-}{9.65 \times 10^4 \text{ C}} = 9.3 \times 10^{-3} \text{ mol e}^-$

Amount of products expected:

$9.3 \times 10^{-3} \text{ mol e}^- \cdot \dfrac{1 \text{ mol I}_2}{2 \text{ mol e}^-} \cdot \dfrac{253.8 \text{ g I}_2}{1 \text{ mol I}_2} = 1.2 \text{ g I}_2$

$9.3 \times 10^{-3} \text{ mol e}^- \cdot \dfrac{2 \text{ mol OH}^-}{2 \text{ mol e}^-} \cdot \dfrac{1 \text{ mol KOH}}{1 \text{ mol OH}^-} \cdot \dfrac{56.11 \text{ g KOH}}{1 \text{ mol KOH}} = 0.52 \text{ g KOH}$

$9.3 \times 10^{-3} \text{ mol e}^- \cdot \dfrac{1 \text{ mol H}_2}{2 \text{ mol e}^-} \cdot \dfrac{2.02 \text{ g H}_2}{1 \text{ mol H}_2} = 9.4 \times 10^{-3} \text{ g H}_2$

54.   a. Electrolysis of KBr (aq):
        Reduction of K⁺ $E° = -2.925 \text{ V}$
        Reduction of H_2O $= -0.83 \text{ V}$
So H_2O will be reduced to H_2 at the cathode (in preference to elemental K).

Oxidation of Br⁻ $E° = -1.066$ V
Oxidation of H_2O $= -1.23$ V (+0.6 V overvoltage)

So Br⁻ will be oxidized to Br_2 at the anode.

b. Electrolysis of NaF (molten):
   At the anode: $2 F^- (aq) \rightarrow F_2 (g) + 2 e^-$
   At the cathode: $Na^+ (aq) + e^- \rightarrow Na (s)$

c. Electrolysis of NaF (aq):
   Reduction of Na^+ $E° = -2.71$ V
   Reduction of H_2O $= -0.83$ V

So H_2O will be reduced to H_2 at the cathode.

   Oxidation of F^- $E° = -2.87$ V
   Oxidation of H_2O $= -1.23$ V (+0.6 V overvoltage)

So H_2O will be oxidized to O_2 at the anode.

56.   Elements function as reducing agents in proportion to their tendencies to be oxidized (lose electrons to other elements). The more negative the value of the reduction potential, the better the substance is oxidized. In order of increasing ability as a reducing agent these elements are:

$$Cl^- (+ 1.358) < Ag (+ 0.7994) < Fe (- 0.44) < Na (- 2.714)$$

58.   The combination producing the largest positive $E°$ is $Mg(s) / Mg^{2+}$.
      The cell reaction is:
$$Cl_2 (g) + Mg (s) \rightarrow Mg^{2+} (aq) + 2 Cl^- (aq)$$

$E°_{cell} = E°_{cathode} + E°_{anode} = + 1.360 + 2.37 = + 3.73$ V

60. $E° = 2.04$ V for the lead storage battery.
      To calculate K for the cell reaction, we need to determine n. The number of moles of electrons, n, can be obtained by examining either the anodic or cathodic half-cell reaction. The anodic half-cell reaction is:

$$Pb (s) + SO_4^{2-} (aq) \rightarrow PbSO_4 (s) + 2 e^-$$

$$\log K = \frac{nE^\circ}{0.0592} = \frac{(2 \text{ mol})(2.04 \text{ V})}{0.0592}$$

$$\log K = 68.9 \qquad \text{and} \qquad K = 8 \times 10^{68}$$

62. Examining the standard reduction potentials in Appendix I, we expect that:

a. Fe^{2+} should be oxidized to Fe^{3+}

O_2 would function as an oxidizing agent. Using the potential for:

$$O_2 (g) + 4 H^+ (aq) + 4 e^- \rightarrow 2 H_2O (l) \qquad E^\circ = +1.229 \text{ V}$$

Construct a cell for $Fe^{2+} \rightarrow Fe^{3+} + 1 e^- \qquad\qquad E^\circ = -0.771 \text{ V}$

Note that $E^\circ_{cell} > 0$ so Fe^{2+} should be oxidized to Fe^{3+}.

Similar reasoning applies in parts b & c.

b. Br^- should be oxidized to Br_2.

$$E^\circ_{cell} = (+1.229 \text{ V} - 1.066 \text{ V}) = 0.163 \text{ V}$$

c. I^- should be oxidized to I_2.

$$E^\circ_{cell} = (+1.229 \text{ V} - 0.535 \text{ V}) = 0.694 \text{ V}$$

64. The two reactions which express the overall corrosion of iron to $Fe(OH)_2$:

$$2 (Fe (s) + 2 OH^- (aq) \rightarrow Fe(OH)_2 (s) + 2 e^-) \qquad E^\circ = 0.877 \text{ V}$$

$$O_2 (g) + 2 H_2O (l) + 4 e^- \rightarrow 4 OH^- (aq) \qquad E^\circ = \underline{0.40 \text{ V}}$$

For an overall reaction :

$$2 Fe (s) + 2 O_2 (g) + 2 H_2O (l) \rightarrow 2 Fe(OH)_2 (s) \quad E^\circ = 1.28 \text{ V}$$

To calculate the E_{net} we use the Nernst equation:

$$E_{net} = 1.28 \text{ V} - \frac{0.0592}{4} \log \left(\frac{1}{[O_2]} \right) \quad \text{where } [O_2] \text{ is expressed by its pressure,}$$

$$(0.20 \text{ atm}).$$

$$E_{net} = +1.27 \text{ V}$$

The equilibrium constant for this reaction is:

$$\log K = \frac{nE^\circ}{0.0592} = \frac{(4)(+1.28 \text{ V})}{0.0592} = 86.5 \quad \text{so } K = 3 \times 10^{86}$$

66. For the desired reaction **one mole of zinc** must lose **two moles of electrons**.

The total charge required is: $10. \text{ hr} \cdot \dfrac{3600 \text{ s}}{1 \text{ hr}} \cdot \dfrac{1.5 \text{ amp}}{1} \cdot \dfrac{1 \text{ C}}{1 \text{ amp} \cdot \text{s}} = 5.4 \times 10^4 \text{ C}$

The amount of electrons required is: $5.4 \times 10^4 \text{ C} \cdot \dfrac{1 \text{ mol e}^-}{9.65 \times 10^4 \text{ C}} = 0.56 \text{ mol e}^-$

The amount of zinc required is: $0.56 \text{ mol e}^- \cdot \dfrac{1 \text{ mol Zn}}{2 \text{ mol e}^-} \cdot \dfrac{65.4 \text{ g Zn}}{1 \text{ mol Zn}} = 18 \text{ g Zn}$

68.   The number of coulombs is:

$450 \text{ amps} \cdot 30. \text{ s} = 1.4 \times 10^4 \text{ C}$ [Remember 1 amp $\cdot$ s = 1 C]

$\dfrac{207 \text{ g Pb}}{1 \text{ mol Pb}} \cdot \dfrac{1 \text{ mol Pb}}{2 \text{ mol e}^-} \cdot \dfrac{1 \text{ mol e}^-}{9.65 \times 10^4 \text{ C}} \cdot \dfrac{1.4 \times 10^4 \text{ C}}{1} = 15 \text{ g Pb}$

70.   a. The stoichiometry of the silver/zinc battery indicates a reaction of one mole each of silver oxide, zinc, and water. The mass of one mole of each of these three substances is:

1 mol Ag_2O 231.7 g
1 mol Zn   65.4 g
1 mol H_2O <u>18.0 g</u>
                 315.1 g

The energy associated with the battery is: $\dfrac{0.1 \text{ C}}{1 \text{ s}} \cdot \dfrac{1.59 \text{ V}}{1} = \dfrac{0.159 \text{ V} \cdot \text{C}}{\text{s}}$

Since 1 V $\cdot$ C = 1 Joule, this energy corresponds to 0.159 J/s. [1 watt = 1 J/s]

The energy/gram for the silver/zinc battery is: $\dfrac{0.159 \text{ J/s}}{315.1 \text{ g}} = 5.05 \times 10^{-4}$ watts/gram.

b. Performing the same calculations for the lead storage battery, using a stoichiometric amount for the overall battery reaction:

1 mol Pb   207.2 g
1 mol PbO_2 239.2 g
2 mol H_2SO_4 <u>196.2 g</u>
                642.6 g

The energy associated with the battery is: $\dfrac{0.1 \text{ C}}{1 \text{ s}} \cdot \dfrac{2.0 \text{ V}}{1} = \dfrac{0.20 \text{ V} \cdot \text{C}}{\text{s}}$

The energy/gram for the lead storage battery is: $\dfrac{0.20 \text{ J/s}}{642.6 \text{ g}} = 3.1 \times 10^{-4}$ watts/g

c. The silver/zinc battery produces more energy (and more power)/gram.

72.   a. Kilowatt-hours consumed by the electrolysis cell producing H_2 and F_2 by the reaction:

$$2 \, HF \, (l) \; \rightarrow \; H_2 \, (g) \; + \; F_2 \, (g)$$

operating at 6.0×10^3 amps at 12 V for 24 hrs.

$$6.0 \times 10^3 \text{ amps} \; \cdot \; \frac{1 \text{ C}}{1 \text{ amp} \cdot \text{s}} \; \cdot \; \frac{3600 \text{ s}}{1 \text{ hr}} \; \cdot \; \frac{24 \text{ hr}}{1} \; = \; 5.2 \times 10^8 \text{ coulombs}$$

The number of joules of energy $= (5.2 \times 10^8 \text{ coulombs})(12 \text{ V})(\frac{1 \text{ J}}{\text{V} \cdot \text{C}}) \; = \; 6.2 \times 10^9 \text{ J}$

The power consumed is :

$$6.2 \times 10^9 \text{ J} \; \cdot \; \frac{1 \text{ kwh}}{3.6 \times 10^6 \text{ J}} \; = \; 1.7 \times 10^3 \text{ kwh}$$

b. The amount of HF consumed:

The half-reactions for HF show that <u>each mole of HF</u> will <u>require 1 mole of electrons</u>.

$$5.2 \times 10^8 \text{ C} \; \cdot \; \frac{1 \text{ mol e}^-}{9.65 \times 10^4 \text{ C}} \; \cdot \; \frac{1 \text{ mol HF}}{1 \text{ mol e}^-} = 5.4 \times 10^3 \text{ mol HF}$$

and $5.4 \times 10^3 \text{ mol HF} \; \cdot \; \dfrac{20.0 \text{ g HF}}{1 \text{ mol HF}} \; = \; 1.1 \times 10^5 \text{ g HF}$

The stoichiometry of the equation indicates that one mol H_2 and one mol F_2 are produced for each two mol of HF electrolyzed.

$$5.4 \times 10^3 \text{ mol HF} \; \cdot \; \frac{1 \text{ mol } H_2}{2 \text{ mol HF}} \; \cdot \; \frac{2.02 \text{ g } H_2}{1 \text{ mol } H_2} \; = \; 5.4 \times 10^3 \text{ g } H_2$$

and $5.4 \times 10^3 \text{ mol HF} \; \cdot \; \dfrac{1 \text{ mol } F_2}{2 \text{ mol HF}} \; \cdot \; \dfrac{38.0 \text{ g } F_2}{1 \text{ mol } F_2} \; = \; 1.0 \times 10^5 \text{ g } F_2$

74.   a. Moles of glucose:

$$2.4 \times 10^3 \text{ kcal} \; \cdot \; \frac{4.184 \text{ kJ}}{1 \text{ kcal}} \; \cdot \; \frac{1 \text{ mol glucose}}{2800 \text{ kJ}} \; = \; 3.6 \text{ mol glucose}$$

Moles of oxygen:

$$3.6 \text{ mol glucose} \; \cdot \; \frac{6 \text{ mol } O_2}{1 \text{ mol glucose}} \; = \; 22 \text{ mol } O_2$$

b. In the half-reaction for O_2: $O_2 \, (g) \; + \; 4 \, e^- \; \rightarrow \; 2 \, H_2O \, (l)$

$$22 \text{ mol } O_2 \; \cdot \; \frac{4 \text{ mol e}^-}{1 \text{ mol } O_2} \; = \; 88 \text{ mol e}^-$$

c. Current flowing from combustion of glucose/second.

$$\frac{3.6 \text{ mol glucose}}{24 \text{ hr}} \cdot \frac{1 \text{ hr}}{3600 \text{ s}} \cdot \frac{88 \text{ mol e}^-}{3.6 \text{ mol glucose}} \cdot \frac{9.65 \times 10^4 \text{ C}}{1 \text{ mol e}^-} \cdot \frac{1 \text{ amp} \cdot \text{s}}{1 \text{ C}}$$

$$= 98 \text{ amps}$$

d. 98 C/s $\cdot$ 1.0 V = 98 J/s

Since 1 watt = 1 J/s, the energy expenditure is **98 watts**.

CHAPTER 22: The Chemistry of Hydrogen and the s-Block Elements

3. The balanced reactions are:

Steam and methane: $H_2O\ (g)\ +CH_4\ (g)\ \rightarrow\ 3\ H_2\ (g)\ +\ CO\ (g)$

Steam and petroleum: $H_2O\ (g)\ +\ CH_2\ (l)\ \rightarrow\ 2\ H_2\ (g)\ +\ CO\ (g)$

Steam and coal : $2\ H_2O\ (g)\ +2\ CH\ (s) \rightarrow 3\ H_2\ (g)\ +\ 2\ CO\ (g)$

Hydrocarbon	Mol hydrocarbon	Hydrocarbon(g)	Mol H_2	Hydrogen (g)
Methane	1	16.0	3	6.0
Petroleum	1	14.0	2	4.0
Coal	2	26.0	3	6.0

	Methane	Petroleum	Coal
Mass of Hydrogen / g of hydrocarbon	$\dfrac{6.0}{16.0} = 0.38$	$\dfrac{4.0}{14.0} = 0.29$	$\dfrac{6.0}{26.0} = 0.23$

4. Energy required to convert 10.0 L of H_2 (g) at 25.0 °C and 3.50 atm to H atoms:

The amount of H_2 gas present is:

$$n = \frac{P \cdot V}{R \cdot T} = \frac{(3.50\ \text{atm})(\ 10.0\ \text{L})}{(\ 0.082057\ \frac{L \cdot atm}{K \cdot mol}\ )(\ 298\ \text{K})} = 1.43\ \text{mol}\ H_2$$

The H-H bond energy is 436 kJ/mol, the energy required to separate the molecules into H atoms is :

$$436\ \text{kJ/mol} \cdot 1.43\ \text{mol}\ H_2 = 624\ \text{kJ}$$

This amount of energy would be released upon recombination of the H atoms into H_2 molecules.

6. Volume of 1.0 kg of H_2 (g) at 25 °C and 1.0 atm:

The amount of H_2 corresponding to 1.0×10^3 g H_2 :

$$1.0 \times 10^3\ \text{g}\ H_2\ \cdot\ \frac{1\ \text{mol}\ H_2}{2.02\ \text{g}\ H_2} = 5.0 \times 10^2\ \text{mol}\ H_2\ \text{(to 2 significant figures)}$$

This amount of H_2 would occupy a volume of :

$$V = \frac{nRT}{P} = \frac{(500 \text{ mol } H_2)\,(0.082057 \frac{L \cdot atm}{K \cdot mol})(298 \text{ K})}{1.0 \text{ atm}} = 1.2 \times 10^4 \text{ L}$$

8.   Electrolysis of KCl (aq):

$$2 \text{ KCl (aq)} + 2 H_2O \text{ (l)} \rightarrow 2 \text{ KOH (aq)} + Cl_2 \text{ (g)} + H_2 \text{ (g)}$$

During the electrolysis the chloride ion is oxidized to chlorine (gas) and water is reduced to hydrogen (gas)--in preference to the K^+ ion being reduced to the elemental metal.

Electrolysis of CsI (aq):

$$2 \text{ CsI (aq)} + 2 H_2O \text{ (l)} \rightarrow 2 \text{ CsOH (aq)} + I_2 \text{ (s)} + H_2 \text{ (g)}$$

Iodide ion is oxidized to iodine, and water is reduced to hydrogen.

10.   Very small cations interact with water strongly. Hence the Li^+ strongly interacts with the dipole water . Larger cations have a weaker ion-dipole interaction, and their salts are not as readily hydrated.

13. BeF_2 is a linear molecule with two atoms attached to the central beryllium atom.

$: \overset{\cdot\cdot}{\underset{\cdot\cdot}{F}} — Be — \overset{\cdot\cdot}{\underset{\cdot\cdot}{F}} :$ **The orbital hybridization for beryllium is sp.**

15.   The reaction of magnesium with nitrogen to form magnesium nitride is:

$$3 \text{ Mg (s)} + N_2 \text{ (g)} \rightarrow Mg_3N_2 \text{ (s)}$$

21.   The hydrolysis of the Li^+ cation is responsible for the acidic nature of $LiCl_2$ solutions.

$$[Li (H_2O)_4]^+(aq) + H_2O \text{ (l)} \rightarrow H_3O^+ \text{ (aq)} + [Li^+ (H_2O)_3(OH^-)] \text{ (aq)}$$

23.   The equations for the equilibria associated with $Mg(OH)_2$ and $Ca(OH)_2$ dissolving are:

$$Mg(OH)_2 \text{ (s)} \Leftrightarrow Mg^{2+} \text{ (aq)} + 2 OH^- \text{ (aq)} \quad K_{sp} = 1.5 \times 10^{-11}$$
$$Ca(OH)_2 \text{ (s)} \Leftrightarrow Ca^{2+} \text{ (aq)} + 2 OH^- \text{ (aq)} \quad K_{sp} = 7.9 \times 10^{-6}$$

The overall reaction has Mg^{2+} ions on the left side of the equation and Ca^{2+} on the right side. This can be accomplished by reversing the first equation above. Remember that the K for such a process would be the reciprocal of the original K.

$$Ca(OH)_2 \text{ (s)} \Leftrightarrow Ca^{2+} \text{ (aq)} + 2\,OH^- \text{ (aq)} \qquad K_{sp} = 7.9 \times 10^{-6}$$
$$Mg^{2+} \text{ (aq)} + 2\,OH^- \text{ (aq)} \Leftrightarrow Mg(OH)_2 \text{ (s)} \qquad \frac{1}{K_{sp}} = 6.7 \times 10^{10}$$

For a net reaction: $Ca(OH)_2 \text{ (s)} + Mg^{2+} \text{ (aq)} \Leftrightarrow Ca^{2+} \text{ (aq)} + Mg(OH)_2 \text{ (s)}$

The K associated with this process is $(7.9 \times 10^{-6})(6.7 \times 10^{10}) = 5.3 \times 10^5$

The magnitude of the equilibrium constant indicates that if one allows sea water--containing magnesium ions--to come into contact with lime, $Ca(OH)_2$, an exchange of cations would occur with solid magnesium hydroxide forming. This magnesium hydroxide could be treated with HCl to form $MgCl_2$, and electrolyzed to elemental magnesium.

24.   a. $MgCl_2$ (l) $\xrightarrow{\text{electricity}}$ Mg (s) + Cl_2 (g)

Mass of Mg produced at the cathode:

$$1000. \text{ kg } MgCl_2 \cdot \frac{1 \text{ mol } MgCl_2}{95.211 \text{ g } MgCl_2} \cdot \frac{1 \text{ mol } Mg}{1 \text{ mol } MgCl_2} \cdot \frac{24.305 \text{ g } Mg}{1 \text{ mol } Mg} = 255.3 \text{ kg } Mg$$

Note the absence of a conversion of mass of $MgCl_2$ from kg to grams. Since the answer was requested in units of kg, any conversion to units of grams would have necessitated a conversion back to kg at the end of the calculation. The two conversion factors would cancel each other, and "leaving them out" causes no harm to the integrity of the reasoning (or the answer).

b. At the anode, chlorine is produced. $[2\,Cl^- \rightarrow Cl_2 + 2\,e^-]$

Mass of Cl_2 produced:

$$1000. \text{ kg } MgCl_2 \cdot \frac{1 \text{ mol } MgCl_2}{95.211 \text{ g } MgCl_2} \cdot \frac{1 \text{ mol } Cl_2}{1 \text{ mol } MgCl_2} \cdot \frac{70.906 \text{ g } Cl_2}{1 \text{ mol } Cl_2} = 744.7 \text{ kg } Cl_2$$

c. Faradays used in the process:

The reduction of magnesium requires 2 Faradays per mole: $Mg^{2+} + 2e^- \rightarrow Mg$

$$255.3 \text{ kg Mg} \cdot \frac{1.000 \times 10^3 \text{ g Mg}}{1.0 \text{ kg Mg}} \cdot \frac{1 \text{ mol Mg}}{24.305 \text{ g Mg}} \cdot \frac{2 \text{ F}}{1 \text{ mol Mg}} = 2.100 \times 10^4 \text{ F}$$

The oxidation of chlorine requires 2 Faradays per mole of chlorine: $2 \text{ Cl}^- \rightarrow Cl_2 + 2 e^-$

$$744.7 \text{ kg Cl}_2 \cdot \frac{1.000 \times 10^3 \text{ g Cl}_2}{1.0 \text{ kg Cl}_2} \cdot \frac{1 \text{ mol Cl}_2}{70.906 \text{ g Cl}_2} \cdot \frac{2 \text{ F}}{1 \text{ mol Cl}_2} = 2.100 \times 10^4 \text{ F}$$

The total number of Faradays of electricity used in the process is 2.100×10^4 F.

d. Joules required per mole of magnesium:

$$\frac{8.4 \text{ kwh}}{1 \text{ lb Mg}} \cdot \frac{3.60 \times 10^6 \text{ J}}{1 \text{ kwh}} \cdot \frac{1 \text{ lb Mg}}{454 \text{ g Mg}} \cdot \frac{24.305 \text{ g Mg}}{1 \text{ mol Mg}} = 1.6 \times 10^6 \frac{\text{J}}{\text{mol Mg}}$$

e. The reaction, $MgCl_2$ (s) $\rightarrow$ Mg (s) + Cl_2 (g), represents the reverse of the formation of magnesium chloride from its elements (each in their standard states). From Appendix J, the ΔH for the process is +641.32 kJ/mol or 6.4×10^5 J/mol. The difference between this value and the value calculated above may be attributed to the energy required to melt the $MgCl_2$.

26.   Calculate ΔH°_{rxn}:   3 CaO (s) + 2 Al (s) $\rightarrow$ 3 Ca (s) + Al_2O_3 (s)
       From Appendix J: ΔH_f° Al_2O_3 (s) = - 1675.7 kJ/mol
                           ΔH_f° CaO (s) = - 635.09 kJ/mol

$$\Delta H^\circ_{rxn} = [3 \cdot \Delta H_f^\circ \text{ Ca} + 1 \cdot \Delta H_f^\circ \text{ Al}_2O_3] - [3 \cdot \Delta H_f^\circ \text{ CaO} + 2 \cdot \Delta H_f^\circ \text{ Al}]$$
$$= (0 - 1675.7 \text{ kJ}) - (-1905.3 \text{ kJ} + 0)$$
$$= 229.6 \text{ kJ}$$

30.   The amount of SO_2 that could be removed by 1000. kg of CaO by the reaction:

$$CaO \text{ (s)} + SO_2 \text{ (g)} \rightarrow CaSO_3 \text{ (s)}$$

$$1.000 \times 10^6 \text{ g CaO} \cdot \frac{1 \text{ mol CaO}}{56.079 \text{ g CaO}} \cdot \frac{1 \text{ mol SO}_2}{1 \text{ mol CaO}} \cdot \frac{64.059 \text{ g SO}_2}{1 \text{ mol SO}_2}$$

$$= 1.142 \times 10^6 \text{ g SO}_2$$

31.   a. The solubility of CaF_2 in water:

$$CaF_2 \text{ (s)} \Leftrightarrow Ca^{2+} \text{ (aq)} + 2 F^- \text{ (aq)} \quad K_{sp} = 3.9 \times 10^{-11}$$

Letting x represent the number of moles of CaF_2 that dissolve per liter we obtain:

$$[Ca^{2+}] = x \quad \text{and} \quad [F^-] = 2x$$

Substituing into the K_{sp} expression: $(x)(2x)^2 = 3.9 \times 10^{-11}$ and solving for x:

$x = 2.1 \times 10^{-4}$ mol CaF_2 /L ; with a molar mass of 78.1 g, the mass of calcium fluoride per liter is:

$$\frac{78.1 \text{ g } CaF_2}{1 \text{ mol } CaF_2} \cdot \frac{2.1 \times 10^{-4} \text{ mol } CaF_2}{1 \text{ L}} = 1.7 \times 10^{-2} \text{ g } CaF_2.$$

b. The mass of CaF_2 to prepare 1.0×10^6 liters of a solution containing $[F^-] = 2.0 \times 10^{-5}$ M

$$\frac{2.0 \times 10^{-5} \text{ mol } F^-}{L} \cdot \frac{1.0 \times 10^6 \text{ L}}{1} \cdot \frac{1 \text{ mol } CaF_2}{2 \text{ mol } F^-} \cdot \frac{78.08 \text{ g } CaF_2}{1 \text{ mol } CaF_2} = 780 \text{ g } CaF_2$$

32.   Reaction scheme:

Clue

1.   1.00 g A + heat $\rightarrow$ B + gas (P = 209 mm; V = 450 mL; T = 298 K)
      white solid   white solid

2.   Gas (from 1) + $Ca(OH)_2$ (aq) $\rightarrow$ C (s)
                                            white solid

3.   Aqueous solution of B is basic (turns red litmus paper blue)

4.   B (aq) + HCl (aq) + heat $\rightarrow$ D
                                   white solid

5.   Flame test for B: green flame

6.   B (aq) + H_2SO_4 (aq) $\rightarrow$ E
                                   white solid

Clue 5 indicates that **B** is a barium salt.

Clue 2 suggests that the gas evolved in Clue 1 is CO_2, and that **C** would be $CaCO_3$.

Heating of carbonates liberates CO_2 (g).

Compound **B** is a metal oxide (Clue 3), and probably has the formula BaO.

Clues 3 and 5 suggest the oxide reacts with HCl and H_2SO_4 to form $BaCl_2$ (compound **D**) and $BaSO_4$ (compound **E**) respectively.

Since **B** is most likely BaO, compound **A** must be $BaCO_3$.

One gram of $BaCO_3$ (Molar mass 197) corresponds to 5.06×10^{-3} mol $BaCO_3$.
Compare this amount of substance to the amount of gas liberated when substance **A** is heated. Substitution of data from clue 1 yields:

$$n = \frac{(209 \text{ mmHg}) (0.450 \text{ L})}{(62.4 \frac{\text{L} \cdot \text{mmHg}}{\text{K} \cdot \text{mol}}) (298 \text{ K})} = 5.06 \times 10^{-3} \text{ mol gas}$$

This is the quantity of CO_2 anticipated from the thermal decomposition of $BaCO_3$.

CHAPTER 23: Metals, Metalloids, and Nonmetals
Periodic Groups 3A and 4A

7. a. BCl_3 (g) + 3 H_2O (l) $\rightarrow$ $B(OH)_3$ (aq) + 3 HCl (aq)

   b. $\Delta H°_{rxn}$ = [1 • $\Delta H°_f$ B(OH)$_3$ + 3 • $\Delta H°_f$ HCl] - [1 • $\Delta H°_f$ BCl$_3$ + 3 • $\Delta H°_f$ H$_2$O]

   = (- 968.92 kJ - 501.48 kJ) - (- 408 kJ - 857.49 kJ)

   = (- 1470.40 kJ) - (- 1265.49 kJ)

   = - 205 kJ

9. In the reaction of sodium borohydride with iodine, **iodine is reduced** to the iodide ion and **sodium borohydride is oxidized**--forming molecular hydrogen and diborane. The equation is:

   $2 NaBH_4$ (s) + I_2 (s) $\rightarrow B_2H_6$ (g) + 2 NaI (s) + H_2 (g)

11. Lewis structure for B_2Cl_4

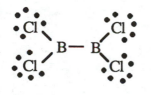

The hybridization of the boron atom is sp^2. The geometry around each boron atom should be trigonal planar, with the atoms attached to the boron atoms separated by angles of approximately 120°.

13. The equation for the production of boron carbide:

   $4 BCl_3$ (g) + 6 H_2 (g) + C(graphite) $\rightarrow$ B_4C (s) + 12 HCl (g)

The mass of B_4C obtainable from 5.45 L of BCl$_3$ (at 26.5 °C and 456 mmHg)
From the ideal gas law:

$$456 \text{ mmHg} • \frac{1 \text{ atm}}{760 \text{ mmHg}} = 0.600 \text{ atm}$$

and substituting :

$$n = \frac{P•V}{R•T} = \frac{(0.600 \text{ atm})(5.45 \text{ L})}{(0.082057 \frac{L•atm}{K•mol})(299.7 \text{ K})} = 0.133 \text{ mol BCl}_3$$

$$0.133 \text{ mol BCl}_3 • \frac{1 \text{ mol B}_4\text{C}}{4 \text{ mol BCl}_3} • \frac{55.26 \text{ g B}_4\text{C}}{1 \text{ mol B}_4\text{C}} = 1.84 \text{ g B}_4\text{C}$$

The mass of H_2 needed: $0.133 \text{ mol } BCl_3 \cdot \dfrac{6 \text{ mol } H_2}{4 \text{ mol } BCl_3} \cdot \dfrac{2.02 \text{ g } H_2}{1 \text{ mol } H_2} = 0.403 \text{ g } H_2$

15.   The aluminum ion is relatively small. The fluoride ion is among the smallest of the anions, and it is possible for six of them to fit around the Al^{3+}. The larger halide ions are sterically prevented from gathering around the aluminum cation.

17.   Estimate the radius of the lead atom if the density of the face-centered cubic solid is 11.342 g/cm^3. Since there are 4 atoms of lead per unit cell, we can calculate the mass of the unit cell:

$$\frac{4 \text{ atoms Pb}}{1 \text{ unit cell}} \cdot \frac{207.2 \text{ g Pb}}{1 \text{ mol Pb}} \cdot \frac{1 \text{ mol Pb}}{6.022 \times 10^{23} \text{ atom Pb}} = 1.376 \times 10^{-21} \text{ g/unit cell}$$

The volume of the unit cell is :

$$\text{Volume} = \frac{\text{Mass}}{\text{Density}} = \frac{1.376 \times 10^{-21} \text{ g/unit cell}}{11.342 \text{ g/cm}^3} = 1.213 \times 10^{-22} \text{ cm}^3$$

and since $\text{Volume} = (\text{length})^3$

$1.213 \times 10^{-22} \text{ cm}^3 = (\text{length})^3$ and length $= 4.951 \times 10^{-8} \text{ cm}$

In a face-centered cubic arrangement, the lead atoms along the edge (length) do not touch. They do, however, touch through the diagonal of a face. The face diagonal represents (4 • atomic radii). The geometry is such that :

$$(4 \cdot \text{atomic radii})^2 = 2 \cdot (\text{length})^2$$

[See Chapter 13 for a discussion of this equation.]

$$\text{or } 4 \cdot \text{atomic radii} = (2)^{0.5} \cdot \text{length}$$

$$\text{atomic radii} = \frac{1.414 \cdot \text{length}}{4} = \frac{1.414 \cdot 4.951 \times 10^{-8} \text{ cm}}{4}$$

$$= 1.750 \times 10^{-8} \text{ cm or } 175.0 \text{ pm}$$

19. $2 Al (s) + 2 NaOH (aq) + 6 H_2O (l) \rightarrow 2 NaAl(OH)_4 (aq) + 3 H_2 (g)$

Volume of H_2 (in mL) produced when 13.2 g of Al react:

$$13.2 \text{ g Al} \cdot \frac{1 \text{ mol Al}}{26.98 \text{ g Al}} \cdot \frac{3 \text{ mol } H_2}{2 \text{ mol Al}} = 0.734 \text{ mol } H_2$$

$$V = \frac{(0.734 \text{ mol } H_2)(0.082057 \frac{L \cdot atm}{K \cdot mol})(295.7 \text{ K})}{735 \text{ mm Hg} \cdot \frac{1 \text{ atm}}{760 \text{ mmHg}}} = 18.4 \text{ L}$$

or 18.4×10^3 mL

21.   Equations for $Ga(OH)_3$ dissolving in HCl and NaOH:

$$Ga(OH)_3 (s) + 3 HCl (aq) \rightarrow GaCl_3 (aq) + 3 H_2O (l)$$
$$Ga(OH)_3 (s) + NaOH (aq) \rightarrow Na[Ga(OH)_4] (aq)$$

Volume of 0.0112 M HCl needed to react with 1.25 g $Ga(OH)_3$:

$$1.25 \text{ g Ga(OH)}_3 \cdot \frac{1 \text{ mol Ga(OH)}_3}{120.7 \text{ g Ga(OH)}_3} \cdot \frac{3 \text{ mol HCl}}{1 \text{ mol Ga(OH)}_3} \cdot \frac{1 \text{ L}}{0.0112 \text{ mol HCl}}$$

= 2.77 L HCl   or 2770 mL HCl

23.   a. $Cl_3Al + : Cl^- \rightarrow [Cl_3Al:Cl]^-$

The chloride ion functions as a Lewis base by donating an electron pair. The complex ion formed would have **tetrahedral geometry** of the chlorine atoms around the aluminum atom. Aluminum would utilize **sp^3 hybridization**.

25.   Since the CCl_4 molecule is heavier than the CF_4 molecule, the boiling point for CCl_4 is expected to be higher than that for CF_4. London forces increase with increasing mass.

27.   a. Reduction of PbO and CaO with C:

$$CaO (s) + C (s) \rightarrow Ca (s) + CO (g)$$
$$\Delta G°_{rxn} = [\Delta G°_f Ca (s) + \Delta G°_f CO (g)] - [\Delta G°_f CaO (s) + \Delta G°_f C (s)]$$
$$= (0 - 137.168 \text{ kJ}) - (- 604.03 \text{ kJ} + 0) = + 466.86 \text{ kJ}$$

$PbO\ (s)\ +\ C\ (s)\ \rightarrow\ Pb\ (s)\ +\ CO\ (g)$

$\Delta G^\circ_{rxn} = [\Delta G^\circ_f\ Pb\ (s)\ +\ \Delta G^\circ_f\ CO\ (g)] - [\Delta G^\circ_f\ PbO\ (s)\ +\ \Delta G^\circ_f\ C\ (s)\ ]$

$= (0 - 137.168\ kJ) - (-187.89\ kJ + 0)\ =\ +50.72\ kJ$

b. The reduction of PbO by C will be more feasible on a commercial scale, since the temperature at which ΔG° would become negative (i.e. process spontaneous) would be much lower than that for CaO.

29.   Amount of PbO introduced by 227,250 tons of $Pb(C_2H_5)_4$:

$$227{,}250\ \text{tons Pb}(C_2H_5)_4\ \bullet\ \frac{1\ \text{mol Pb}(C_2H_5)_4}{323.449\ \text{g Pb}(C_2H_5)_4}\ \bullet\ \frac{1\ \text{mol PbO}}{1\ \text{mol Pb}(C_2H_5)_4}$$

$$\bullet\ \frac{223.200\ \text{g PbO}}{1\ \text{mol PbO}}\ =\ 156{,}820\ \text{tons PbO}$$

31.   a. $Si\ (s)\ +\ 2\ CH_3Cl\ (g)\ \rightarrow\ (CH_3)_2SiCl_2\ (g)$

b. Stoichiometric amount of CH_3Cl :

$$2.65\ \text{g Si}\ \bullet\ \frac{1\ \text{mol Si}}{28.09\ \text{g Si}}\ \bullet\ \frac{2\ \text{mol CH}_3Cl}{1\ \text{mol Si}}\ =\ 0.189\ \text{mol CH}_3Cl$$

$$P\ =\ \frac{(0.189\ \text{mol CH}_3Cl)\ (0.082057\ \frac{L \bullet atm}{K \bullet mol})\ (297.7\ K)}{5.60\ L}\ =\ 0.823\ \text{atm}$$

c. Mass of $(CH_3)_2SiCl_2$ produced:

$$0.0943\ \text{mol Si}\ \bullet\ \frac{1\ \text{mol }(CH_3)_2SiCl_2}{1\ \text{mol Si}}\ \bullet\ \frac{129.1\ \text{g }(CH_3)_2SiCl_2}{1\ \text{mol }(CH_3)_2SiCl_2}\ =\ 12.2\ \text{g }(CH_3)_2SiCl_2$$

33.   a. Compare the electron dot structures of CO and CN^- :

$:C\equiv O:$ $[:C\equiv N:]^-$ The two species are isoelectronic, and share the common features of a triple bond and two lone pairs of electrons.

b. The electron dot structure for cyanamide:

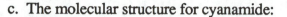

c. The molecular structure for cyanamide:

The geometry around the central N is tetrahedral, consistent with **sp³** hybridization. The anticipated bond angle for this N is approximately 109°. The C atom, having two groups attached, utilizes **s p** hybridization with a resulting bond angle for N-C-N of 180°.

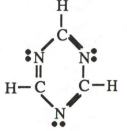

35.   a. A proposed structure for the cyclic trimer (HCN)₃:

b. Energy for the trimerization reaction:

Bonds broken: 3 • C≡N bonds 3 • 887 kJ/mol = <u>2661 kJ</u>

                               Energy input: 2661 kJ

Bonds formed: 3 • C-N bonds 3 • 305 kJ/mol =   915 kJ

                    3 • C=N bonds 3 • 615 kJ/mol = <u>1845 kJ</u>

                         Energy released: 2760 kJ

          Energy change = 2661 kJ - 2760 kJ = - 99 kJ

37.   The volume of hydrogen (measured at 26 °C and 745 mmHg):

$$0.56 \text{ g Al} \cdot \frac{1 \text{ mol Al}}{27.0 \text{ g Al}} \cdot \frac{3 \text{ mol H}_2}{2 \text{ mol Al}} = 0.031 \text{ mol H}_2$$

$$V = \frac{(0.031 \text{ mol H}_2)(0.082057 \frac{\text{L} \cdot \text{atm}}{\text{K} \cdot \text{mol}})(299 \text{ K})}{745 \text{ mm Hg} \cdot \frac{1 \text{ atm}}{760 \text{ mmHg}}} = 0.779 \text{ L}$$

                                        or 780 mL (to 2 significant figures)

CHAPTER 24: The Chemistry of the Nonmetals: Periodic Groups 5A Through 7A and the Rare Gases

2. a. $3 \, Mg \, (s) + N_2 \, (g) \rightarrow Mg_3N_2 \, (s)$

   b. $P_4 \, (s) + 3 \, KOH \, (aq) + 3 \, H_2O \, (l) \rightarrow PH_3 \, (g) + 3 \, KH_2PO_2 \, (aq)$

   c. $CH_4 \, (g) + H_2O \, (g) \rightarrow CO \, (g) + 3 \, H_2 \, (g)$
      Δ, Ni

   d. $HNO_3 \, (aq) + KOH \, (aq) \rightarrow KNO_3 \, (aq) + H_2O \, (l)$

   alkali
   e. $2 \, NH_3 \, (aq) + OCl^- \, (aq) \rightarrow N_2H_4(aq) + Cl^- \, (aq) + H_2O \, (l)$
      gelatin

   $250°C$
   f. $NH_4NO_3 \, (s) \rightarrow N_2O \, (g) + 2 \, H_2O \, (g)$

   g. $NaNO_2 \, (aq) + HCl \, (aq) \rightarrow NaCl \, (aq) + HNO_2 \, (aq)$

   h. $4 \, NH_3 \, (g) + 5 \, O_2 \, (g) \rightarrow 4 \, NO \, (g) + 6 \, H_2O \, (g)$

   i. $3 \, Cu \, (s) + 8 \, HNO_3 \, (aq) \rightarrow 3 \, Cu(NO_3)_2 \, (aq) + 4 \, H_2O \, (l) + 2 \, NO \, (g)$

   j. $P_4O_{10} \, (s) + 6 \, H_2O \, (l) \rightarrow 4 \, H_3PO_4 \, (aq)$

4. a. The reaction of hydrazine with dissolved oxygen:

   $$N_2H_4 \, (l) + O_2 \, (aq) \rightarrow N_2 \, (g) + 2 \, H_2O \, (l)$$

   b. Mass of hydrazine to consume the oxygen in 3.00×10^4 L of water:

   $$3.00 \times 10^4 \, L \cdot \frac{3.08 \, cm^3 \, O_2}{0.100 \, L} \cdot \frac{1 \, mol \, O_2}{22400 \, cm^3} \cdot \frac{1 \, mol \, N_2H_4}{1 \, mol \, O_2} \cdot \frac{32.05 g N_2H_4}{1 \, mol \, N_2H_4}$$
   $\uparrow$ at STP

   $$= 1320 \, g \, N_2H_4$$

7.   The half equations:

$$N_2H_5^+ \text{ (aq)} \rightarrow N_2 \text{ (g)} + 5\,H^+ \text{ (aq)} + 4\,e^-$$

$$IO_3^- \text{ (aq)} \rightarrow I_2 \text{ (s)}$$

are balanced according to the procedure in Chapter 5 to give:

$$5\,N_2H_5^+ \text{ (aq)} + 4\,IO_3^- \text{ (aq)} \rightarrow 5\,N_2 \text{ (g)} + 2\,I_2 \text{ (s)} + 12\,H_2O \text{ (l)} + H^+ \text{ (aq)}$$

E° for the reaction is: E° for the hydrazine equation (oxidation) $= +\ 0.23$ V

E° for the iodate equation (reduction) $= \underline{+\ 1.195}$ V

$E^\circ{}_{cell} = +\ 1.43$ V

8.   For the reaction:

$$H_2N\text{-}N(CH_3)_2 \text{ (l)} + 2\,N_2O_4 \text{ (l)} \rightarrow 3\,N_2 \text{ (g)} + 4\,H_2O \text{ (g)} + 2\,CO_2 \text{ (g)}$$
(DMH)

Mass of N_2O_4 to react with 4.5 tons of DMH:

$$4.5 \text{ T DMH} \cdot \frac{1 \text{ mol DMH}}{60.1 \text{ g DMH}} \cdot \frac{2 \text{ mol } N_2O_4}{1 \text{ mol DMH}} \cdot \frac{92.0 \text{ g } N_2O_4}{1 \text{ mol } N_2O_4} = 14 \text{ T } N_2O_4$$

Mass of N_2 generated:

$$4.5 \text{ T DMH} \cdot \frac{1 \text{ mol DMH}}{60.1 \text{ g DMH}} \cdot \frac{3 \text{ mol } N_2}{1 \text{ mol DMH}} \cdot \frac{28.0 \text{ g } N_2}{1 \text{ mol } N_2} = 6.3 \text{ T } N_2$$

Mass of H_2O generated:

$$4.5 \text{ T DMH} \cdot \frac{1 \text{ mol DMH}}{60.1 \text{ g DMH}} \cdot \frac{4 \text{ mol } H_2O}{1 \text{ mol DMH}} \cdot \frac{18.0 \text{ g } H_2O}{1 \text{ mol } H_2O} = 5.4 \text{ T } H_2O$$

Mass of CO_2 generated:

$$4.5 \text{ T DMH} \cdot \frac{1 \text{ mol DMH}}{60.1 \text{ g DMH}} \cdot \frac{2 \text{ mol } CO_2}{1 \text{ mol DMH}} \cdot \frac{44.0 \text{ g } N_2O_4}{1 \text{ mol } CO_2} = 6.6 \text{ T } CO_2$$

12.   a. The structure for N_2O_3 indicates two NO bonds which have a bond order of 1.5 and one N-O bond which has bond order of 2 (and is therefore shorter).

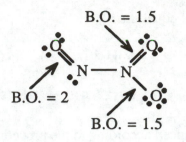

B.O. = 1.5

B.O. = 2

B.O. = 1.5

b. For the reaction: N_2O_3 (g) $\Leftrightarrow$ NO (g) + NO_2 (g), $\Delta H° = 40.5$ kJ/mol and

$\Delta G° = -1.59$ kJ/mol

Calculate $\Delta S°$:

$$\Delta G° = \Delta H° - T \Delta S°$$
$$-1.59 \text{ kJ/mol} = 40.5 \text{ kJ/mol} - (298 \text{ K}) \Delta S°$$
$$-42.09 \text{ kJ/mol} = -(298 \text{ K}) \Delta S°$$
$$0.141 \frac{kJ}{K \cdot mol} = \Delta S°$$
$$141 \frac{J}{K \cdot mol} = \Delta S°$$

Calculate K_p

$$\Delta G° = -RT \ln K_p$$
$$-1.59 \times 10^3 \text{ J/mol} = -(8.314 \text{ J/K} \cdot \text{mol})(298 \text{ K}) \ln K_p$$
$$\frac{-1.59 \times 10^3 \text{ J/mol}}{-(8.314 \text{ J/K} \cdot \text{mol})(298 \text{ K})} = \ln K_p$$
$$1.90 = K_p$$

Calculate $\Delta H°_f$ for N_2O_3:

$$\Delta H°_{rxn} = [1 \cdot \Delta H°_f NO + 1 \cdot \Delta H°_f NO_2] - [1 \cdot \Delta H°_f N_2O_3]$$
$$40.5 \text{ kJ} = (90.25 \text{ kJ} + 33.18 \text{ kJ}) - \Delta H°_f N_2O_3$$
$$-82.9 \text{ kJ} = -\Delta H°_f N_2O_3$$
$$82.9 \text{ kJ} = \Delta H°_f N_2O_3$$

14.   a. Relation of P_2O_5 to H_3PO_4 Molar mass of P_2O_5 : 141.94

Molar mass of H_3PO_4: 98.00

To have equal number of P atoms, we need two molecules of H_3PO_4 for each molecule of P_2O_5:

$$\frac{2 \cdot H_3PO_4}{1 \cdot P_2O_5} = \frac{196.00}{141.94} = 1.3800$$

b. 11.7×10^6 T P_2O_5 $\cdot \dfrac{1.3800 \text{ T } H_3PO_4}{1 \text{ T } P_2O_5}$ $= 16.1 \times 10^6$ T H_3PO_4

$\hspace{5cm}$ (16.1 million tons)

c. $\dfrac{\$300}{1 \text{ T } P_2O_5} \cdot \dfrac{1.00 \text{ T } P_2O_5}{1.3800 \text{ T } H_3PO_4} \cdot 1 \text{ T } H_3PO_4 = \$217$

15. Approximate pH of 0.010 M Na_3PO_4:

Hydrolysis of the PO_4^{3-} ion may be represented:

$$PO_4^{3-} \text{ (aq)} + H_2O \text{ (l)} \Leftrightarrow HPO_4^{2-} \text{ (aq)} + OH^- \text{ (aq)}$$

$$K_b = \frac{[HPO_4^{2-}][OH^-]}{[PO_4^{3-}]} = \frac{1.0 \times 10^{-14}}{3.6 \times 10^{-13}} \xleftarrow{} \frac{K_w}{K_a \text{ for } HPO_4^{2-}}$$

$$= \frac{[x][x]}{(0.010 - x)} = 2.8 \times 10^{-2}$$

$K_b \cdot 100 > 0.010$ so it will be necessary to use the quadratic formula or successive approximations to solve this equation.

$x = 7.8 \times 10^{-3} = [OH^-]$ so pOH = 2.11 and pH = (14.00 - 2.11) = 11.89

17. A possible preparation for $CaHPO_4$:

$$Ca(OH)_2 \text{ (s)} + H_3PO_4 \text{ (aq)} \rightarrow CaHPO_4 \text{ (s)} + 2 H_2O \text{ (l)}$$

19. A possible reaction when base is added to an aqueous solution of P_4 (s)

$$P_4 \text{ (s)} + 3 H_2O \text{ (l)} + 3 OH^- \text{ (aq)} \rightarrow 3 H_2PO_2^- \text{ (aq)} + PH_3 \text{ (g)}$$

22. A balanced equation for the reaction of arsenic with nitric acid:

$$As \text{ (s)} + HNO_3 \text{ (aq)} \rightarrow H_2 \text{ (g)} + H_3AsO_3 \text{ (aq)} + NO \text{ (g)}$$

Balancing this equation by inspection is a task at best. Using the methods outlined in Chapter 5:

$\hspace{3cm} As \rightarrow H_3AsO_3 \hspace{4cm} HNO_3 \rightarrow NO$

$3 H_2O + As \rightarrow H_3AsO_3 \hspace{3.5cm} HNO_3 \rightarrow NO + 2 H_2O$

$3 H_2O + As \rightarrow H_3AsO_3 + 3 H^+ \hspace{2cm} 3 H^+ + HNO_3 \rightarrow NO + 2 H_2O$

$3 H_2O + As \rightarrow H_3AsO_3 + 3 H^+ + 3 e^- \hspace{0.5cm} 3 H^+ + HNO_3 + 3 e^- \rightarrow NO + 2 H_2O$

Note that the number of electrons gained and lost is equal, so we add the equations:

$$3\ H_2O + As \rightarrow H_3AsO_3 + 3\ H^+ + 3\ e^-$$
$$3\ H^+ + HNO_3 + 3\ e^- \rightarrow NO + 2\ H_2O$$

Removing duplications (in water, hydrogen ions, and electrons) we get:

$$H_2O + As + HNO_3 \rightarrow H_3AsO_3 + NO$$

23.   Balance the following equations:

These are easily balanced by inspection. A hint about a good place to start follows each balanced equation.

a. $2\ KClO_3\ (s) \rightarrow 2\ KCl\ (s) + 3\ O_2\ (g)$

Noting the odd number of oxygens in $KClO_3$ and the even number in O_2, one can equalize these by using coefficients of 2 and 3 respectively.

b. $2\ H_2S\ (g) + 3\ O_2\ (g) \rightarrow 2\ H_2O\ (g) + 2\ SO_2\ (g)$

Noting that one mole of water and one mole of sulfur dioxide require a total of 3 oxygen atoms, a coefficient of 3/2 for O_2 is needed. To provide <u>integral</u> coefficients for all substances, multiply the equation by two.

c. $2\ Na\ (s) + O_2\ (g) \rightarrow Na_2O_2\ (s)$

The mandate for at least two Na's in the product drives this balancing procedure.

d. $K\ (s) + O_2\ (g) \rightarrow KO_2\ (g)$

e. $2\ ZnS\ (s) + 3\ O_2\ (g) \rightarrow 2\ ZnO\ (s) + 2\ SO_2\ (g)$

An odd **total** number of oxygen atoms on the right with an even number on the left suggests that the coefficient of either ZnO or SO_2 will have to be even. Note that Zn and S appear in the reactants combined--indicating the the coefficient of ZnO and SO_2 will have to be equal.

f. $SO_2\ (g) + H_2O\ (l) \rightarrow H_2SO_3\ (aq)$

25.   For the reaction: O_2 (g) + 4 H_3O^+ (aq) + 4 e$^-$ $\rightarrow$ 2 H_2O (l) , $E° = 1.229$ V

a. Calculate E when pH = 1.00 and P_{oxy} = 1.0 atm :

$$E = E° - \frac{0.0592}{n} \log Q \text{ where } Q = \frac{1}{[O_2][H_3O^+]^4}$$

$$= 1.229 - \frac{0.0592}{4} \log\left(\frac{1}{[1.0][1.00 \times 10^{-1}]^4}\right)$$

$$= 1.229 - 0.0592 = 1.170 \text{ V}$$

b. Calculate E when pH = 1.00 and P_{oxy} = 0.22 atm :

$$E = E° - \frac{0.0592}{n} \log Q \text{ where } Q = \frac{1}{[O_2][H_3O^+]^4}$$

$$= 1.229 - \frac{0.0592}{4} \log\left(\frac{1}{[0.22][1.00 \times 10^{-1}]^4}\right)$$

$$= 1.229 - \frac{0.0592}{4} \cdot (4.55 \times 10^4)$$

$$= 1.229 - 0.0689 \quad \text{or}$$

$$= 1.160 \text{ V}$$

28.   Hydrogen peroxide functioning as a Bronsted acid:

$$H_2O_2 \text{ (aq)} + H_2O \text{ (l)} \Leftrightarrow HO_2^- \text{ (aq)} + H_3O^+ \text{ (aq)}$$

pH of 0.10 M H_2O_2 (assuming loss of one proton)

$$K_a = \frac{[HO_2^-][H_3O^+]}{[H_2O_2]} = 1.78 \times 10^{-12}$$

The magnitude of the K_a tells us that the peroxide concentration at equilibrium is almost equal to its initial value. So we can substitute into the K_a expression:

$$\frac{x^2}{(0.10 - x)} = 1.78 \times 10^{-12}$$

and simplifying the denominator to 0.10 we obtain: x = $[H_3O^+] = 4.2 \times 10^{-7}$

and the pH = 6.37

30.   The metallic character in the Periodic Table decreases from left to right in a period.

Oxides in order of increasing basicity : $BrO_2 < As_2O_3 < GeO_2 < Ga_2O_3 < CaO < K_2O$

31.   CaO is more basic than BeO. The metallic character of elements increases down a group, hence the oxide of calcium is more basic.

Hydrolysis of SrO: $SrO \ (s) + H_2O \ (l) \rightarrow Sr(OH)_2 \ (aq)$

34.   a. Allowable release of SO_2: (0.3 %)

$$2.00 \times 10^3 \ T \ H_2SO_4 \cdot \frac{1 \ mol \ H_2SO_4}{98.07 \ g \ H_2SO_4} \cdot \frac{1 \ mol \ SO_2}{1 \ mol \ H_2SO_4} \cdot \frac{64.06 \ g \ SO_2}{1 \ mol \ SO_2}$$

$$\cdot \frac{0.003 \ T \ SO_2 \ released}{1.00 \ T \ SO_2 \ produced} = 3.92 \ T \ SO_2$$

b. Mass of $Ca(OH)_2$ to remove 3.92 T SO_2:

$$3.92 \ T \ SO_2 \cdot \frac{1 \ mol \ SO_2}{64.06 \ g \ SO_2} \cdot \frac{1 \ mol \ Ca(OH)_2}{1 \ mol \ SO_2} \cdot \frac{74.09 \ g \ Ca(OH)_2}{1 \ mol \ Ca(OH)_2}$$

$$= 4.53 \ T \ Ca(OH)_2$$

36.   Balance the following equations:

a. $UF_4 \ (s) + F_2 \ (g) \rightarrow UF_6 \ (s)$

b. $2 \ Br^- \ (aq) + Cl_2 \ (aq) \rightarrow 2 \ Cl^- \ (aq) + Br_2 \ (aq)$

c. $2 \ I^- \ (aq) + Br_2 \ (aq) \rightarrow 2 \ Br^- \ (aq) + I_2 \ (aq)$

d. $Br^- \ (aq) + AgNO_3 \ (aq) \rightarrow AgBr \ (s) + NO_3^- \ (aq)$

e. $Cl_2 \ (g) + 2 \ OH^- \ (aq) \rightarrow OCl^- \ (aq) + Cl^- \ (aq) + H_2O \ (l)$

38.   The reaction is identical to that of study question 36-c. In it, bromine will displace iodide ion from solution, producing elemental iodine.

41.   The oxidation involved may be written: $2 \ F^- \ (aq) \rightarrow F_2 \ (g) + 2e^-$ [1 mol F_2/2 mol e^-]

$$\frac{38.00 \text{ g } F_2}{1 \text{ mol } F_2} \cdot \frac{1 \text{ mol } F_2}{2 \text{ mol } e} \cdot \frac{1 \text{ mol } e}{96500 \text{ C}} \cdot \frac{1 \text{ C}}{1 \text{ A} \cdot \text{s}} \cdot \frac{5.00 \times 10^3 \text{ A}}{1} \cdot \frac{3600 \text{ s}}{1 \text{ hr}}$$

$$\cdot \frac{24 \text{ hr}}{1} \cdot \frac{1 \text{ T } F_2}{9.08 \times 10^5 \text{ g } F_2} = 9.37 \times 10^{-2} \text{ T } F_2$$

46.   The ClO_2 molecule has a total of 19 electrons. It must have at least one unpaired electron.

48.   Lewis dot structures and molecular structures:

a. I_3^- : This anion contains 22 electrons (7 for each I and 1 more for the negative change.

The molecular structure for this ion involves the placement of three lone pairs of electrons in the equatorial plane, with the other two iodine atoms occupying the axial positions. Hence while the structural geometry is trigonal bipyramidal, the molecular geometry is linear.

b. $BrCl_2^-$: This ion is isoelectronic with the I_3^- ion shown above.

Note that the structural geometry is trigonal bipyramidal, and the molecular geometry linear, as seen for the I_3^- ion above.

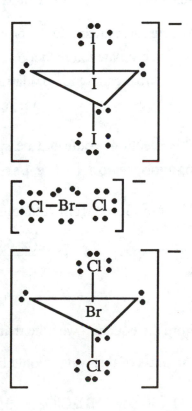

c. ClF_2^+: This cation has a total of 20 electrons (7 each from the three halogen atoms - 1 electron lost to form the cation).

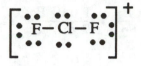

The structural geometry for this cation is tetrahedral (with four major groups attached). The ion has a "molecular geometry" which is bent-- like water.

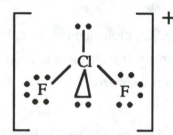

49.   One can oxidize aqueous Mn^{2+} (1.0 M) to aqueous MnO_4^- with bromate ion if the ΔG° for the process is negative-- $\Delta G^\circ < 0$ implies spontaneity. Since $\Delta G^\circ = - n F E^\circ$, the ΔG° will be negative if (and only if) E° is positive. [Remember n and F are always positive.] From Appendix I:

$$Mn^{2+} (aq) + 4 H_2O \rightarrow MnO_4^- (aq) + 8 H^+ (aq) + 5 e^- \qquad E^\circ = \quad -1.51 \text{ V}$$

Adding this reaction to the bromate reaction potential (1.495 V) gives a voltage of:

        bromate ion reduction $E^\circ = 1.495$ V

        manganese ion oxidation $E^\circ = \underline{-1.51}$ V

               net voltage =   - 0.02 V

This reaction would not be spontaneous--i.e.the **bromate ion will not oxidize the manganese ion to the permanganate ion.**

51.   Water produced from 700. tons of NH_4ClO_4:

$$700. \text{ T } NH_4ClO_4 \cdot \frac{9.08 \times 10^5 \text{ g } NH_4ClO_4}{1 \text{ T } NH_4ClO_4} \cdot \frac{1 \text{ mol } NH_4ClO_4}{117.5 \text{ g } NH_4ClO_4}$$

$$= 5.41 \times 10^6 \text{ mol } NH_4ClO_4$$

From the equation we note that 2 mol of NH_4ClO_4 gives rise to 4 mol of H_2O.

The mass of this water would be:

$$5.41 \times 10^6 \text{ mol } NH_4ClO_4 \cdot \frac{4 \text{ mol } H_2O}{2 \text{ mol } NH_4ClO_4} \cdot \frac{18.02 \text{ g } H_2O}{1 \text{ mol } H_2O} \cdot \frac{1 \text{ T } H_2O}{9.08 \times 10^5 \text{ g } H_2O}$$

$$= 215 \text{ T } H_2O$$

Grams of oxygen produced:

$$5.41 \times 10^6 \text{ mol NH}_4\text{ClO}_4 \cdot \frac{2 \text{ mol O}_2}{2 \text{ mol NH}_4\text{ClO}_4} \cdot \frac{32.00 \text{ g O}_2}{1 \text{ mol O}_2} = 1.73 \times 10^8 \text{ g O}_2$$

Tons of Aluminum needed: $4 \text{ Al (s)} + 3 \text{ O}_2 \text{ (g)} \rightarrow 2 \text{ Al}_2\text{O}_3 \text{ (s)}$

$$5.41 \times 10^6 \text{ mol NH}_4\text{ClO}_4 \cdot \frac{2 \text{ mol O}_2}{2 \text{ mol NH}_4\text{ClO}_4} \cdot \frac{4 \text{ mol Al}}{3 \text{ mol O}_2} \cdot \frac{26.98 \text{ g Al}}{1 \text{ mol Al}}$$

$$\cdot \frac{1 \text{ T Al}}{9.08 \times 10^5 \text{ g Al}} = 214 \text{ T Al}$$

Tons of Al$_2$O$_3$ formed:

$$5.41 \times 10^6 \text{ mol O}_2 \cdot \frac{2 \text{ mol Al}_2\text{O}_3}{3 \text{ mol O}_2} \cdot \frac{102.0 \text{ g Al}_2\text{O}_3}{1 \text{ mol Al}_2\text{O}_3} \cdot \frac{1 \text{ T Al}}{9.08 \times 10^5 \text{ g Al}}$$

$$= 405 \text{ T Al}_2\text{O}_3$$

55.   Constructing a cell with the oxidation of Mn^{2+} to MnO_4^- (Data from Appendix I):

xenon oxide reduction	$E° = 2.10$ V
manganese ion oxidation	$E° = -1.51$ V
net voltage	$E° = 0.59$ V

The positive voltage indicates that XeO_3 is capable of oxidizing Mn^{2+} to MnO_4^-

To determine if XeO_3 can oxidize F^- to F_2 , a similar cell is "constructed":

xenon oxide reduction	$E° = 2.10$ V
fluoride ion oxidation	$E° = -2.87$ V
net voltage	$E° = -0.77$ V

XeO_3 is not capable of oxidizing F^- to F_2.

CHAPTER 25: The Transition Elements and Their Chemistry

10.   Electron configurations for:

         a. Y^{3+} [Kr]   diamagnetic

         b. Rh^{3+} [Kr] $4d^6$ paramagnetic

         c. Ce^{4+} [Xe]   diamagnetic

         d. Pt^{4+} [Xe] $4f^{14}\,5d^6$ paramagnetic

         e. Y^{2+} [Ar] $3d^3$ paramagnetic

         f. U^{4+} [Rn] $5f^2$ paramagnetic

13.   The roasting of chalcopyrite, $CuFeS_2$, may be represented:

$$2\,CuFeS_2\ (s) + 3\,O_2\ (g) \rightarrow 2\,CuS\ (s) + 2\,FeO\ (s) + 2\,SO_2\ (g)$$

The amount of SO_2 produced when one ton of ore is roasted:

$$\frac{908\ kg\ ore}{1} \cdot \frac{1\ mol\ ore}{183.5\ g\ ore} \cdot \frac{2\ mol\ SO_2}{2\ mol\ ore} \cdot \frac{64.06\ g\ SO_2}{1\ mol\ SO_2} \cdot \frac{1.0\ T\ SO_2}{908\ kg\ SO_2}$$

$$= 0.35\ T\ SO_2$$

15.   a. $FeTiO_3\ (s) + 2\,HCl\ (aq) \rightarrow FeCl_2\ (aq) + TiO_2\ (s) + H_2O\ (l)$

     b. To combine the equations, multiply the equation in part (a) by two and add:

$$2\,FeTiO_3\ (s) + 4\,HCl\ (aq) \rightarrow 2\,FeCl_2\ (aq) + 2\,TiO_2\ (s) + 2\,H_2O\ (l)$$

$$2\,FeCl_2\ (aq) + 2\,H_2O\ (l) + \frac{1}{2}O_2\ (g) \rightarrow Fe_2O_3\ (s) + 4\,HCl\ (aq)$$

$$2\,FeTiO_3\ (s) + \frac{1}{2}O_2\ (g) \rightarrow 2\,TiO_2\ (s) + Fe_2O_3\ (s)$$

The net equation shows no HCl, indicating that HCl is recovered in the second step.

     c. Mass of iron(III) oxide from one ton of ilmenite:

$$\frac{9.08 \times 10^5\ g\ FeTiO_3}{1} \cdot \frac{1\ mol\ FeTiO_3}{151.7\ g\ FeTiO_3} \cdot \frac{1\ mol\ Fe_2O_3}{2\ mol\ FeTiO_3} \cdot \frac{159.7\ g\ Fe_2O_3}{1\ mol\ Fe_2O_3}$$

$$= 4.78 \times 10^5\ g\ Fe_2O_3$$

17.   Classify each of the following as monodentate or multidentate:

        a. CH_3NH_2 monodentate (lone pair on N)

        b. Br^- monodentate (lone pair on Br)

        c. ethylenediamine   bidentate (lone pairs on terminal N atoms)

        d. N_3^- monodentate (lone pair on a N atom)

        e. $C_2O_4^{2-}$ bidentate (lone pairs on terminal O atoms)

        f. $H_3C-C\equiv N$ monodentate (lone pair on N)

        g. phenanthroline   bidentate (lone pairs on N atoms)

19. $[Ni(en)(NH_3)_3(H_2O)]^{2+}$ The compound is positively charged, as all the ligands attached to the Ni^{2+} ion are neutral.

21.   Formulas for:

        a. dichlorobis(ethylenediamine)nickel (II) $Ni(en)_2Cl_2$ [where en = $H_2NCH_2CH_2NH_2$]

        b. potassium tetrachloroplatinate (II) $K_2[PtCl_4]$

24.   Names for the coordination complexes in Figure 25.12:

Formula	Name
$[Co(NH_3)_5H_2O]Cl_3$	pentaammineaquacobalt(III) chloride
$K_3[Fe(CN)_6]$	potassium hexacyanoferrate(III)
$Cr(CO)_6$	hexacarbonylchromium(0)
$K_3[Fe(C_2O_4)_3]$	potassium tris(oxalato)ferrate(III)
$[Co(H_2NCH_2CH_2NH_2)_3]I_3$	tris(ethylenediamine)cobalt(III) iodide

26.

Name	Formula
a. hydroxopentaaquairon(III) ion	$[Fe(H_2O)_5(OH)]^{2+}$
b. potassium tetracyanonickelate(II)	$K_2[Ni(CN)_4]$
c. potassium diaquabis(oxalato)chromate(III)	$K[Cr(C_2O_4)_2(H_2O)_2]$

28.   Geometric isomers of:

a.   Pd(NH₃)₄Cl₂

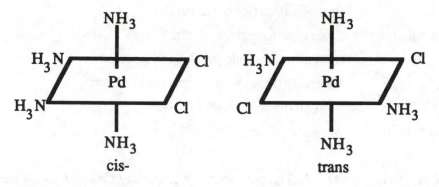

cis-   trans

b.   Pt(NH₃)₂(NCS)(Br)

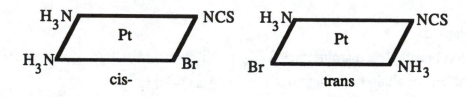

cis-   trans

c.   Co(NH₃)₃(NO₂)₃

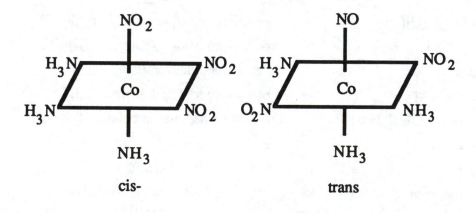

cis-   trans

30.   Decide if the underlined atom is or is not a chiral center:

a. $\underline{C}H_2Cl_2$ Not chiral   has two H and two Cl's attached
b. $H_2N\text{-}\underline{C}H(CH_3)\text{-}COOH$ chiral
c. $Cl\text{-}\underline{C}H(OH)\text{-}CH_2Cl$ chiral

326

32.   The four isomers of $[Co(en)(NH_3)_2(H_2O)Cl]^{2+}$:

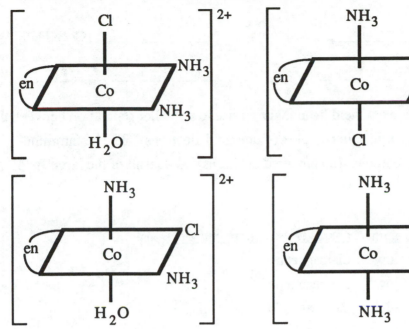

The ligand ethylenediamine is represented in these drawings by the symbolism shown to the right:

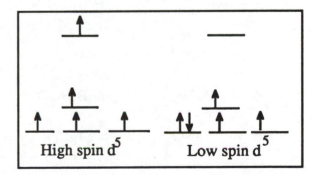

34.   Electron configurations with possible <u>high and low</u> spin complexes: d^5, d^6, d^7, d^8. The high and low spin complex for a d^5 case is shown here. The Ni $^{2+}$ ion (d^8) is shown in your text. In most cases however, the high spin case does not exist, owing to the large separation of the uppermost orbital from the lower orbitals.

327

36.   Electron configurations for Co^{3+}:   (d^6 configuration)

The fluoro- ligand is a weak field ligand, giving rise to a smaller separation between the two d sublevels, and a high spin complex (4 unpaired electrons). The hexammine-complex is a low spin complex (no unpaired electrons), as a result of the large field strength for the ammine ligand.

41.

complex	spin	character	unpaired electrons
a. $[Fe(CN)_6]^{4-}$	low	diamagnetic	
b. $[Cr(en)_3]^{3+}$	low	paramagnetic	3
c. $[MnF_6]^{4-}$	high	paramagnetic	5
d. $[Cu(phen)_3]^{2+}$	low	paramagnetic	1

Electron configuration:

a. Fe^{2+} _____ _____   b. Cr^{3+} _____ _____

                ⇅   ⇅   ⇅   ↑   ↑   ↑

c. Mn^{2+} ↑   ↑   d. Cu^{2+} ⇅   ↑

              ↑   ↑   ↑   ⇅   ⇅   ⇅

43.   A + $BaCl_2$ ppt ($BaSO_4$)   ⇒ A = $[Co(NH_3)_5Br]SO_4$

        B + $AgNO_3$ ppt ($AgBr$)   ⇒ B = $[Co(NH_3)_5SO_4]Br$

45.   The reaction:

          $2\ FeTiO_3\ (s) + 7\ Cl_2\ (g) + 6\ C\ (s) \rightarrow 2\ TiCl_4\ (l) + 2\ FeCl_3\ (s) + 6\ CO\ (g)$

is an oxidation-reduction reaction. [Examine, for example, the oxidation state of chlorine before (0) and after (-1) the reaction.]

Mass of Cl_2 and C for 25.0 kg of ilmenite:

$$\frac{25.0 \text{ kg FeTiO}_3}{1} \cdot \frac{1 \text{ mol FeTiO}_3}{151.7 \text{ g FeTiO}_3} \cdot \frac{7 \text{ mol Cl}_2}{2 \text{ mol FeTiO}_3} \cdot \frac{70.91 \text{ g Cl}_2}{1 \text{ mol Cl}_2}$$

$$= 40.9 \text{ kg Cl}_2$$

$$\frac{25.0 \text{ kg FeTiO}_3}{1} \cdot \frac{1 \text{ mol FeTiO}_3}{151.7 \text{ g FeTiO}_3} \cdot \frac{6 \text{ mol C}}{2 \text{ mol FeTiO}_3} \cdot \frac{12.01 \text{ g C}}{1 \text{ mol C}} = 5.94 \text{ kg C}$$

Mass of titanium produced:

$$\frac{25.0 \text{ kg FeTiO}_3}{1} \cdot \frac{1 \text{ mol FeTiO}_3}{151.7 \text{ g FeTiO}_3} \cdot \frac{1 \text{ mol Ti}}{1 \text{ mol FeTiO}_3} \cdot \frac{47.9 \text{ g Ti}}{1 \text{ mol Ti}} = 7.89 \text{ kg Ti}$$

47.   Amount of $KMnO_4$ used:

$$0.0173 \frac{\text{mol KMnO4}}{\text{L}} \cdot 0.0124 \text{ L} = 2.15 \times 10^{-4} \text{ mol KMnO}_4$$

In acidic solution MnO_4^- is reduced to Mn^{2+}, a change of 5 moles of electrons/mole of permanganate ion.

$$2.15 \times 10^{-4} \text{ mol MnO}_4^- \cdot \frac{5 \text{ mol e}^-}{1 \text{ mol MnO}_4^-} = 1.07 \times 10^{-3} \text{ mol e}^-$$

The amount of uranyl (VI) nitrate present is:

$$0.213 \text{ g UO}_2(\text{NO}_3)_2 \cdot \frac{1 \text{ mol UO}_2(\text{NO}_3)_2}{394.0 \text{ g UO}_2(\text{NO}_3)_2} = 5.41 \times 10^{-4} \text{ mol UO}_2(\text{NO}_3)_2$$

Calculation of the ratio of mol e^-/mol $UO_2(NO_3)_2$ yields:

$$\frac{1.07 \times 10^{-3} \text{ mol e}^-}{5.41 \times 10^{-4} \text{ mol UO}_2(\text{NO}_3)_2} = 1.98 \text{ or } 2 \text{ mol e}^-/\text{mol UO}_2(\text{NO}_3)_2$$

If two (2) moles of electrons are transferred per mole of uranyl nitrate, the oxidation must have been from U^{4+} to U^{6+}.

Using the method outlined in Chapter 5, the balanced net ionic equation for the oxidation may be written:

$$6 \text{ H}_2\text{O (l)} + 5 \text{ U}^{4+} \text{(aq)} + 2 \text{ MnO}_4^- \text{(aq)} \rightarrow 5 \text{ UO}_2^{2+}\text{(aq)} + 2 \text{ Mn}^{2+}\text{(aq)} + 4 \text{ H}_3\text{O}^+\text{(aq)}$$

49.   The ions $[Cr(CN)_6]^{4-}$ and $[Cr(SCN)_6]^{4+}$ contain the chromium (II) ion. The electron configuration for this ion is: $[Ar]3d^4$. The hexacyano complex would have (with two unpaired electrons) the configuration:

$$\underline{\uparrow\downarrow} \quad \underline{\uparrow} \quad \underline{\uparrow}$$

This is a low spin complex, with two unpaired electrons. The hexathiocyano- complex, with 4 unpaired electrons, would have the configuration:

$$\underline{\uparrow} \quad \underline{\quad}$$
$$\underline{\uparrow} \quad \underline{\uparrow} \quad \underline{\uparrow}$$

This high spin complex along with the low spin cyano- complex indicates that SCN^- is a **weaker field ligand** than CN^-.

51.   The isomers (and their three mirror images) of the copper complex are:
(The glycinate ligand is shown as:)

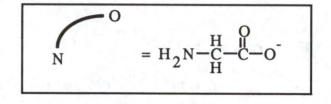

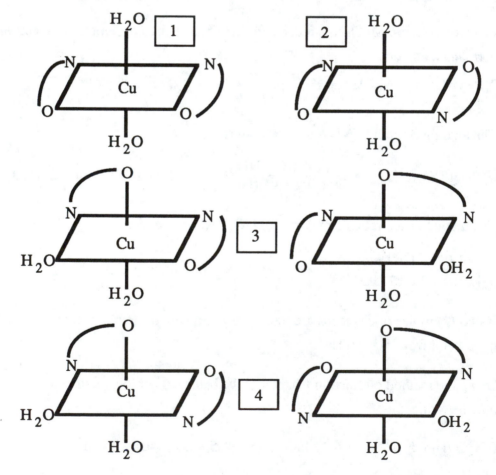

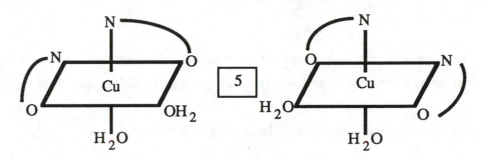

5

53.   Compound: Cr 19.51 % ; C 39.92 %; H_2O 40.57 %

Determine the number of moles of each atom present: (Assume 100 grams of compound.)

$$19.51 \text{ g Cr} \cdot \frac{1 \text{ mol Cr}}{51.996 \text{ g Cr}} = 0.3752 \text{ mol Cr atoms}$$

$$39.92 \text{ g Cl} \cdot \frac{1 \text{ mol Cl}}{35.453 \text{ g Cl}} = 1.126 \text{ mol Cl atoms}$$

$$40.57 \text{ g } H_2O \cdot \frac{1 \text{ mol } H_2O}{18.015 \text{ g } H_2O} = 2.252 \text{ mol } H_2O \text{ molecules}$$

The ratio of moles of atoms are:

$$\frac{1.126 \text{ mol Cl atoms}}{0.3752 \text{ mol Cr atoms}} = \frac{3.00 \text{ mol Cl atoms}}{1 \text{ mol Cr atoms}}$$

$$\frac{2.252 \text{ mol } H_2O \text{ molecules}}{0.3752 \text{ mol Cr atoms}} = \frac{6.00 \text{ mol } H_2O \text{ molecules}}{1 \text{ mol Cr atoms}}$$

The chlorine atoms are in the outer sphere of the complex, since they precipitate upon the addition of silver ions. The formula for the compound must be: $[Cr(H_2O)_6]Cl_3$,

hexaaquachromium(III) chloride.

The structure of this complex would be:

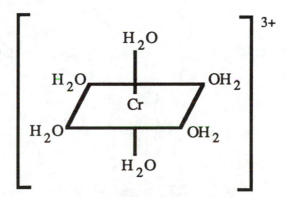

C HAPTER 26: Organic Chemistry

1. The name indicates the number of carbons (oct), the degree of branching (straight-chain : normal or n) and the degree of saturation (-ane). For C_8H_{18} the name is n-octane.

3. The name for the group $CH_3CH_2CH_2CH_2-$ is n-butyl.

5. The structures for the nine isomers of heptane are:

$CH_3-CH_2-CH_2-CH_2-CH_2-CH_2-CH_3$

n-heptane

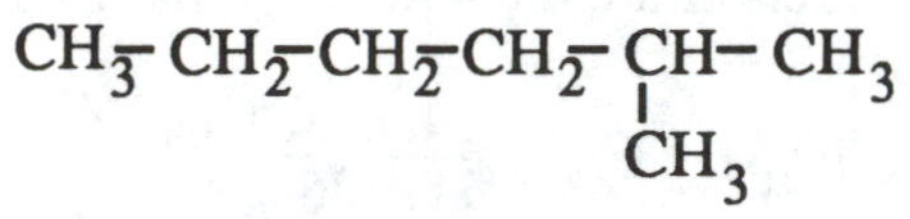

$CH_3-CH_2-CH_2-CH_2-\underset{\underset{\displaystyle CH_3}{|}}{CH}-CH_3$

2-methylhexane

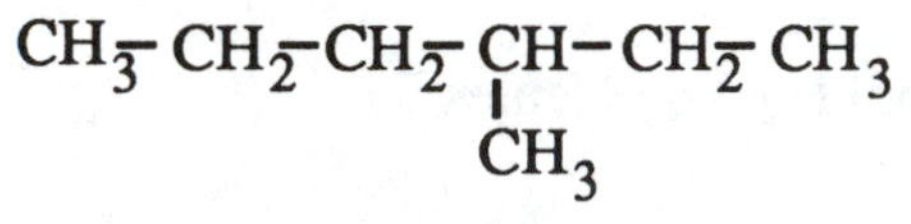

$CH_3-CH_2-CH_2-\underset{\underset{\displaystyle CH_3}{|}}{CH}-CH_2-CH_3$

3-methylhexane

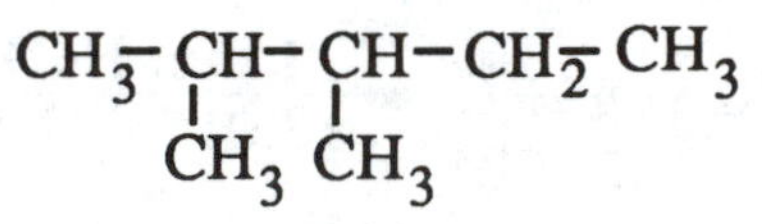

$CH_3-\underset{\underset{\displaystyle CH_3}{|}}{CH}-\underset{\underset{\displaystyle CH_3}{|}}{CH}-CH_2-CH_3$

2,3-dimethylpentane

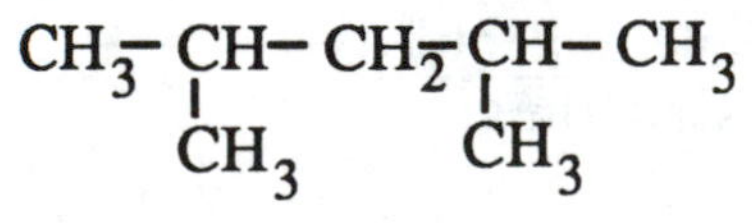

$CH_3-\underset{\underset{\displaystyle CH_3}{|}}{CH}-CH_2-\underset{\underset{\displaystyle CH_3}{|}}{CH}-CH_3$

2,4-dimethylpentane

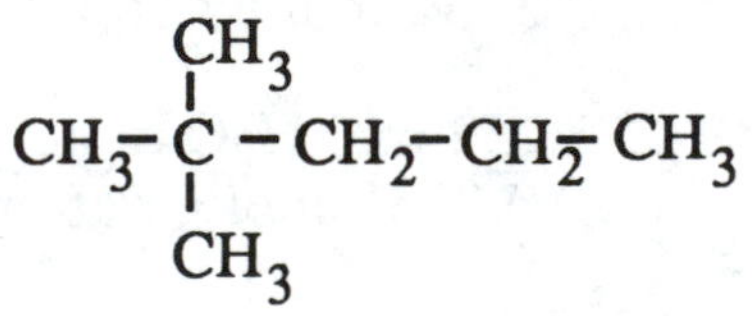

$CH_3-\overset{\overset{\displaystyle CH_3}{|}}{\underset{\underset{\displaystyle CH_3}{|}}{C}}-CH_2-CH_2-CH_3$

2,2-dimethylpentane

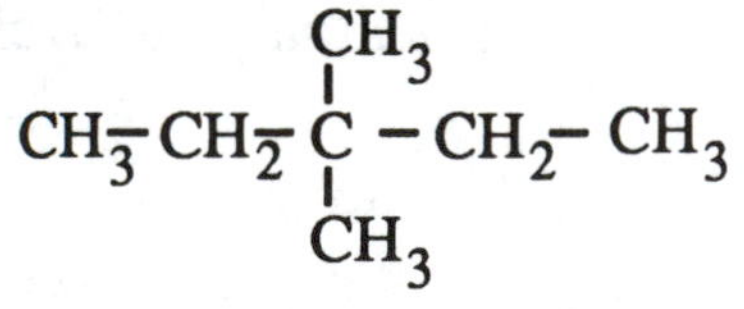

$CH_3-CH_2-\overset{\overset{\displaystyle CH_3}{|}}{\underset{\underset{\displaystyle CH_3}{|}}{C}}-CH_2-CH_3$

3,3-dimethylpentane

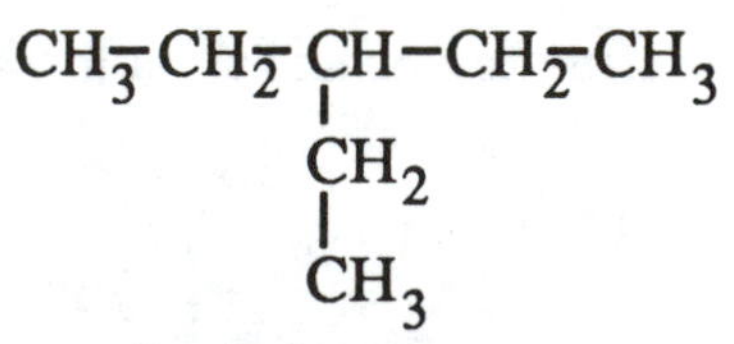

$CH_3-CH_2-\underset{\underset{\underset{\underset{\displaystyle CH_3}{|}}{CH_2}}{|}}{CH}-CH_2-CH_3$

3-ethylpentane

$CH_3-\overset{\overset{\displaystyle CH_3}{|}}{\underset{\underset{\displaystyle CH_3}{|}}{C}}-\overset{\overset{\displaystyle CH_3}{|}}{CH}-CH_3$

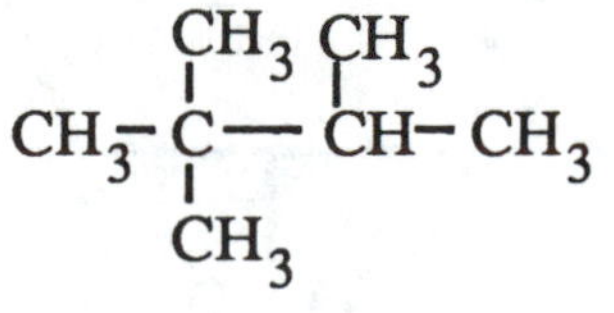

2,2,3-trimethylbutane

7.   a. 2,3-dimethylbutane;   Longest chain has 4 carbons. The two methyl groups (di) are
                                      located on carbons 2 and 3.

       b. 2-methylpentane;   Longest chain has 5 carbons. The methyl group is on the
                                      second carbon (numbered right to left).

       c. 4,4-dimethylheptane;   Longest chain has 7 carbons. The two methyl groups (di) are
                                      both located on carbon 4.

       d. 3-methylhexane;   Longest chain has 6 carbons. The methyl group is located on
                                      carbon 3 (numbered right to left).

9.   a. 2,2-dimethylhexane   b. 3,3-diethylpentane

$$CH_3-\underset{\underset{CH_3}{|}}{\overset{\overset{CH_3}{|}}{C}}-CH_2-CH_2-CH_2-CH_3$$

$$CH_3-CH_2-\underset{\underset{CH_2-CH_3}{|}}{\overset{\overset{CH_2-CH_3}{|}}{CH}}-CH_2-CH_3$$

       c. 2-methyl-3-ethylheptane   d. isobutane

$$CH_3-\overset{\overset{CH_3}{|}}{CH}-\underset{\underset{CH_2-CH_3}{|}}{CH}-CH_2-CH_2-CH_2-CH_3$$

$$CH_3-\overset{\overset{CH_3}{|}}{CH}-CH_3$$

10.   a. $CH_3CH=CHCH_3$ Double bond $\rightarrow$ alkene
       b. $CH_3-O-CH_2CH_3$ Oxygen bonded to two carbons $\rightarrow$ ether
       c. $C_6H_5C\equiv CCH_3$ Triple bond $\rightarrow$ alkyne
       d. $CH_3CH_2NH_2$ Nitrogen bonded to carbon $\rightarrow$ amine
       e. $CH_3CO_2CH_2CH_2CH_3$ Alkyl group bonded to carboxylate group $\rightarrow$ ester
       f. $CH_3CH_2CH_2C(O)CH_3$ Carbonyl group attached to two alkyl groups $\rightarrow$ ketone
       g. $CH_3CH_2C(OH)HCH_2CH_3$ OH group bonded to alkane chain $\rightarrow$ alcohol
       h. $CH_3CH_2CH_2CO_2H$ Carboxylic acid group $\rightarrow$ acid
       i. $(CH_3CH_2)_2CHCH_2C(O)H$ Carbonyl group attached to an alkyl group and a H $\rightarrow$
                                      aldehyde

12.   The electron dot structure for acetaldehyde may be shown as:

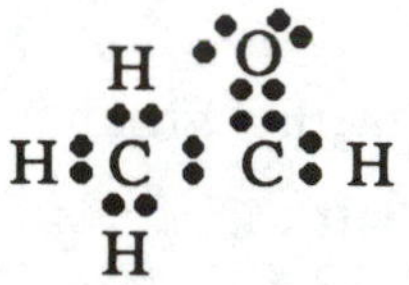

The bond angle for the C-C-O bond would be 120° since the hybridization of the C (attached to O) is sp^2 and that of the methyl carbon is sp^3.

14.   Electron dot for acetonitrile:

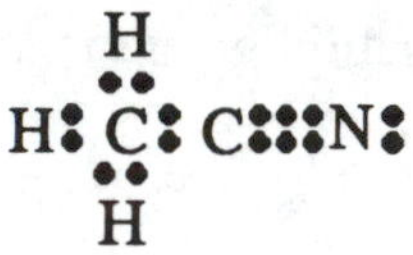

The C-C-N bond angle is 180° (two groups are attached to the intermediate C atom).

The hybridization of the cyano C will be **sp**.

15.   a. ethanol or ethyl alcohol   b. 2-methyl-2-propanol or **tert**-butyl alcohol

       c. l-butanol or **n**-butyl alcohol   d. 2-methyl-2-butanol

17.   a. Ethanol is a **primary** alcohol.   b. 2-methyl-2-propanol is a **tertiary** alcohol.

       c. l-butanol is a **primary** alcohol.   d. 2-methyl-2-butanol is a **tertiary** alcohol.

20.   a. $CH_3(CH_2)_2\text{-}CH_2OH + KMnO_4$ (aq) $\rightarrow$ $CH_3(CH_2)_2\text{-}COOH$

       b. $CH_3(CH_2)_2\text{-}CH_2OH + PCC \rightarrow CH_3(CH_2)_2\text{-}CHO$

$$\underset{\displaystyle\overset{\textstyle OH}{|}}{}$$

       c. $CH_3CH_2\overset{\displaystyle\overset{OH}{|}}{C}HCH_3 + KMnO_4$ (aq) $\rightarrow$ $CH_3CH_2\overset{\displaystyle\overset{O}{\|}}{C}CH_3$

       d. $CH_3CH_2\overset{\displaystyle\overset{OH}{|}}{C}HCH_3 + HBr \rightarrow CH_3CH_2\overset{\displaystyle\overset{Br}{|}}{C}HCH_3$

       e. $CH_3CH_2CH_2CH_2OH \xrightarrow{\ H_2SO_4,\ 180°C\ } CH_3CH_2CH{=}CH_2$ and depending upon conditions

       an ether could be formed by condensation of two molecules of the alcohol:

$$(CH_3CH_2CH_2CH_2)_2O$$

       f. $CH_3CH_2OH + Na \rightarrow CH_3CH_2O^-\ Na^+$

22.   **Cis-trans** isomers are possible for alkenes if either of the doubly-bonded carbons has different atoms. For the figure shown below, A and X must be different <u>and</u> B and Y must be different.

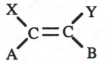

The linear nature of carbon-carbon triple bonds removes the possibility of cis-trans isomerism in alkynes.

24.   a. $CH_3CH_2CH=CH_2 + Br_2 \rightarrow CH_3CH_2CHBrCH_2Br$

     b. $CH_3CH_2CH=CH_2 + HBr \rightarrow CH_3CH_2CHBrCH_3$

     c. $CH_3CH_2CH=CH_2 + H_2 \rightarrow CH_3CH_2CH_2CH_3$

     d. $CH_3CH_2CH=CH_2 + H_2O \rightarrow CH_3CH_2CHOHCH_3$

27.   a. Preparation of 1,2-dibromobutane from an alkene:

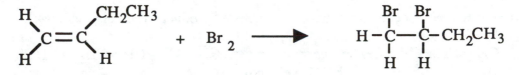

     b. Preparation of 2-bromobutane from an alkene:

     c. Preparation of 2-butanol from an alkene:

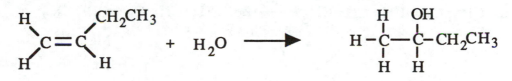

29.   The preparation of 2-butanone from 1-butanol may be accomplished by:

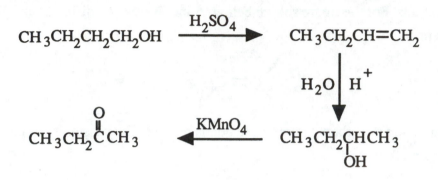

$$CH_3CH_2CH_2CH_2OH \xrightarrow{H_2SO_4} CH_3CH_2CH=CH_2$$

$$H_2O \downarrow H^+$$

$$CH_3CH_2\overset{O}{\overset{\|}{C}}CH_3 \xleftarrow{KMnO_4} CH_3CH_2\underset{OH}{CHCH_3}$$

33.   Isopropylbenzene may be prepared in the following manner:

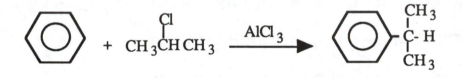

35.   For dehydrohalogenation to occur, a molecule must possess a **halogen atom** on a carbon adjacent to a carbon possessing at least one **hydrogen atom**.

37.   Types of reaction needed to produce the requested products from alkyl halides:

a. $CH_3{-}CH_2{-}\underset{Cl}{CH}{-}CH_3 \xrightarrow[\text{KOH in ethanol}]{\text{elimination}} CH_3{-}CH_2{-}CH=CH_2 + HCl$

b. $CH_3{-}CH_2{-}CH_2{-}CH_2Cl \xrightarrow[\text{NaOH, }H_2O]{\text{substitution}} CH_3{-}CH_2{-}CH_2{-}CH_2OH + NaCl$

c. $CH_3{-}CH_2{-}CH_2{-}CH_2Br + HOCH_3 \xrightarrow{\text{substitution}} CH_3{-}CH_2{-}CH_2{-}CH_2OCH_3 + HBr$

d. $CH_3{-}CH_2{-}CH_2{-}\underset{Br}{CH}{-}CH_3 \xrightarrow[\text{substitution}]{Mg} CH_3{-}CH_2{-}CH_2{-}\underset{\underset{Br}{\overset{|}{Mg}}}{CH}{-}CH_3$

$$CH_3CH_2CH_2CH_2CH_3 \xleftarrow{H_2O}$$

38.   The preparation of di-n-propyl ether may be accomplished by the reaction scheme below:

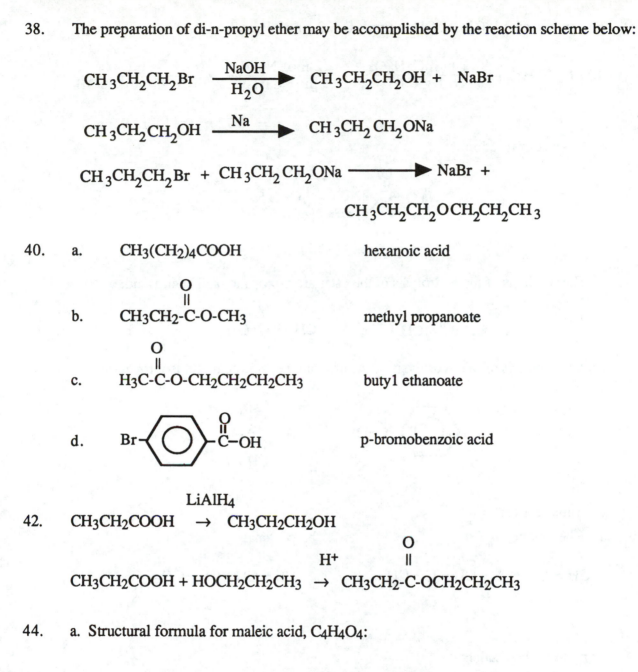

$$CH_3CH_2CH_2Br \xrightarrow[H_2O]{NaOH} CH_3CH_2CH_2OH + NaBr$$

$$CH_3CH_2CH_2OH \xrightarrow{Na} CH_3CH_2CH_2ONa$$

$$CH_3CH_2CH_2Br + CH_3CH_2CH_2ONa \longrightarrow NaBr +$$

$$CH_3CH_2CH_2OCH_2CH_2CH_3$$

40.   a. $CH_3(CH_2)_4COOH$ hexanoic acid

                         O
                         ‖

      b. $CH_3CH_2\text{-}C\text{-}O\text{-}CH_3$ methyl propanoate

                   O
                   ‖

      c. $H_3C\text{-}C\text{-}O\text{-}CH_2CH_2CH_2CH_3$ butyl ethanoate

      d.   Br—⟨O⟩—C—OH   p-bromobenzoic acid

                   LiAlH$_4$

42. $CH_3CH_2COOH \rightarrow CH_3CH_2CH_2OH$

$$CH_3CH_2COOH + HOCH_2CH_2CH_3 \xrightarrow{H^+} CH_3CH_2\text{-}C\text{-}OCH_2CH_2CH_3$$

44.   a. Structural formula for maleic acid, $C_4H_4O_4$:

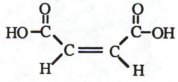

b. Volume of 0.130 M NaOH to titrate 0.522 g maleic acid:

$$0.522 \text{ g } C_4H_4O_4 \cdot \frac{1 \text{ mol } C_4H_4O_4}{116.1 \text{ g } C_4H_4O_4} \cdot \frac{2 \text{ mol NaOH}}{1 \text{ mol } C_4H_4O_4} \cdot \frac{1 \text{ L}}{0.130 \text{ mol NaOH}}$$

$$= 6.92 \times 10^{-2} \text{ L or } 69.2 \text{ mL}$$

c. Product of titration:

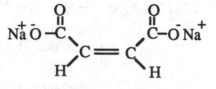

46.   a. The products of the hydrolysis of the ester are n-propanol and acetic acid:

$$CH_3CH_2CH_2OH$$

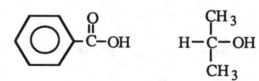

b. The products of the hydrolysis of the ester are benzoic acid and isopropanol:

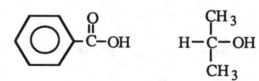

48.   Formulas for :

a. 2-hexanone

$$CH_3\text{--}CH_2\text{--}CH_2\text{--}CH_2\text{--}\underset{\underset{O}{\|}}{C}\text{--}CH_3$$

b. pentanal

$$CH_3\text{--}CH_2\text{--}CH_2\text{--}CH_2\text{--}\underset{\underset{O}{\|}}{C}\text{--}H$$

c. m-chlorobenzaldehyde

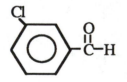

50.   a. The product of oxidation is propanoic acid:

$$CH_3CH_2\overset{\overset{\displaystyle O}{\|}}{C}-OH$$

b. The reduction product is n-propanol:

$$CH_3CH_2CH_2\,OH$$

c. The reduction yields 2-butanol

$$CH_3CH_2\overset{\overset{\displaystyle OH}{|}}{C}H\,CH_3$$

d. The oxidation of 2-butanone with potassium permanganate gives <u>no reaction</u>.

e. The reaction with methyl magnesium iodide and subsequent acid hydrolysis yields 2-pentanol:

$$CH_3^--CH_2^--CH_2^--\overset{\overset{\displaystyle}{|}}{C}H-CH_3$$
$$\underset{\displaystyle OH}{}$$

f. The reaction with methyl magnesium iodide yields 2-methyl-2-butanol:

$$CH_3CH_2\overset{\overset{\displaystyle OH}{|}}{\underset{\underset{\displaystyle CH_3}{|}}{C}}CH_3$$

52.   2-Pentanol may be prepared by the following synthesis:

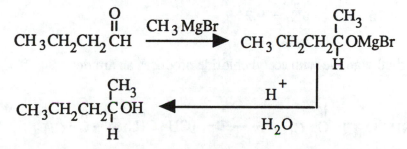

For a preparation of 2-pentanol from a ketone see question 50 (e).

339

54.   The preparation of n-butanol from n-propanol :

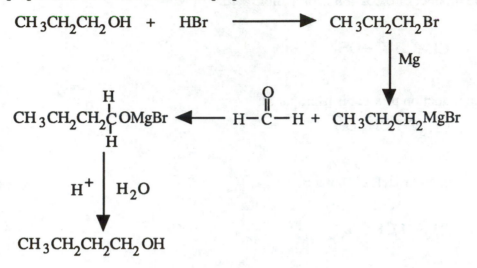

$$CH_3CH_2CH_2OH \; + \; HBr \longrightarrow CH_3CH_2CH_2Br$$

57.   The pH of 0.2 M pyridine :

The equilibrium for pyridine in water may be written:

$$C_5H_5N + H_2O \Leftrightarrow C_5H_5NH^+ + OH^- \quad \text{and } K_b = 1.5 \times 10^{-9}$$

$$K_b = \frac{[C_5H_5NH^+][OH^-]}{[C_2H_5N]} = 1.5 \times 10^{-9}$$

Substituting 0.20 M for the concentration of C_5H_5N, and multiplying both sides of the equation by 0.20 (Since $K_b \cdot 100 << [C_5H_5N]$), you obtain:

$$[OH^-]^2 = 3.0 \times 10^{-10} \quad \text{and} \quad [OH^-] = 1.7 \times 10^{-5}$$

$$pOH = 4.76 \qquad pH = 9.24$$

59.   The reaction of diethylamine with acetyl chloride produces an amide:

$$(CH_3CH_2)_2NH + Cl-\overset{\overset{\displaystyle O}{\|}}{C}-CH_3 \longrightarrow (CH_3CH_2)_2N-\overset{\overset{\displaystyle O}{\|}}{C}-CH_3$$

62.   a. Isomers for C_3H_8O:

Compound	Name	Class
CH₃CH₂CH₂-OH	1-propanol	primary alcohol
CH₃CH-CH₂ │ OH	2-propanol	secondary alcohol
CH₃-O-CH₂CH₃	methyl ethyl ether	ether

    b. Aldehyde with molecular formula C_4H_8O:

       CH₃CH₂CH₂CHO   butanal   aldehyde

    Ketone with molecular formula C_4H_8O:

              O
              ‖
       CH₃CH₂CCH₃   butanone   ketone

64.   Synthesis for methyl salicylate:

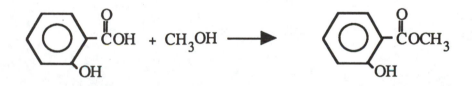

Synthesis for acetylsalicylic acid:

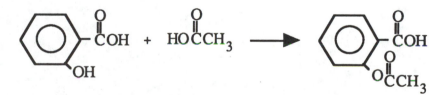

66.   The preparation of 2,2-dimethylpropanol from tert-butyl bromide:

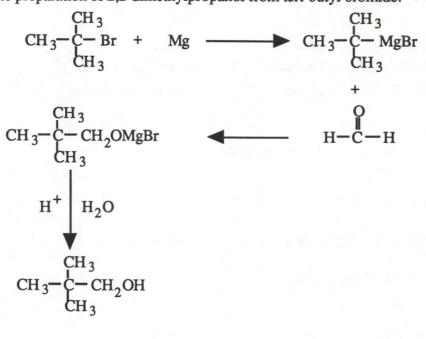

CHAPTER 27: Polymers: Natural and Synthetic Macromolecules

1. The structure of glyceraldehyde shows that the molecule possesses a chiral center. The central C atom is attached to four different groups: H, OH, CH_2OH, and CHO.

5. Dipeptides formed on mixing alanine and serine:

    ala-ala

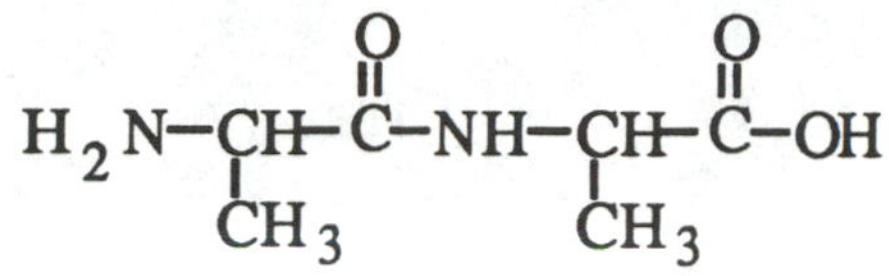

    ser-ser

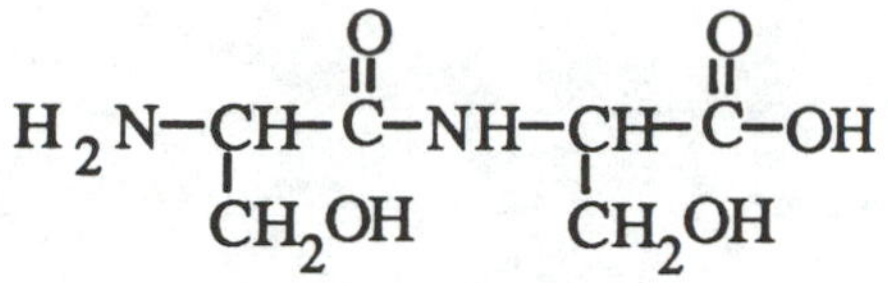

    ala-ser

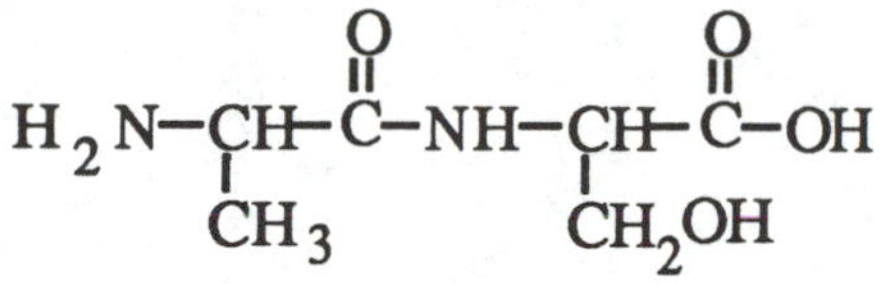

    ser-ala

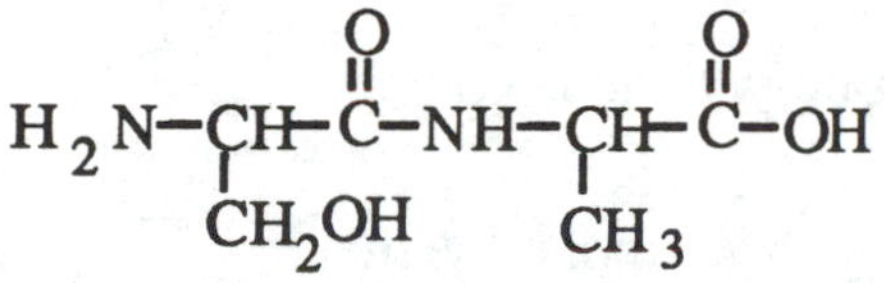

7. The primary structure of a protein refers to the order of amino acids in the peptide chain. The secondary structure results from hydrogen bonding interactions between neighboring protein moieties.

9. Fibrous proteins: wool, hair, skin, beaks, nails and claws
   Globular proteins : albumin, hemoglobin and myoglobin

11. Carbohydrates have the general structure: $C_x(H_2O)_y$.
    Glyceraldehyde's formula, $CH(OH)(CHO)CH_2OH$, may be written $C_3(H_2O)_3$.

13. a. Monosaccharides: fructose, galactose, glucose, ribose and mannose
    b. Disaccharides: cellobiose, lactose, maltose and sucrose
    c. Polysaccharides: cellulose, glycogen and starch (amylose and amylopectin)

15.   The helical structure of DNA arises from the geometry of the carbon atoms in the backbone of the polymer and the phosphate groups (tetrahedral). The conformation of the nitrogen bases creates the possibility of H-bonding. See Figure 27.11 in your text.

17.   Adenine doesn't pair with guanine due to a "mismatch" in the number of hydrogen bonding groups in the two amino acids. Adenine possesses one N-H group capable of hydrogen bonding while guanine has two.

19.   The structure for polyvinylacetate:

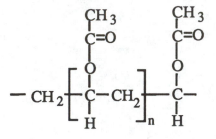

21.   Moles of ethylene to make a 150 g polyethylene bottle:

$$150 \text{ g ethylene} \cdot \frac{1 \text{ mol ethylene}}{28.1 \text{ g ethylene}} = 5.3 \text{ mol ethylene}$$

23.   With respect to their elasticity, polymers are classified as elastomers, fibers, and plastics.

Elastomers are noted for their ability to return to their original shape following a deformation. Neoprene and natural rubber are two well-known elastomers.

Fibers are more rigid than elastomers or plastics and consist of semisynthetic fibers (cellulose acetate), synthetic fibers(nylon, rayon,dacron), and natural fibers (cotton, silk).

Plastics are intermediate in their ability to suffer deformation reversibly. Plastics are further classified as to their response to heat.

Thermoplastic polymers are those which soften on heating. Examples include polyethylene, polystyrene, polyvinyl chloride (PVC), and polyurethane.

Thermosetting polymers are those which do not soften upon heating. Heating will degrade these polymers. Perhaps the best known example of a thermosetting polymer is Bakelite.

25. Polymer formation:

Addition: Teflon from tetrafluoroethylene

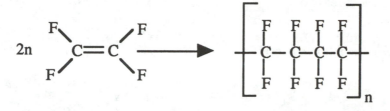

Condensation: Nylon from acid chlorides and diamines:

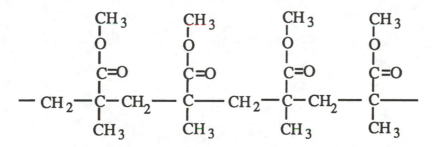

27. A polymethyl methacrylate chain containing four monomeric units:

$$
\begin{array}{cccc}
CH_3 & CH_3 & CH_3 & CH_3 \\
| & | & | & | \\
O & O & O & O \\
| & | & | & | \\
C=O & C=O & C=O & C=O \\
| & | & | & | \\
-CH_2-C-CH_2-C-CH_2-C-CH_2-C- \\
| & | & | & | \\
CH_3 & CH_3 & CH_3 & CH_3
\end{array}
$$

29. Mass of $CHCl_3$ and HF to make 1.0 kg of Teflon:

$$1.0 \times 10^3 \text{ g Teflon} \cdot \frac{1.0 \times 10^3 \text{ g } C_2F_4}{1.0 \times 10^3 \text{ g Teflon}} \cdot \frac{1 \text{ mol } C_2F_4}{100. \text{ g } C_2F_4} \cdot \frac{2 \text{ mol } CHClF_2}{1 \text{ mol } C_2F_4}$$

$$\cdot \frac{1 \text{ mol } CHCl_3}{1 \text{ mol } CHClF_2} = 20. \text{ mol } CHCl_3$$

and this amount of CHCl$_3$ has a mass of:

$$20. \text{ mol CHCl}_3 \cdot \frac{119 \text{ g CHCl}_3}{1 \text{ mol CHCl}_3} = 2.4 \times 10^3 \text{ g CHCl}_3$$

From the reaction CHCl$_3$ + 2 HF →..................

We note that 40. mol HF are necessary to react with 20. mol CHCl$_3$.

This amount of HF will have a mass of:

$$40. \text{ mol HF} \cdot \frac{20. \text{ g HF}}{1 \text{ mol HF}} = 8.0 \times 10^2 \text{ g HF}$$